New Visions on Form and Growth

Fingered Growth, Dendrites, and Flames

Pierre Pelcé

Université de Provence, Marseille

Translated by

Jasna Brujić and Laurent Costier

University of Cambridge

OXFORD
UNIVERSITY PRESS

OXFORD
UNIVERSITY PRESS

Great Clarendon Street, Oxford OX2 6DP

Oxford University Press is a department of the University of Oxford.
It furthers the University's objective of excellence in research, scholarship,
and education by publishing worldwide in

Oxford New York

Auckland Bangkok Buenos Aires Cape Town Chennai
Dar es Salaam Delhi Hong Kong Istanbul Karachi Kolkata
Kuala Lumpur Madrid Melbourne Mexico City Mumbai Nairobi
São Paulo Shanghai Taipei Tokyo Toronto

Published in the United States
by Oxford University Press Inc., New York

Translation of Théorie des formes de croissance by Pierre Pelcé originally published in French by EDP Sciences, Les Ulis, and CNRS Éditions, Paris 2000

Aidé par le ministère français chargé de la culture

First published in English 2004

A catalogue record for this title is available from the British Library

Library of Congress Cataloging in Publication Data

(Data available)

ISBN 0 19 852701 2

10 9 8 7 6 5 4 3 2 1

Typeset using the French publisher's files by Laurent Costier and Jasna Brujić
Printed in Great Britain
on acid-free paper by
Biddles Ltd., King's Lynn
<http://www.biddles.co.uk>

Preface

The present book is closely related, firstly by its title, to the book of D'Arcy Thompson entitled *On Growth and Form*, unifying a large number of patterns in growth from inanimate to living matter. Both texts take into consideration dimensional analysis, surface tension, changes in the reference frame, spirals, crystals, liquid crystals, analogies between different morphologies and biological analogues.

They differ, however, in that this book employes a quantitative approach to describe the mechanisms of growth, the equations which govern their dynamics, the theoretical basis of physics out of equilibrium, as well as growth laws which are thus derived. Such an in-depth analysis has limited the number of growth patterns under consideration mainly to patterns observed in inanimate matter, which are now well understood. These patterns, including the simplest to the most complicated fingered growth, viscous fingers, crystalline dendrites, flames and electrolytic aggregates belong to very different fields in physics: hydrodynamics, elasticity, physical chemistry and electrochemistry. The universal behaviour of their growth dynamics is therefore even more striking.

Remarkably, it seems as though this unification could somehow be applied to cellular forms by 'tip growth', discussed in the last chapter. If that is the case, this book could be considered as a useful guide to biology students researching this topic.

I would like to thank Claude Godrèche for suggesting that I should write this book. I also thank the numerous coworkers and collaborators cited here who have participated in the compilation of this review. I am grateful to Jean-Paul Pahin for his participation in including the figures in this book. Last, but not least, I thank my wife and children, as well as my family, for their patience and for the time which I could not spend with them during the writing of this book.

Pierre PELCÉ
January 2000

Contents

Pattern formation and growth rate 203

Stability of a curved shape 255

Evolution of growth forms 287

CHAPTER 1

Introduction

1.1 ABOUT PATTERNS IN GROWTH

Many natural shapes are designed during their growth. Thus, if their rigidity is high enough their final state reflects the different mechanisms responsible for their development. Two classes arise: the first one corresponds to a state where the shape mimics the inhomogeneities of the matter of the given object. In this case the growth is mainly controlled by the structure of the object itself, independent of the state of the surrounding medium. The second one is more ornamented and results from growth instabilities of regular shapes. In the latter case the constituent matter can be homogeneous and isotropic. However, if shape heterogeneities occur, which are merely triggered by fluctuations in the surrounding media, they are amplified by the growth dynamics. In general they are more sensitive to the state of the surrounding medium and in particular to the presence of gradients, which allow close groups of objects to influence their mutual growth dynamics.

By natural shapes we consider not only inanimate matter such as viscous fingering, crystalline and electrochemical dendrites and flames, but also living organisms such as algae, filamentous mushrooms and neurons. The study of the properties of inanimate matter is generally very distinct from that of living matter. Nevertheless, it seems that it would be helpful to consider them together in the limit of pattern in growth, uniting physical sciences and biology. In fact, it has already been admirably presented in D'Arcy Thompson's book, *On Growth and Form*, where many analogies between inert and living growth forms are mentioned. These analogies are often qualitative and therefore limited to stating similarities between morphologies from different origins. For instance, the shape of droplets falling into another fluid is compared to the shape of some jellyfish and the crystalline structures to cytoskeletons of radiolarians. These forms must be explained in a familiar framework for physicists since, as mentioned by D'Arcy Thompson, '*Now the state, including the shape or form, of a portion of matter is the resultant of a number of forces, which represent or symbolise the manifestations of various kinds of energy; and it is obvious, accordingly, that a great part of physical science must be understood or taken for granted as the necessary preliminary to the discussion on which we are engaged*'.

More profound analogies concerning growth mechanisms could arise, leading to similar growth laws. A first step in this direction, described in the present work, is to notice such universality among the growth forms in inanimate matter (Langer 1980, Pelcé 1986, Kessler *et al.* 1988, Langer 1989). Such a quantitative study requires the elaboration of physical models of growth. These models define a set of equations with solutions that are growth forms themselves. Some growth laws are deduced from them and are compared to experimental results. The same approach should be applicable to biological forms since, as quoted by D'Arcy Thompson again, '*we want to see how, in some cases at least, the forms of living things, and of the parts of living things, can be explained by physical considerations, and to realise that in general no organic form exist save such as are in conformity with physical and mathematical laws*'. But this time, we should maybe restrict the growth laws to relations between orders of magnitude since the mathematical detail required by the equations of the models is completely hidden by the complexity of the systems studied. Indeed, as stated by Jacob (1970), '*Between the world of living and the world of inanimate, there is a difference of complexity, not of nature*'.

The morphogenesis is programmed by the genes. Yet the way they act on the shape is hardly known and two extreme hypotheses can be envisaged. The first one is a permanent control by the genes on the shape. The second one would only imply the action of the genes in a first step, during the synthesis of the different components of the cell. Then these components would organise themselves according to the standard laws of physics and chemistry. The mechanisms of inanimate growth could take an important part in the morphogenesis of living organisms.

Inanimate matter is often in thermodynamic equilibrium. This equilibrium is described by the laws of thermodynamics of Joule, Carnot and Clausius and by the statistical physics of Boltzman, Gibbs and Maxwell. In a transient regime, such as during crystal growth, gradients of different substances occur in the liquid and solid phases maintaining the growth form. The creation of gradients by a dendrite is well described by the thermodynamics of systems slightly out of equilibrium. On the contrary, living matter is always out of equilibrium. A cell without gradients is a dead cell. The issue at stake is to find out whether or not the physics of out of equilibrium systems, as for the crystalline dendrite, can be applied to determine its shape and dimensions. As stated by Schrödinger (1944), '*Living beings, to whom the laws of physics apply, involve other unknown laws of physics yet but which will be part and parcel of this field of science once revealed*'. To solve this question, we must first check that the growth forms of inanimate matter satisfy the laws of out of equilibrium systems, selecting a few characteristic systems such as a viscous fluid displacing a less viscous fluid in a Hele–Shaw cell, a crystalline dendrite growing in a supercooled or oversaturated fluid, a premixed flame spreading in a tube or a metallic electrode growing in an electric field. Each of these systems generates structures with forms, dimensions and growth rates in good agreement with the results obtained by this law of out of equilibrium systems. It even applies to rather complicated systems involving many parameters, such as flames and dendrites. The same approach can be applied for the morphogenesis of living organisms: selecting a few simple systems in well-controlled culture conditions, applying the laws of out-of-equilibrium systems at that moment, and comparing the theoretical results to the experimental morphologies obtained, just as we do in the last chapter.

Let us consider for instance the rather simple experiment during which a viscous liquid is first introduced in a Hele–Shaw cell – a space delimited by two glass windows with a gap far smaller than the transversal dimensions. If air is injected into one extremity of the cell an air–water interface is created, moving at the velocity of the fluids. Rather than staying planar the interface first loses its shape with cells of well-defined widths. Later some cells tend to move quicker than others until there remains only an air finger of stationary shape moving at a constant speed, as shown in figure 1.1 (Saffman and Taylor 1958). The velocity of the finger increases with the imposed pressure gradient until the finger breaks into smaller ones at a critical speed, determined by the experimental conditions.

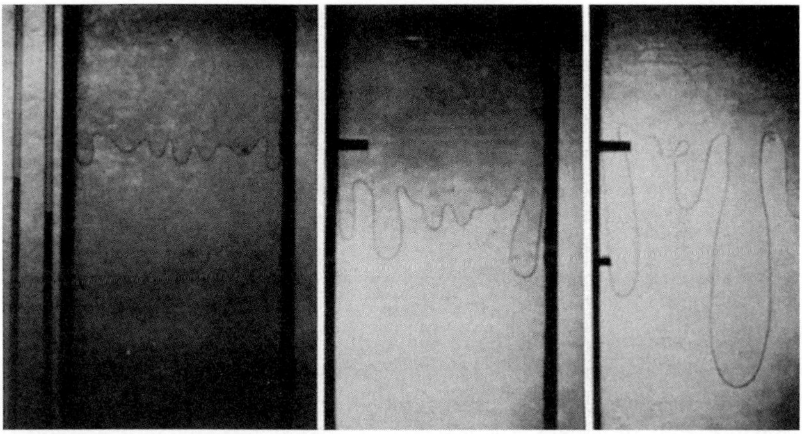

Figure 1.1 Snapshot of a distorted water–air interface in a Hele–Shaw cell (Saffman and Taylor, 1958).

When a solid germ is introduced into a supercooled liquid it has a spherical shape at first and its radius increases at a given rate depending on the supercooling of the liquid and on the interfacial tension between the liquid and the solid phases. Starting from a critical radius, again dependent on the degree of supercooling of the liquid, lobes and petals appear, the number of which depends on the underlying symmetry of the crystal (for instance six petals grow in each of the main crystallographic directions in the case of succinonitrile, which belongs to a class of transparent materials with cubic symmetry). As shown in figure 1.2 (Huang and Glicksman 1981) the lobes turn into dendrites: crystalline tips grow at a constant rate and induce a secondary branching at the borders.

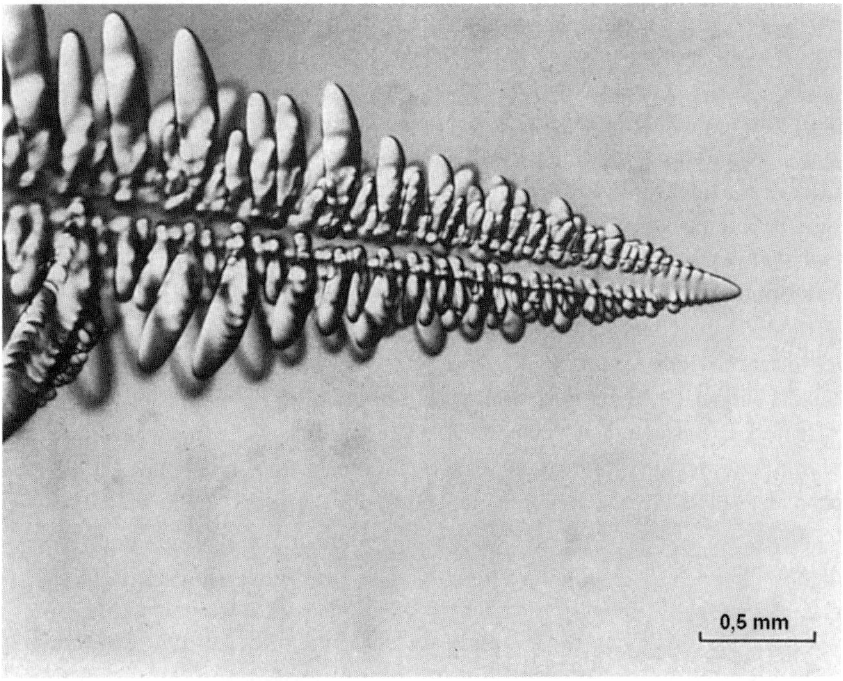

Figure 1.2 Dendrite of succinonitrile (Huang Glicksman, 1981).

Let us consider then a reactive gaseous mixture, combustible at room temperature, and let us significantly raise the temperature at a given point. Since the combustion is highly exothermic, the reaction will spread owing to the diffusion of heat towards the fresh mixture and owing to the activation of the reaction by the increase in the temperature of the gas. The activation energy of this kind of reaction is generally far larger than kT. The rate of the reaction is very sensitive to the temperature and the reaction takes place in a very small region: the flame front. This region separates the fresh from the burnt gases and spreads at a given speed with respect to the fresh gases. As shown in figure 1.3 (Pelcé-Savornin *et al.* 1988), when the flame propagates in a vertical tube a very stable curved shape can be observed.

Starting from a zinc sulphate solution, thin layers of zinc begin to grow at the interface between the solution and n-butyl acetate by electrodeposition (Matsushita

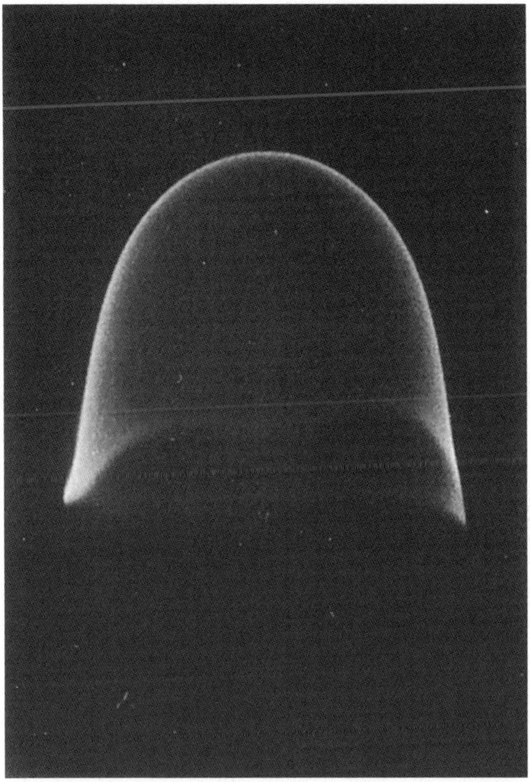

Figure 1.3 Curved flame dynamically stabilised in a tube. The reactive gases flow downwards, with the fresh and burnt ones respectively above and below (Pelcé-Savornin *et al.* 1988).

et al. 1984). A radial electric field is kept in the solution between a carbon electrode in the centre and an annular zinc anode at the borders. As shown in figure 1.4, if the applied field is moderate (several mV) thin layers of zinc with an extremely jagged shape appear at the surface of the solution. This growth form features many branches, which can be classified into primary, secondary, tertiary, etc. The structure of a higher order branch is the same as the structure of a primary branch, giving the metallic aggregate the property of an invariant scaling law. It is a fractal of a well-defined dimension, $D \approx 1.7$.

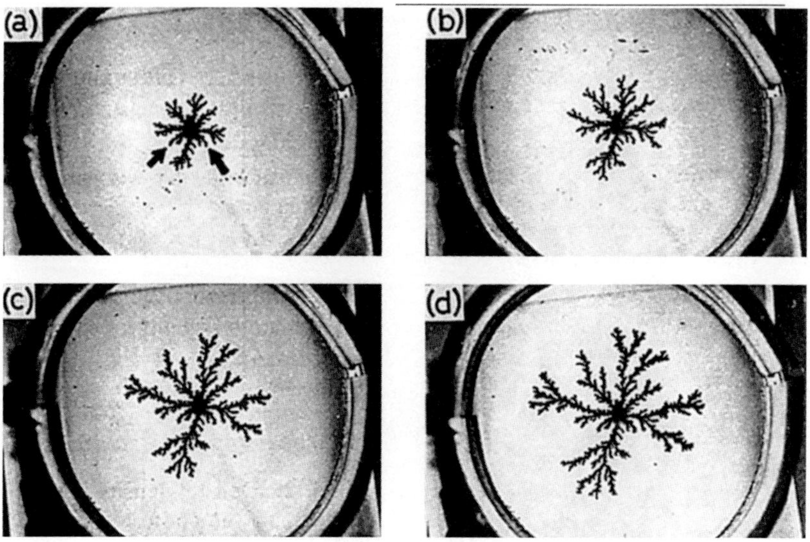

Figure 1.4 Metallic aggregate formed by electrodeposition (Matsushita *et al.*, 1984).

1.2 OUT-OF-EQUILIBRIUM FORMS

These growth forms develop in out-of-equilibrium media. Gradients fixed by the external medium guide the growth. Hence in most of the cases growth forms are not the results of a variational law minimising a free energy. They are solutions of equations which set the growth dynamics. When the driving force of the growth is suppressed these forms generally undergo a relaxation to equilibrium forms. Nevertheless if the growth kinetics are very slow as in the case of faceted crystals, or if the material which the object is made of is rigid enough as in the case of a quenched metallic structure, growth forms can be observed for a long time. Thus, the fundamental laws which set the growth of these forms arise from the physics of out of equilibrium systems.

The general theory of out of equilibrium media concerning inanimate matter is rather old, since it started last century with the Navier equations (1822) and Stokes equations (1845) for fluids, and with the Fourier equation of heat diffusion (1822). These equations, presented in the sixth volume of the *Course of Theoretical Physics* by Landau and Lifshitz (edited in 1987), are the bases of out-of-equilibrium thermodynamics (Glansdorff and Prigogine 1971, Prigogine 1996), summarised by the rate of entropy production, by linear relationships between thermodynamic forces and flux and by properties of symmetry between kinetic constants discovered by

Onsager (1931). These equations have rather robust solutions, dissipative structures and stationary states resulting from a balance between the energy favoured by external conditions and its dissipation by viscous and diffusive processes. The growth forms hereby introduced are a few examples of these systems. This theory should somehow be extended to living organisms and should therefore explain part of the phenomena observed during the morphogenesis.

The examples stated above constitute a particular class of out-of-equilibrium media. The shape of the objects and organisms under consideration is a borderline between the inside and the outside part, in which fields develop according to the appropriate dynamics. Boundary conditions have to be satisfied at the interface for these fields. They are determined by conservation equations of different physical quantities which govern the growth and the physical properties of the interface itself. For example, in the viscous fingering introduced at the beginning, theinterface between two fluids of different viscosities changes under the influence of the motion of the fluids. In the case of crystals the solid–liquid interface interacts with the temperature field or the concentration field of impurities released by the phase transition. During the combustion of the premixed gas a thin region of gas where the reaction is taking place interacts with the temperature field, the concentration field of reagents and the velocity field of fresh and burnt gases. For the electro-deposition of metal atoms on an electrode, the deposition rate of the atoms on the metallic surface is influenced by the ionic concentration, the electric potential and the electro-osmotic based flows.

Solving these examples often involves complicated mathematical methods. Yet we can simply obtain the relevant orders of magnitude if we restrict ourselves to a dimensional analysis. They are often satisfactory for rather complicated systems.

1.3 A CERTAIN COMPLEXITY

We can easily see that the systems become progressively more complex from viscous fingering to crystals, from crystals to flames and electrolytic deposits and further to biological forms. This trend accounts for the choice of the examples described above. The first example is the simplest one, both theoretically and experimentally. The construction of the model – the evaluation of the relevant physical quantities and of the parameters which govern the dynamics of viscous fingering – is rather ancient. Apart from subtle physical–chemistry related problems such as the contact between a moving fluid and a wall, the standard model is now well established. The motion of the fluids has to be described according to the Navier–Stokes equations, interacting with an interface which intervenes dynamically mostly by the effect of its surface energy. Most of the viscous fingering is relatively easy to set up experimentally. The ideal configuration is probably the experiment by Saffman and Taylor (1958) where a fluid displaces a more viscous fluid in a Hele–Shaw cell and fully wets the cell wall. Since both the modelling of the viscous fingering and its experimental set-up are easy, they naturally appear as template growth forms from which we are inspired to understand the forms generated by more complicated dynamics.

The study of crystalline growth is slightly more complicated. The structure of the liquid–solid interface made of the same material accounts for the theoretical and experimental complexity. Contrary to the previous example where the interface separates two immiscible fluids and is displaced under the motion of the two fluids, the liquid–solid interface moves by the transition from one phase to the other, where both phases are likely to stay at rest. Kinetic effects related to the phase transition, which are as yet not fully understood, can come into play. Even if these effects are negligible the problem of the structure of the interface must be carefully considered. The interface may be rough even at equilibrium and change shape like an interface between two fluids, or it may be smooth and feature planar facets on a macroscopic scale. The dependence of the surface energy on the crystalline direction is therefore important to distinguish between one or the other behaviour. Another source of complexity in a real experiment is the abundance of the fields present: a temperature field, a concentration field of impurities and a velocity field of the fluid set in motion by convective effects for instance.

Combustion is also complicated, but for different reasons. It does not feature two phases in contact but just one gaseous phase in two different chemical states. The interface is then a thin region where the exothermic chemical reaction takes place. The complexity of the phenomenon arises from the quantity of intermediate reagents involved in the reaction on the one hand and, as in the previous case, from the abundance of the fields involved in the combustion process on the other hand: a temperature field, a concentration field of reagents, a velocity field of the fresh and burnt gas set in motion by the change in the density of gases through the flame front. Also, the experimental conditions are not easy to master. Large temperature gaps have to be controlled over a very small scale, not to mention all the problems related to the boundary conditions on the walls and the parasite convective effects which can alter the intrinsic dynamics of the flames.

The level of complexity of electrodeposition is of the same order. Electrodeposition shares common properties with dendritic growth in that they both feature the growth of a condensed phase; it also shares common properties with combustion because of the surface chemical reactions and the electro-osmotic or convective fluid flows (Fleury 1998). Moreover, the dynamics of charged particles plays a role, coupling ionic concentrations and the electric field. As for combustion, the kinetics introduce Arrhenius-type exponential laws which are very sensitive to the changes in certain parameters. This sensitivity plays an important role in the diversity of the observed growth regimes.

Apart from the trivial case of viscous fingering, the systems considered are rather complicated and involve a lot of control parameters. The modelling of complicated systems implies the use of gross assumptions. We begin by retaining only several parameters, a priori the most important ones, and the corresponding growth forms and growth rates are determined according to those few parameters alone. The discrepancies with the experimental results lead to the introduction of secondary phenomena in the theory and to the selection of extra control parameters. Hence, contrary to the quantitative laws of physics such as the law of gravity or the wavelength of a corpuscle as a function of its momentum, a growth law may be expressed by many different analytical forms according to the domain of control parameters considered. The general case can only be expressed by numerical analysis in the form of graphs and curves.

1.4 EQUATIONS OF CONSERVATION

These are the general equations of fluid mechanics. They are brilliantly presented in the sixth volume of *Course of Theoretical Physics* by Landau and Lifshitz (1987), but we will introduce them again below to make this book easier to read. They reflect the conservation of quantities such as the mass of the fluid, its energy, its momentum and the number of molecules of the solute.

The conservation of mass means that the rate of mass change of the fluid inside a control volume V comes from the quantity of fluid crossing the boundary of V per unit time alone,

$$\frac{\partial}{\partial t}\int_V dV\rho = -\int_S dS\rho\vec{w} = \int_V dV\vec{\nabla}\cdot(\rho\vec{w}), \tag{1.1}$$

where ρ is the mass density and $\vec{w}$ the velocity field of the fluid. We deduce:

$$\frac{\partial}{\partial t}\rho + \vec{\nabla}\cdot(\rho\vec{w}) = 0. \tag{1.2}$$

The first law of thermodynamics appears when we write the rate of energy change of the control volume, $\partial/\partial t(\rho e + \rho w^2/2)$, where e is the internal energy per unit mass of fluid. First the variation of kinetic energy:

$$\frac{\partial(\rho w^2)}{\partial t} = \frac{w^2}{2}\frac{\partial\rho}{\partial t} + \rho\vec{w}\frac{\partial\vec{w}}{\partial t}. \tag{1.3}$$

Considering first an ideal fluid the fluid dynamics satisfies the Euler equation

$$\rho\left(\frac{\partial\vec{w}}{\partial t}+\vec{w}\cdot\vec{\nabla}\vec{w}\right) = -\vec{\nabla}p \tag{1.4}$$

where p is the pressure. Using the thermodynamic relation $dh = Tds + dp/\rho$, where h and s are the enthalpy and the entropy per mass unit of fluid respectively, and equations (1.2) and (1.4), we obtain from equation (1.3)

$$\frac{\partial(\rho w^2)}{\partial t} = -\frac{w^2}{2}\vec{\nabla}\cdot(\rho\vec{w})-\rho\vec{w}\cdot\vec{\nabla}\left(h+\frac{w^2}{2}\right)+\rho T\vec{w}\cdot\vec{\nabla}s. \tag{1.5}$$

Similarly,

$$\frac{\partial(\rho e)}{\partial t} = h\frac{\partial\rho}{\partial t}+\rho T\frac{\partial s}{\partial t} = -h\vec{\nabla}\cdot(\rho\vec{w})-\rho T\,\vec{w}\cdot\vec{\nabla}s, \tag{1.6}$$

where we have used the relation $d(\rho e) = ed\rho + \rho\, de = hd\rho + \rho T\, ds$, and the condition of adiabaticity $ds/dt = 0$ of ideal fluids. Adding equations (1.5) and (1.6), we deduce the conservation of energy

$$\frac{\partial\left(\rho e+\frac{\rho w^2}{2}\right)}{\partial t} = -\vec{\nabla}\cdot\left(\rho\vec{w}\left(\frac{w^2}{2}+h\right)\right). \tag{1.7}$$

For the conservation of momentum, we express

$$\frac{\partial(\rho w_i)}{\partial t} = \rho\frac{\partial w_i}{\partial t}+\frac{\partial\rho}{\partial t}w_i \tag{1.8}$$

which we can group together using equations (1.2) and (1.4) :

$$\frac{\partial(\rho w_i)}{\partial t} = -\frac{\partial\Pi_{ik}}{\partial x_k}. \tag{1.9}$$

where $\Pi_{ik} = p\delta_{ik}+\rho w_i\, w_k$ is the tensor of the density of momentum flux.

In the absence of a diffusion current the concentration of a solution remains constant during its displacement, that is:

$$\frac{dc}{dt} = \frac{\partial c}{\partial t}+\vec{w}\cdot\vec{\nabla}c = 0. \tag{1.10}$$

Therefore using the law of conservation of mass of the fluid, determined by equation (1.2), it follows that:

$$\frac{d(\rho c)}{dt} = \frac{\partial(\rho c)}{\partial t}+\vec{\nabla}\cdot(\rho c\,\vec{w}) = 0. \tag{1.11}$$

If the fluid is no longer considered ideal, the preceding quantities are still conserved, but the corresponding fluxes are implemented using an additional flux of an irreversible nature. Their expression can be deduced in a general form, from the law of increasing entropy. Hence we can write down the equations of energy conservation for the solution and momentum in the form:

$$\frac{\partial(\rho e)}{\partial t} + \vec{\nabla} \cdot \left(\rho \vec{w}\left(h + \frac{w^2}{2}\right) - (\vec{w} \cdot \vec{\sigma}') + \vec{j}_q\right) = 0$$

$$\frac{\partial(\rho c)}{\partial t} + \vec{\nabla} \cdot (\rho \vec{w} c + \vec{j}_s) = 0 \tag{1.12}$$

$$\frac{\partial(\rho w_i)}{\partial t} + \frac{\partial}{\partial x_k}(\rho \delta_{ik} + \rho w_i w_k - \sigma'_{ik}) = 0 ,$$

where $\vec{j}_q$ and $\vec{j}_s$ are the irreversible flux arising from heat and the molecules of the solute, respectively, and σ'_{ik} is the viscous tensor of the constraints. Their expression, as a function of temperature gradients, chemical potential and velocity gradients, is as yet undetermined. It will become evident from the general principles revealed in the following paragraph.

Integrating these conservation equations over a small volume encompassing a portion of the interface, the transformation of a volume integral of the divergence of a vector field integral into a surface integral of the flux of the same field leads to equality of the normal components of these vectors on both sides of the interface.

1.5 THE SECOND LAW OF THERMODYNAMICS

The equilibrium state of an isolated thermodynamic system, characterised by the variables x_1, x_2, ..., x_n, is determined by the state of maximum entropy $S(x_1, x_2, \ldots, x_n)$. The quantities $X_i = \partial S / \partial x_i$ are therefore zero at equilibrium. If the system is perturbed from its equilibrium state, the variables x_i become time dependent. Moreover, if the perturbation from equilibrium is small we can expect the change, which is always small near equilibrium, to be predicted by the derivatives dx_i/dt, expressed in the form of a linear combination of quantities X_i. We can therefore write for each i

$$\frac{dx_i}{dt} = \gamma_{ik} X_k . \tag{1.1}$$

A first constraint on the coefficients γ_{ik} can be obtained by expressing the entropy variation of the system with respect to time. We therefore have:

$$\frac{dS}{dt} = X_i \frac{dx_i}{dt} = \gamma_{ik} X_i X_k . \tag{1.2}$$

Since this derivative must be positive, the coefficients γ_{ik} must have a sign such that the quadratic form of the last member of equation (1.2) is positive. A second constraint, which is much less obvious, uses the symmetry of the equations of motion of each particle in the system with respect to time. Under these conditions, it can be shown (Onsager 1931) that the kinetic coefficients γ_{ik} are symmetric, $\gamma_{ki} = \gamma_{ik}$.

If the law of the evolution of a thermodynamic system close to equilibrium considered in this generalisation appears relatively simple, its application to real systems is less so. First of all, the variables x_i and their conjugate quantities X_i should be identified. So, we express the change in the total entropy of the system in the form (1.2), or even in the case of a continuous medium, the expression of the derivative $\partial/\partial t \int \rho s \, dV$. In the latter, we depart from the expression for the change in energy *e* and the enthalpy *h* by the unit mass as a function of their respective variables, namely

$$\begin{aligned} de &= T ds + \frac{p}{\rho^2} d\rho + \mu dc \\ dh &= T ds + \frac{1}{\rho} dp + \mu dc. \end{aligned} \tag{1.3}$$

By taking the partial derivative with respect to time of the first of the above equations and the gradient of the second, we then obtain using the law of conservation of mass,

$$\rho T\left(\frac{\partial s}{\partial t} + \vec{w} \cdot \vec{\nabla} s\right) = \frac{\partial(\rho e)}{\partial t} + \vec{\nabla} \cdot (\rho \vec{w}\, h) - \rho\mu\left(\frac{\partial c}{\partial t} + \vec{w} \cdot \vec{\nabla} c\right) - \vec{w} \cdot \vec{\nabla} p \tag{1.4}$$

We then use the conservation relations (1.12) in paragraph 1.4 to obtain:

$$\rho T\left(\frac{\partial s}{\partial t} + \vec{w} \cdot \vec{\nabla} s\right) = -\vec{\nabla} \cdot \vec{j}_q + \mu \vec{\nabla} \cdot \vec{j}_s + \sigma'_{ik} \frac{\partial w_i}{\partial x_k} . \tag{1.5}$$

From equation (1.5), we can deduce

$$\frac{\partial}{\partial t} \int \rho s \, dV = \int dV\left(s \frac{\partial \rho}{\partial t} + \rho \frac{\partial s}{\partial t}\right) = \int dV\left(-\frac{1}{T} \vec{\nabla} \cdot \vec{j}_q + \frac{\mu}{T} \vec{\nabla} \cdot \vec{j}_s + \frac{\sigma'_{ik}}{T} \frac{\partial w_i}{\partial x_k}\right) . \tag{1.6}$$

Upon integrating by parts the last member of equation (1.6) we obtain the law of the growth of entropy in the form

$$\frac{\partial}{\partial t} \int \rho s \, dV = -\int dV \frac{(\vec{j}_q - \mu \vec{j}_s) \cdot \vec{\nabla} T}{T^2} - \int dV \frac{\vec{j}_s \cdot \vec{\nabla} \mu}{T} + \int dV \frac{\sigma'_{ik}}{T} \frac{\partial w_i}{\partial x_k} . \tag{1.7}$$

The derivatives dx_i/dt are therefore naturally $\vec{j}_q - \mu \vec{j}_s$, $\vec{j}_s$ and σ'_{ik} is to be expanded as a function of $\vec{\nabla} T/T^2$, $\vec{\nabla} \mu/T$ and $(\partial w_i/\partial x_k)/T$. Following the Curie symmetry principle we can only link together tensors of the same order (Curie 1908). We then write (De Groot and Mazur 1969, Caroli *et al.* 1992)

$$\vec{j}_q - \mu\vec{j}_s = -\gamma_{qs}\frac{\vec{\nabla}\mu}{T} - \gamma_{qq}\frac{\vec{\nabla}T}{T^2}$$

$$\vec{j}_s = -\gamma_{ss}\frac{\vec{\nabla}\mu}{T} - \gamma_{sq}\frac{\vec{\nabla}T}{T^2} \tag{1.8}$$

$$\sigma'_{ik} = -\frac{\gamma_{ww}}{T}\left(\frac{1}{2}\left(\frac{\partial w_i}{\partial x_k} + \frac{\partial w_k}{\partial x_i}\right) - \frac{1}{3}\delta_{ik}\vec{\nabla}\cdot\vec{w}\right) - \frac{\gamma'_{ww}}{T}\delta_{ik}\vec{\nabla}\cdot\vec{w}.$$

The constraints on the kinetic coefficients mentioned at the beginning of the paragraph are therefore that γ_{qq}, γ_{ss}, γ_{ww} and γ'_{ww} are positive and that $\gamma_{qs} = \gamma_{sq}$. Let us suppose that the flux of solute is zero ($j_s = 0$) in a fluid at rest ($\vec{w} = 0$). From equation (1.8) we deduce the Fourier law of heat conduction $\vec{j}_q = -\lambda\vec{\nabla}T$, where $\lambda = (\gamma_{qq} - \gamma_{sq}^2/\gamma_{ss})/T^2$. From the expression of the viscous tensor of stresses

$$\sigma'_{ik} = \eta\left(\frac{\partial w_i}{\partial x_k} + \frac{\partial w_k}{\partial x_i} - \frac{2}{3}\delta_{ik}\vec{\nabla}\cdot\vec{w}\right) + \zeta\delta_{ik}\vec{\nabla}\cdot\vec{w} \tag{1.9}$$

we deduce that the coefficients of viscosity η and ζ are directly proportional to γ_{ww} and γ'_{ww} by $\eta = \gamma_{ww}/(2T)$, $\zeta = \gamma'_{ww}/T$.

In practice it is more convenient to use the variables T and c. We therefore express the gradient of the chemical potential in the form

$$\vec{\nabla}\mu = \left(\frac{\partial\mu}{\partial c}\right)_{p,T}\vec{\nabla}c + \left(\frac{\partial\mu}{\partial T}\right)_{p,c}\vec{\nabla}T + \left(\frac{\partial\mu}{\partial p}\right)_{c,T}\vec{\nabla}p. \tag{1.10}$$

By defining the coefficients of diffusion $D = (\gamma_{ss}/\rho T)(\partial\mu/\partial c)_{p,T}$ and thermal diffusion $k_T D = (T/\rho)((\gamma_{ss}/T)(\partial\mu/\partial T)_{c,p} + \gamma_{qs}/T^2)$, we can rewrite the flux in the form

$$\vec{j}_s = -\rho D\left(\vec{\nabla}c + \frac{k_T}{T}\vec{\nabla}T + \frac{k_p}{p}\vec{\nabla}p\right)$$

$$\vec{j}_q - \mu\vec{j}_s = \left(k_T\left(\frac{\partial\mu}{\partial c}\right)_{p,T} - T\left(\frac{\partial\mu}{\partial T}\right)_{p,c}\right)\vec{j}_s - \lambda\vec{\nabla}T. \tag{1.11}$$

The law of the increase in entropy is simply expressed in the form

$$\frac{\partial}{\partial t}\int \rho s dV = \int dV \frac{\lambda\left(\vec{\nabla} T\right)^2}{T^2} + \int dV \left(\frac{\partial \mu}{\partial c}\right)_{P,T} \frac{\vec{j}_s^{\,2}}{D \rho T}$$
$$+ \int dV \left(\frac{\eta}{2T}\left(\frac{\partial w_i}{\partial x_k} + \frac{\partial w_k}{\partial x_i} - \frac{2}{3}\delta_{ik}\vec{\nabla} \cdot \vec{w}\right)^2 + \frac{\zeta}{T}\left(\vec{\nabla} \cdot \vec{w}\right)^2\right) \tag{1.12}$$

implying that the coefficients λ, D, η and ζ are positive, since the stability of a thermodynamic system implies $(\partial \mu / \partial c)_{p,\mathrm{T}}$ is positive.

Having determined the flux as a function of the gradients, the equations for the temperature, the concentration of a solute, or the velocity field of a fluid, can now be obtained starting from the equations of conservation in paragraph 1.4. They will be determined in chapter 5.

1.6 DIMENSIONAL ANALYSIS AND SELF-SIMILARITY

The dimensional analysis of model systems is the stage in which the understanding of the physical phenomena is realistically considered. The rule is to list all the realistic parameters of the model and to determine their dimension. We consider as parameters not only those explicitly contained within the equations but also those which play a role in the boundary conditions; the time and the spatial coordinates can intervene just as much as the parameters. For example, in the system MLTθ (mass, length, time, temperature) a coefficient of diffusion is written as $[D] = \mathrm{L}^2/\mathrm{T}$. Let us suppose that a given physical quantity depends on n of these parameters. We determine the k quantities of mutually independent dimensions so that the other $n-k$ parameters are expressed as a rational fraction of the preceding k. Let Π_1, Π_2, ..., Π_{n-k} be their ratios. The Π theorem of Buckingham indicates that the physical quantity adimensionalised by the correct combination of k quantities of independent dimensions is expressed in terms of $n-k$ dimensionless numbers Π_i.

Let us consider a bubble of air which evacuates the viscous fluid contained in a Hele–Shaw cell. The parameters are the viscosity of the fluid $[\eta] = \mathrm{M}/\mathrm{LT}$, the surface tension $[\gamma] = \mathrm{M}/\mathrm{T}^2$, the thickness of the fluid layer $[b] = \mathrm{L}$ and the flow $[Q] = \mathrm{L}^2/\mathrm{T}$. For a small Reynolds number the fluid inertia is negligible and the volumetric mass of the fluid does not enter into the equations explicitly. The only dimensionless numbers which can be introduced are then R/b, Qt/b^2 and $\eta Q/\gamma b$, where R is the radius of the bubble and t is the time. The law of the radius of the bubble as a function of time should therefore be of the form

$$R(t) = bf\left(\frac{Qt}{b^2}, \frac{\eta Q}{\gamma b}\right). \tag{1.1}$$

Next, let us consider the growth of a dendrite at a constant velocity U in a supercooled liquid. The parameters are the thermal diffusion coefficient, $[D] = \mathrm{L}^2/\mathrm{T}$, $[Q] = \mathrm{M}/\mathrm{LT}^2$, the latent heat given off per unit volume of solid, $[\mathrm{T}_s] = \theta$ and $[\mathrm{T}_\infty] = \theta$, respectively, the temperatures of fusion of the planar interface and the temperature of the liquid at long distances away from the crystal, $[c_p] = \mathrm{M}/\mathrm{LT}^2\theta$ the specific heat per unit volume of solid, and $[\gamma] = \mathrm{M}/\mathrm{T}^2$ the liquid–solid surface tension. The dimensionless numbers formed are then $DQ/U\gamma$, $c_p\mathrm{T}_s/Q$ and $\Delta = c_p(T_s - T_\infty)/Q$ the supercooling of the fluid. The velocity of the crystal U must therefore be of the form

$$U = \frac{DQ}{\gamma} f(c_p T_s/Q, \Delta). \tag{1.2}$$

Now, let us consider the case of a planar flame propagating at a constant velocity U. The parameters are: $[D] = \mathrm{L}^2/\mathrm{T}$ the thermal diffusion coefficient, $[Q] = \mathrm{M}/\mathrm{LT}^2$ the latent heat given off during the reaction per unit mass of the reaction mixture, $[c_p] = \mathrm{M}/\mathrm{LT}^2\theta$ the specific heat per unit mass of the reaction mixture, $[\mathrm{T}_0] = \theta$ the temperature of the fresh gases, $[E] = \mathrm{ML}^2/\mathrm{T}^2$, the activation energy of the reaction, $[R] = \mathrm{ML}^2/\mathrm{T}^2\theta$ the ideal gas constant, $[Y]$ the mass fraction of the reactants in the fresh gas, and $[\tau] = \mathrm{T}$ and the average time elapsed between two collisions of the gas molecules. The dimensionless numbers relevant to the propagation phenomenon are $U^2\tau/D$, Y, E/RT_b, where $T_b = T_0 + Q/c_p$ is the temperature of the burnt gases, and $\beta = E(T_b - T_0)/RT_b^2$ the reduced activation energy. The velocity of the flame propagation should therefore be of the form

$$U = \sqrt{\frac{D}{\tau}}\, f\left(Y, \frac{E}{RT_b}, \beta\right). \tag{1.3}$$

As in other physical problems, the notion of self-similarity is essential in the problems of growth and of form. A phenomenon is said to be self-similar if its characteristics at a given moment can be determined by its same characteristics at another moment by a similarity transformation. Let us consider, for example, the temperature field $T(x, t)$ in an unevenly heated medium. The field would be called self-similar if it could be put into the form $T(x, t) = t^\alpha \theta\,(x/t^\beta)$. In this type of solution, space and time are found to be simply related. The coefficients α and β are said to be the exponents of similarity. For example, the temperature field of the fluid in which a crystal of a stationary form is growing at a constant velocity U is of the form $T(x - Ut, y)$. If new variables $\zeta = \log x$ and $\tau = \log t$ are introduced, the temperature field can be written as $T(\log(\zeta/\tau^U), y)$.

In order to determine this type of solution we introduce the presupposed form of the field into the equations of evolution. The exponents of similarity then appear as the eigenvalues of the problem. The resolution of the problem is found to be simplified by the fact that space and time can be regrouped into a single variable. Instead of an equation of partial derivatives we integrate an ordinary differential equation. The exponents of similarity are determined as a function of other dimensionless numbers of the problem. Several possible cases can arise (Barenblatt and Zeldovich 1972).

The simplest and the rarest case is that of the so-called first kind. The dimensional analysis unambiguously indicates what the exponent of similarity should be. It generally occurs when a single dimensionless number is important. For example, in the case of the air bubble considered in this paragraph, $R(t)$ grows indefinitely with time, as does the first dimensionless number Qt/b^2. We should therefore expect a dependence of the form $R(t) = b(Qt/b^2)^{\alpha}$ at long times. If the thickness b is not an essential lengthscale in the problem, it should disappear from the relation, and $R(t) = (Qt)^{1/2}$. Thus $\alpha = 1/2$, and we verify afterwards that a self-similar solution for a velocity potential in $r/(Qt)^{1/2}$ is really the solution to the problem.

The most widespread case, especially in complex systems, is that of the second kind. The exponent of similarity is a non-linear eigenvalue which can only be determined at the same time as the solution, thus said to be an eigenfunction. For example, viscous fingering in a Hele–Shaw cell can only propagate in a stationary way if its relative size is larger than $1/2$ (McLean and Saffman 1981). The velocity U of a needle-like crystal appears as an eigenvalue of an integrodifferential equation for the stationary form of the interface. It is remarkable to note that by only considering the dimensionless numbers $c_p T_s/Q$ and Δ, we cannot find the solution of a self-similar problem raised in this way. On the other hand, a discrete number of solutions and therefore of growth velocities can be found if an additional parameter such as the coefficient of anisotropy of the surface tension ε is taken into account, even if it is of a very small order of magnitude (Kessler *et al.* 1986). So, we then arrive at the main question probed by the self-similar solutions, mainly in complex systems where there are numerous dimensionless numbers : which are the principal dimensionless numbers allowing the determination of the exponents of similarity? Even if some of them are of a small order of magnitude, they can turn out to be essential in the determination of the exponents of similarity or the growth velocity.

Beyond the simple uses, self-similar solutions are important because as long as they are stable they describe the behaviour at long times of solutions to problems set under arbitrary initial conditions. Therefore they are often associated with robust growth forms (a Saffman–Taylor finger, a dendrite, a curved flame).

1.7 THE ROLE OF FLUCTUATIONS

The above equations describe the average motion and deformations of the medium under consideration. However, due to the thermodynamic fluctuations of these media their diverse physical quantities (such as temperature, concentration of the solute, fluid viscosity, shape of the interface) fluctuate around a mean value. Hence the solutions of these equations only make sense if the variations of all the physical quantities associated with them are large compared to the amplitude of thermal fluctuations. This usually only happens at the very beginning of growth, when the deformations at the interface are of a small amplitude, of the order of their magnitude. If they do not appear at the linear level, where only the amplification rates are determined, they appear as initial conditions for the dynamic instability. So, even if the growth rate of instability of a perturbation mode is greater than that of another, the latter will only be observed if its initial condition is not too small compared to the former. The microscopic dynamics of the fluctuations may play an important role in shaping the observed morphologies because if the unstable dynamics develop for long enough, the growth form will macroscopically reflect them (Zeldovich *et al.* 1981, Bensimon *et al.* 1986, Pelcé 1986, Langer 1989).

They could appear much more subtly during the determination of certain growth forms, as in a Saffman–Taylor finger or a dendrite. Under certain parameter regimes, when the growth rate imposed on the form is different from those compatible with the stationary growth, the tip presents a discontinuity in the slope of the amplitude which is exponentially small (Kessler *et al.* 1986, Ben Amar and Pomeau 1986, Combescot *et al.* 1986). Under the usual conditions of dendritic growth this discrepancy is clearly comparable to the amplitude of a fluctuation of the shape of the interface. This observation weakens the predictability of the exact values obtained from the deterministic equations of motion.

When the fluctuations are large it is no longer possible to describe the growth by the dynamics of a continuous medium. A growth form which is representative of this case is the diffusion-limited aggregation (Witten and Sander 1981). We come back to the microscopic description of Brownian motion, in which the probability density is satisfied by the diffusion equation (Chandrasekhar 1943), and we attach the particles to the aggregate using simple rules. The resulting patterns, stochastic by construction and very jagged, are in fact fractal. These structures, although a little artificial in that they arise from computer simulations, could bring us closer to a number of very jagged growth forms seen in nature, by epitaxy under weak flux or within electrolyte aggregates (Vicsek 1992, Barabasi and Stanley 1995).

The measure of the fluctuation amplitude of a medium is therefore of utmost importance in the characterisation of pattern formation in growth. In general, the materials under investigation are slightly out-of-equilibrium, in a sense that their fluctuations are always the same as at equilibrium and therefore the theory of thermal fluctuations is sufficient to determine them (*Course of Theoretical Physics: Statistical Physics* by Landau and Lifshitz). However, it is difficult to determine the limits of its validity. For instance, a light scattering experiment in front of a moving liquid–solid interface by Güttinger and coworkers reveals an anomalous scattering of the light even though the growth rate of the crystal is not even a few μm/s .

1.8 GENERAL QUESTIONS

How can we determine the patterns, the dimensions and the growth rate of these objects, as well as possible transitions between different patterns?

One must first determine the mechanisms which control the observed pattern formation in growth.

What is the origin of viscous fingering, crystalline microstructures, cellular flames and jagged structures of electrolytic aggregates? Apart from being the most important question, it is also the most time-consuming. In undertaking such a broad question the essential contributors to the research are generally very specialised. They come from specific sectors of applied science: flow of porous media for viscous fingers (Hill 1952), metallurgy for crystalline structures (Flemings 1974, Kurz and Fisher 1989, Billia and Trivedi 1993), combustion for flames (Strehlow 1984), batteries for electrolytic aggregates (Bockris and Reddy 1970). These mechanisms use the same basic physics and chemistry which we will explore in chapters 2 – 4.

Once the basic mechanisms have been determined we can formulate more quantitative questions. The inanimate structures studied here arise from instabilities of simpler structures. As discussed in chapter 5, regular geometries such as planar, circular and spherical are possible, yet unstable. They therefore degenerate into more complex forms, which grow according to their degree of stability. The analysis of the linear stability of simple forms shown in chapter 6 gives much more information about the nature of the mechanisms which come into play. The determination of the limits of stability, in other words the curve fitted to control parameters which separate the stable and the unstable domains, the critical perturbation wavelengths which model the form at the first instance and the analysis of the linear growth rate of perturbations as a function of their wave number, or azimuthal number (the dispersion relation), gives validity to the adopted mechanisms.

Then comes the problem of robust structures that develop following primary instabilities: fingers, lobes, dendrites, and as many shapes as are extended at a given velocity. The velocity field, the temperature, the ionic concentration and the electric potential self-organise in a way that now generates a polarity with an approximately constant growth velocity of tips. Chapters 7 and 8 derive this growth velocity as a function of the control parameters.

In chapter 9 we investigate the question of their stability. As we have previously mentioned, these forms are robust and therefore much more stable than the simple forms at the start. Even if a secondary branching is often present at the sides of the extended form, reminiscent of the initial primary instability, its tip is still a robust structure for a wide range of control parameters. Hence, the interpretation of a break-up of a Saffman–Taylor finger, or a flame made up of two or more parts when its width becomes too large, or a secondary branching of dendrites, have repercussions on the general understanding of pattern formation in growth. According to the systems under investigation, there could be other instabilities which manifest themselves as a non-stationary motion of the tip.

As we show in chapter 10, the analysis of the evolution of the growth forms can only be done numerically due to the mathematical difficulties of the growth models. The simulations of viscous fingers, dendrites and flames reproduce similar patterns to those observed experimentally. A difference may arise, however, from the different nature of the fluctuations which are introduced into the simulations. The origin of these artificial fluctuations is the discretisation of equations. In the range of parameters where the pattern is very sensitive to the fluctuations they can lead to quantitatively different results to those observed in nature.

In chapter 11, we take on the analysis of stochastic patterns. The analysis of these complex structures is purely numerical at present. These patterns are generally characterised by a fractal dimension, the probability density and the exponents of the scaling relations. The theoretical part of the chapter is limited to diffusion limited aggregates, insisting on analogies with the fingers and dendrites discussed in previous chapters.

Finally, in chapter 12, we discuss the applications of the developed concepts of pattern formation in the growth of inert matter to biological forms growing from the tips. These systems coincide with certain unicellular algae, filament fungi, and even neurons, which thus initiate a framework for the development of a quantitative theory of morphogenesis.

For the sake of coherence we favour the analysis of unique patterns, which are at the heart of the problem of morphogenesis. The more difficult problem of interactions between growth forms, such as the movement of a stream of droplets in a Hele–Shaw cell, the late coalescence of crystalline germs, the dynamics of cellular interfaces (Flesselles *et al.* 1991, Cross and Hohenberg 1993), the coalescence of fractal aggregates (Weitz and Oliveira 1984, Jullien and Botet 1987), or the effects of the groups observed in the fertilised eggs of a Fucus, is just as important and remains topical. However, certain aspects of the study of interactions are beyond the scope of this book, but play a part in understanding the more general problem of out-of-equilibrium systems.

BIBLIOGRAPHY

[1] A.L. Barabasi, H.E. Stanley, Fractal Concepts in Surface Growth, Cambridge University Press, 1995.

[2] G.I. Barenblatt, Ya.B. Zeldovich, Self-Similar Solutions as Intermediate Asymptotics, *Ann. Rev. Fluid. Mech*, **4**, p. 285, 1972.

[3] M. Ben Amar, Y. Pomeau, 'Theory of Dendritic Growth in a Weakly Undercooled Melt', *Europhys. Lett.*, **2** (4), p. 307, 1986.

[4] D. Bensimon, L.P. Kadanoff, S. Liang, B.I. Shraiman, C. Tang, 'Viscous Flow in Two Dimensions', *Rev. Mod. Phys.*, **58**, p. 977, 1986.

[5] B. Billia, R. Trivedi, Handbook of Crystal Growth, Elsevier science publishers, Amsterdam, vol. 1, chap. 14, 1993.

[6] J.O.M. Bockris, A.K.N. Reddy, Modern Electrochemistry, Plenum/Rosetta edition, 1970.

[7] B. Caroli, C. Caroli, B. Roulet, Instabilities of Planar Solidification Fronts, in Solids far from Equilibrium, C. Godrèche ed., Cambridge University Press, Alea Saclay, 1992.

[8] S. Chandrasekhar, 'Stochastic Problems in Physics and Astronomy', *Rev. Mod. Phys.*, **15**, p. 1, 1943.

[9] R. Combescot, T. Dombre, V. Hakim, Y. Pomeau, A. Pumir, 'Shape Selection of Saffman–Taylor Fingers', *Phys. Rev. Lett.*, **56**, p. 2036, 1986.

[10] M.C. Cross, P.C. Hohenberg, 'Pattern Formation Outside of Equilibrium', *Rev. Mod. Phys.*, **65**, p. 864, 1993.

[11] P. Curie, Œuvres, Paris, Gauthier-Villars, 1908. Reedition, éditions des archives contemporaines, Gordon and Breach Science publishers, 1984.

[12] D'Arcy Wentworth Thompson, On Growth and Form, first published in 1917, abridged edition, Cambridge University Press, 1961, translated into French by D. Teyssié, Forme et croissance, Seuil, 1994.

[13] S.R. De Groot, P. Mazur, Non Equilibrium Thermodynamics, North Holland, Amsterdam, 1969.

[14] M.C. Flemings, Solidification Processing, McGraw Hill, 1974.

[15] J.M. Flesselles, A.J. Simon, A. Libchaber, 'Dynamics of One-Dimensional Interfaces: an Experimentalist's View', *Adv. Phys.*, **40**, p. 1, 1991.

[16] V. Fleury, Arbres de pierre, Flammarion, 1998.

[17] J. Fourier (1822) Théorie analytique de la chaleur, 1822, reedited by the éditions Jacques Gabay 1988.

[18] P. Glansdorff, I. Prigogine, Thermodynamic Theory of Structure, Stability and Fluctuations, Wiley-Interscience, 1971.

[19] H. Güttinger, J.H. Bilgram, W. Känzig, 'Dynamic Light Scattering at the Ice Water Interface during Freezing', *J. Phys. Chem. Solids*, **40**, p. 55, 1979.

[20] S. Hill, 'Channelling in Packed Columns', *Chem. Eng. Sci.*, **1**, p. 247, 1952.

[21] S.C. Huang and M.E. Glicksman, 'Fundamentals of Dendritic Solidification – I. Steady State Tip Growth; II. Development of Sidebranch Structure', *Acta Metall.*, **29**, p. 701, 1980.

[22] F. Jacob, La logique du vivant, Paris, Gallimard, 1970.

[23] R. Jullien, R. Botet, Aggregation and Fractal Aggregates, Word Scientific, Singapore, 1987.

[24] D. Kessler, J. Koplik, H. Levine, 'Steady State Dendritic Crystal Growth', *Phys. Rev.*, **A33**, p. 3352, 1986.

[25] D. Kessler, J. Koplik, H. Levine, 'Pattern Selection in Fingered Growth Phenomena', *Adv. Phys.*, **37**, p. 255, 1988.

[26] W. Kurz, D.J. Fisher, Fundamentals of Solidification, third edition, Trans Tech publications, 1989.

[27] L.D. Landau, E.M. Lifshitz, Course of Theoretical Physics, volume 5, Statistical Physics (2nd revised and enlarged edition), Oxford, Pergamon, 1969.

[28] L.D. Landau, E.M. Lifshitz, Course of Theoretical Physics, volume 6, Fluid Mechanics (2nd English edition, revised), Oxford, Pergamon, 1987.

[29] J.S. Langer, 'Instabilities and Pattern Formation in Crystal Growth', *Rev. Mod. Phys.*, **52**, pp. 1, 1980.

[30] J.S. Langer, 'Dendrites, Viscous Fingers, and the Theory of Pattern Formation', *Science*, **243**, p. 1150, 1989.

[31] J.W. McLean, P.G. Saffman, 'The Effect of Surface Tension on the Shape of Fingers in a Hele–Shaw Cell', *J. Fluid. Mech.*, **102**, p. 455, 1981.

[32] M. Matsushita, M. Sano, Y. Hayakawa, H. Honjo, Y. Sawada, 'Fractal Structures of Zinc Metal Leaves Grown by Electrodeposition', *Phys. Rev. Lett.*, **53**, p. 286, 1984.

[33] C.L.M.H. Navier, Mémoire sur les lois du mouvement des fluides, Mémoires de l'académie des sciences, **6**, p. 389, 1922.

[34] L. Onsager, *Phys. Rev.*, **37**, p. 405, 1931.

[35] P. Pelcé, 'Dynamique des fronts courbés', thesis in Physical Sciences, university of Provence, 1986, and, 'Introduction', in Dynamics of Curved Fronts, Academic Press, New York, 1988.

[36] C. Pelcé-Savornin, J. Quinard, G. Searby, 'The Flow Field of a Curved Flame Propagating Freely Upwards', *Combust. Sci. and Tech.*, **58**, p. 337, 1988.

[37] I. Prigogine, Introduction à la thermodynamique des processus irréversibles, Éditions Jacques Gabay, 1996.

[38] P.G. Saffman, G.I. Taylor (1958), The Penetration of a Fluid into a Porous Medium or Hele-Shaw Cell Containing a More Viscous Liquid, The scientific papers of G.I. Taylor, **4**, p. 311, 1958, Cambridge University Press, 1971.

[39] E. Schrödinger, What is life? Mind and Matter, Cambridge University Press, 1944.

[40] G.G. Stokes, 'On the Theories of the Internal Friction of Fluids in Motion', *Camb. Trans.*, **8**, p. 287, 1845.

[41] R.A. Strehlow, Combustion Fundamentals, McGraw Hill, New York, 1984.

[42] T. Vicsek, Fractal Growth Phenomena, second edition, World Scientific, 1992.

[43] D.A. Weitz, M. Oliveiria, 'Fractal Structures formed by Kinetic aggregation of aqueous gold colloids', *Phys. Rev. Lett.*, **52**, p. 1433, 1984.

[44] T. A. Witten, L.M. Sander, 'Diffusion-Limited Aggregation, a Kinetic Critical Phenomenon', *Phys. Rev. Lett.*, **47**, p. 1400, 1981.

[45] Ya.B. Zeldovich, A.G. Istratov, N.I. Kidin, V.B. Librovich, 'Flame Propagation in Tubes: Hydrodynamics and Stability', 1980, in Selected Works of Ya.B. Zeldovich, Vol. I Chemical Physics and Hydrodynamics, p. 459, Princeton University Press, 1992.

CHAPTER 2

Surface tension

2.1 SURFACE TENSION

The origin of interfacial tension is the interaction energy between the molecules in the two media present. The first approach by Laplace involves calculating the energy required to divide two parts of an infinite fluid into two parts with a planar separation surface. Let $V(\vec{r})$ be the interaction potential between two molecules of the fluid. This potential is supposed to be attractive and represents short-range interactions. The interaction energy of a volume element of a fluid of density ρ with a semi-infinite space of the same fluid at a distance z (figure 2.1) is

$$W(z) = \rho^2 \int_{z' > z} V(\vec{r}) dx dy dz' . \tag{2.1}$$

The interaction energy per unit surface area between two semi-infinite spaces, z apart is then

$$U(z) = \int_z^{+\infty} W(z') dz' . \tag{2.2}$$

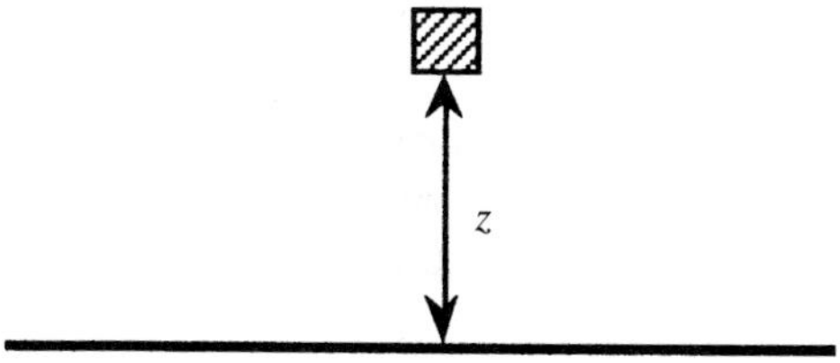

Figure 2.1 Sketch representing the interaction between a volume element of fluid and a semi-infinite space.

We deduce the energy required to create an interface of unit surface area,

$$\gamma = \frac{1}{2}(U(\infty) - U(0)) = -\frac{\rho^2}{2}\int_0^{+\infty} zdz \int V(\vec{r})dxdy$$
$$= -\frac{\rho^2}{8}\int rV(\vec{r})d\vec{r}. \tag{2.3}$$

Since the force is attractive, the potential energy V is negative, the surface tension γ is positive and has the same dimension as an energy per unit surface area $(\gamma = \mathrm{M}/\mathrm{T}^2)$.

To calculate the interfacial tension γ_{ij} between two phases i and j two semi-infinite spaces have to be created from the phases i and j and have to be reassembled thereafter. γ_{ij} will then include three terms: one term from equation (2.3) for each phase i and j and a term corresponding to the interaction between the two phases. We can deduce the expression of the interfacial tension γ_{ij}

$$\gamma_{ij} = -\frac{\rho_i^2}{8}\int rV_{ii}(\vec{r})d\vec{r} - \frac{\rho_j^2}{8}\int rV_{ii}(\vec{r})d\vec{r} + \frac{\rho_i\rho_j}{4}\int rV_{ij}(\vec{r})d\vec{r}\ , \tag{2.4}$$

where the last term is calculated in a similar way as in equation (2.3). Even if this last term is negative the interfacial tension is always positive since the integrals of the interaction energies can be written as a product k_ik_j.

To obtain the order of magnitude of an interfacial tension we must reconsider the nature of the interaction energy $V(\vec{r})$. As far as the liquid and vapour phases are concerned the potential is governed by Van der Waals forces at long range. The order of magnitude of V_{ij} is then $-e^2a_0^5/4\pi\varepsilon_0 d^6$ where d is the intermolecular distance, a_0 the Bohr radius and e the charge of the electron. Thus the surface tension γ is of the order of $(a_0/d)^5(e^2/4\pi\varepsilon_0 d)/d^2$. If $d \approx 4a_0$ and given that $e = 1.6\times10^{-19}$ C, $\varepsilon_0 = 8.854\times10^{-12}\ \mathrm{C^2/J\cdot m}$, $e^2/4\pi\varepsilon_0 = 2.3\times10^{-28}\ \mathrm{J\cdot m}$, we obtain $\gamma \approx 5\times10^{-2}\ \mathrm{J/m^2}$. In the case of an interface with a solid phase the surface tension can be far higher since the cohesive energy of a solid phase can result from ionic or covalent bonds. Table 2.1 reports the values of some interfacial tensions between non-miscible fluids, between the liquid and vapour phases of the same material and between the hexatic and isotropic phases of a liquid crystal.

	air/water	water/oil	air/oil	SCN	PVA	Al	Cu	Het/iso
γ $\mathrm{J/m^2}$	7.2×10^{-2}	1.5×10^{-2}	3×10^{-2}	8.9×10^{-3}	2.8×10^{-3}	9.3×10^{-2}	1.8×10^{-1}	5.2×10^{-4}

Table 2.1 Numerical values of interfacial tensions.

As previously mentioned interfacial tension can be defined as the energy required to create an interface of unit surface area. It is clear that this definition is meaningless if the two blocks are separated by a distance exceeding the interaction range of the Van der Waals forces. In that case, if we want to rate the energy of a thin liquid film we have to reconsider the energy per unit surface area of two blocks z apart, as defined by equation (2.2). Thus for a fluid film of thickness h lying on a solid substrate we introduce the notion of disjoining pressure $\Pi(h)$ (Deryagin 1940) related to the energy per unit surface area of film.

$$U(h) = \int_h^{+\infty} \Pi(\xi)d\xi . \tag{2.5}$$

Consider the potential $V(r) \approx 1/r^6$, already used in the integrals (2.1) and (2.2), leading to an energy $U(h)$ per unit surface area, proportional to $d^2\gamma/h^2$ and then to a disjoining pressure $\Pi(h)$ proportional to $d^2\gamma/h^3$ where d is of the same order as the intermolecular distance. For ranges larger than the characteristic length of light absorption in the medium the effects of retarded potentials have to be taken into account and $\Pi(h)$ is inversely proportional to h^4 (Landau and Lifshitz 1969). This energy is generally positive and the Van der Waals forces tend to thicken the film.

The presence of adsorbed molecules at the surface of a fluid – surfactants – modifies the properties of the interface. In particular the interfacial tension becomes a function of the concentration of these molecules. Amphiphilic molecules are an important class of surfactants. They consist of a hydrophilic polar head – ionic or not – and a hydrophobic tail made of a carbon backbone. They have a high molecular weight in general. When they are in an aqueous solution they preferentially gather together at the free surface, the polar head being in the water and the hydrophobic tail pointing towards the outside. Let μ_l and μ_s be the chemical potentials of the solute in the volume and at the surface of the fluid, respectively. They are determined by the help of the corresponding concentrations by the relations

$$\mu_l = \mu_l^0 + kT\mathrm{Log}c_l, \quad \mu_s = \mu_s^0 + kT\mathrm{Log}c_s, \tag{2.6}$$

so that at thermodynamic equilibrium, when the chemical potentials are equal, the ratio of the concentrations between the surface and the volume of the fluid is

$$\frac{c_s}{c_l} = \exp((\mu_l^0 - \mu_s^0)/kT). \tag{2.7}$$

Since $\mu_l^0 > \mu_s^0$ the surface concentration is far higher than the volume concentration. Thus, if a surfactant at low concentration is progressively added to a volume of water, almost all the surfactant is concentrated at the free surface of the fluid. At concentrations higher than the critical micellar concentration (CMC) the monomers contained in the volume of the fluid can gather and form spherical aggregates called micelles. They have a lower energy per monomer since all the polar heads of the monomers are turned towards the outside and therefore interact directly with the fluid whereas the hydrophobic tails in the centre of the aggregate do not have any interaction with the fluid (Israelachvili 1985). The concentration of monomers at the surface of the fluid saturates and the surface tension hardly depends on the concentration of surfactant anymore.

2.2 THE LAPLACE LAW

If the separation surface between the two media is curved the pressures on either side of this surface are different. To determine this pressure difference we write down the condition for thermodynamic equilibrium between the two interacting media. Let us assume first that medium 1 is fully contained within medium 2 which is supposed to be at a constant external pressure p_2. The thermodynamic potential to be minimised is then $G = F_1 + F_2 + p_2 V + \gamma S$ where F is the thermodynamic potential at a constant temperature of each phase, V is the total volume and S the separation surface. Hence the condition for thermodynamic equilibrium is written $\delta G = (p_2 - p_1)\delta V_1 + \gamma \delta S = 0$. Let us consider the normal direction to the point of the still surface and let us introduce $\delta\xi$ as the segment of this normal cut by the two surfaces infinitesimally close to each other (figure 2.2a). The change in the elementary surface dS is written as

$$\delta(dS) = \left(\frac{1}{R_1} + \frac{1}{R_2}\right) dS\delta\xi , \tag{2.1}$$

where R_1 and R_2 are the two principal radii of curvature at the given point of the surface. They are assumed to be positive if the surface of the first medium is convex. The condition of equilibrium is expressed as

$$\int dS\delta\xi\left((p_1 - p_2) - \gamma\left(\frac{1}{R_1} + \frac{1}{R_2}\right)\right) = 0 , \tag{2.2}$$

where the integral is extended to the whole unperturbed surface. As this condition must be satisfied whatever the infinitesimal displacement is, the integrand must be null. This leads to the Laplace law:

$$p_1 - p_2 = \gamma\left(\frac{1}{R_1} + \frac{1}{R_2}\right) . \tag{2.3}$$

Hence the pressure of the medium with a convex surface is higher.

Let us assume now that the liquid–vapour interface ends on a planar solid wall. The displacement of the extreme point $\delta\xi(0)$ is linked to a change in the free energy per unit length of contact line $\delta G = (\gamma_{SL} - \gamma_{SV} + \gamma_{LV}\cos\theta)\delta\xi(0)$. The energy minimum is then reached for the equilibrium contact angle θ_e such that $\cos\theta_e = (\gamma_{SV} - \gamma_{SL})/\gamma_{LV}$ (figure 2.2b). It is the boundary condition for an equilibrium structure on the solid wall. As stated above the interfacial tension of an interface with a solid phase has a higher order of magnitude than the interfacial tension of a liquid–solid interface. Nevertheless the equilibrium angle involves the difference between two of these interfacial tensions, and the dominant part, which is deduced from the cohesive energy of the solid, disappears. So, the ratio determining the cosine of the equilibrium angle has an order of magnitude of one. When the absolute value of this ratio is higher than one, equilibrium is impossible and the fluid fully wets the wall.

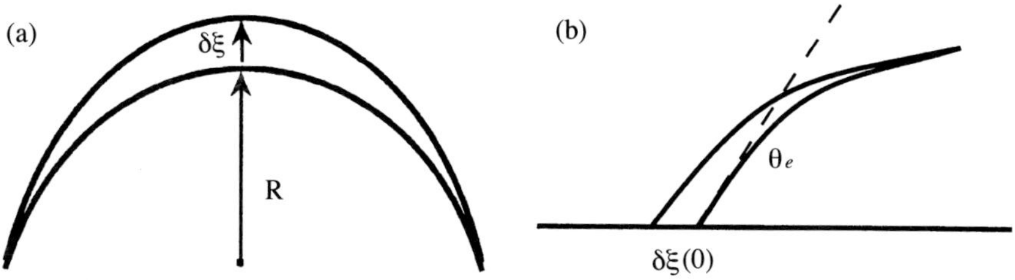

Figure 2.2 (a) Infinitesimal displacement of an interface; (b) infinitesimal displacement with a contact point on a wall.

The Plateau borders (Plateau 1873, D'Arcy Thompson 1917, for instance) are the revolution equilibrium surfaces which an interface between two fluids is likely to adopt in the absence of external forces. In this case the pressure of each medium is constant and the surface satisfies the following equation

$$\frac{1}{R_1} + \frac{1}{R_2} = \text{Cst} . \tag{2.4}$$

There are six possible geometries: the plane, the sphere, the cylinder, the catenoid, the onduloid and the nodoid. Apart from the sphere they all have infinite dimensions. They can be produced experimentally, but truncated and supported by metal frames at the borders.

When the fields are in a gravity field of magnitude g and when the density of the external field is negligible in comparison to the density ρ of the inner field, the pressure p_2 is constant and $p_1 = \text{Cst} - \rho g z$, the z-axis being pointing vertically upwards. Hence the equilibrium condition of the free surface becomes

$$\frac{1}{R_1} + \frac{1}{R_2} + \frac{\rho g}{\gamma} z = \text{Cst} . \tag{2.5}$$

This equation introduces a characteristic length, $l = (\gamma/\rho g)^{1/2}$, called the capillary length, over which the surface bends.

Appendix: Plateau borders for which the internal pressure is higher than the external pressure.

Considering the surfaces of revolution of equation $z(r)$ alone we can write

$$\frac{1}{R_1} = -\frac{d}{dr}\left(\frac{\frac{dz}{dr}}{\sqrt{1+\left(\frac{dz}{dr}\right)^2}}\right), \frac{1}{R_2} = -\frac{\frac{dz}{dr}}{r\sqrt{1+\left(\frac{dz}{dr}\right)^2}}$$

and equation (2.4) can be integrated once to obtain

$$\frac{r\frac{dz}{dr}}{\sqrt{1+\left(\frac{dz}{dr}\right)^2}} = \frac{Cr^2}{2} + C',$$

where C and C′ are constants. Let us introduce r_1 and r_2, the minimum and maximum radii of the surface. Two classes of curves can be obtained depending on whether dz/dr is positive or negative at r_1. In the first case we obtain

$$C = \frac{2}{r_1 + r_2}, \; C' = \frac{r_1 r_2}{r_1 + r_2}$$

and

$$\frac{dz}{dr} = \frac{r^2 + r_1 r_2}{\sqrt{(r^2 - r_1^2)(r_2^2 - r^2)}}$$

the solutions of which are

$$z(r) = \left(r_2\left(E\left(c, \frac{\pi}{2}\right) - E(c, \phi)\right)\right) + \left(r_1\left(F\left(c, \frac{\pi}{2}\right) - F(c, \phi)\right)\right).$$

Here,

$$F(c, \phi) = \int_0^{\phi} \frac{d\phi'}{\sqrt{1 - c^2 \sin^2 \phi'}}, \quad E(c, \phi) = \int_0^{\phi} \sqrt{1 - c^2 \sin^2 \phi'}\, d\phi'$$

are elliptic functions of the first and second orders and

$$c = \sqrt{\frac{r_2^2 - r_1^2}{r_2^2}}, \; \phi = \sin^{-1}\sqrt{\frac{r_2^2 - r^2}{r_2^2 - r_1^2}}.$$

These are onduloid. These curves are similar to sine curves. They oscillate parallel to the z-axis between radii r_1 and r_2 (figure 2.3).

Several limiting cases can be obtained: when $r_1 = r_2$ the surface is a cylinder; when $r_1 = 0$ the surface is a sphere. When r_2 tends to infinity in equation (2.5) we obtain the equation of a small chain

$$r = r_1 \cosh\left(\frac{z}{r_1}\right)$$

The corresponding surface is a catenoid (figure 2.4) and corresponds to the case $C = 0$, which means that the pressure is the same on both sides of the interface.

Figure 2.3 Onduloids.

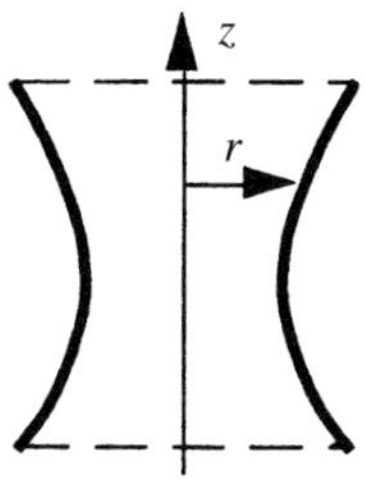

Figure 2.4 Catenoid.

Appendix: Determination of the free surface of a fluid in a gravity field near a planar wall.

As we have seen above, the wall imposes an equilibrium contact angle θ_e at the surface whereas the free surface of the fluid is planar far away from this wall. To determine the profile of the free surface, equation (2.5) has to be integrated with R_1 being infinite and $R_2 = -(1+z'^2)^{3/2}/z''$. We obtain

$$-\frac{\frac{d^2z}{dx^2}}{\sqrt{\left(1+\left(\frac{dz}{dx}\right)^2\right)^{3/2}}}+\frac{z}{l^2}=0$$

and then

$$\frac{1}{\sqrt{1+\left(\frac{dz}{dx}\right)^2}} = 1 - \frac{z^2}{2l^2}$$

since $z = 0$, $dz/dx = 0$ far away from the wall. The shape of the free surface is obtained after a second integration (figure 2.5):

$$x = -l\cosh^{-1}\left(\frac{2l}{z}\right) + 2l\sqrt{1-\frac{z^2}{4l^2}} + x_0$$

where the constant of integration x_0 is determined by the condition $dz/dx(0) = -\cot\theta_e$. In particular the height of the fluid near the wall is $h = l(2(1-\sin\theta_e))^{1/2}$.

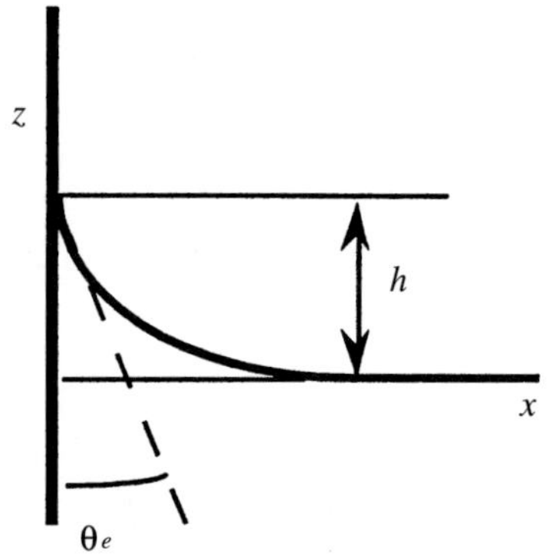

Figure 2.5 Shape of a meniscus near a vertical wall.

Appendix: An interface between two fluids is confined between two parallel plates at a distance b. Determining the pressure jump at the interface as a function of its deformation (Park and Homsy 1985).

We suppose that two fluids fully wet the walls so that the shape of the interface is a semi-cylinder of radius $b/2$ in the xz plane. When the axis of the cylinder bends in the y direction the shape of the interface in the xz plane is also changed, which leads to the creation of an extra capillary effect. Thus the shape of the interface $z = h(x, y)$ satisfies the equation

$$\frac{\partial}{\partial x}\left(\frac{\frac{dh}{dx}}{\sqrt{1+\left(\frac{dh}{dx}\right)^2+\left(\frac{dh}{dy}\right)^2}}\right) + \frac{\partial}{\partial y}\left(\frac{\frac{dh}{dy}}{\sqrt{1+\left(\frac{dh}{dx}\right)^2+\left(\frac{dh}{dy}\right)^2}}\right) = \frac{\Delta P}{\gamma}.$$

We are looking for a weakly curved solution in the y direction, that is, in the form

$$h(x, y) = h_0(x + b/2 - f(\varepsilon y)) + \varepsilon^2 h_2(x, y)$$

where $z = h_0(x)$ is the semicircle of radius $b/2$ and of the equation

$$h_0(x-x_0) = \frac{b}{2}\left(1-\sqrt{1-\frac{4}{b^2}(x-x_0)^2}\right)$$

where the vertex is at $x = f(\varepsilon y)$ and its contact point with the wall is at $x_0 = f(\varepsilon y) - b/2$. We then develop the pressure gap in the form $\Delta P = 2\gamma/b + \varepsilon^2 \Delta P_2$ and we determine h_2 by identifying the terms of order ε^2 in the equation of the interface. If we use the boundary condition on the wall $h_2(f-b/2, y) = 0$ we obtain

$$\Delta P_2 = -\frac{1}{4}\gamma\pi f''.$$

We deduce the pressure of the fluid at the interface

$$p = -\gamma\left(\frac{2}{b}+\frac{\pi}{4}\frac{1}{R}\right)$$

where R is the radius of curvature of the line of the joining vertices.

2.3 INTERFACE STRUCTURES

In the previous chapter we considered an interface between two non-miscible fluids. In this part we will consider an interface between two phases of the same material. The interface can now move not only by the motion of the fluids but above all by the transfer of molecules from one phase to another, with both phases remaining at rest. The structure of an interface in general varies over a few interatomic distances depending on its thickness, and its description cannot be achieved by a thermodynamic approach. On the other hand, near a critical point of a second order phase transition, the order parameter changes over distances far larger than the interatomic distance and a description of continuum mechanics can be used. We write the expression of the free energy of the non-uniform system (Van der Waals 1893, Cahn and Hilliard 1958) as

$$F = \int_V dV\left(f(\rho) + K\left(\vec{\nabla}\rho\right)^2\right). \tag{2.1}$$

In this expression ρ is the order parameter of the phase transition, such as the fluid density if we consider a liquid–vapour transition, $f(\rho)$ is the free energy per unit volume of the homogeneous phase and K is the coefficient of the lowest derivative which takes into account the density variations. As shown in figure 2.6, when the phases coexist, if μ is the chemical potential of the system the function $f-\mu\rho$ has two equal minima at ρ_1 and ρ_2, the densities of the respective phases. The density profile in the interface must then link ρ_1 and ρ_2 along x and ensure a constant chemical potential $\mu = \delta F/\delta\rho$ throughout the system. From the variation of the free energy

$$\delta F = \int_V dV\left(\frac{df}{d\rho}(\delta\rho) + 2K\left(\vec{\nabla}\rho\right)\cdot\left(\vec{\nabla}\delta\rho\right)\right)$$
$$= \int_V dV\delta\rho\left(\frac{df}{d\rho} - 2K\Delta\rho\right) \tag{2.2}$$

we deduce the equation for the order parameter

$$\mu = \frac{df}{d\rho} - 2K\frac{d^2\rho}{dx^2} \tag{2.3}$$

which is also obtained by multiplying by $d\rho/dx$ and after integration

$$f - \mu\rho - K\left(\frac{d\rho}{dx}\right)^2 = 0 \tag{2.4}$$

where the value of $f - \mu\rho$ is fixed at zero at the densities ρ_1 and ρ_2.

The density profile is deduced by

$$x = \int_{\rho_1}^{\rho} \sqrt{\frac{K}{f - \mu\rho}}\, d\rho \,. \tag{2.5}$$

By definition the surface energy is the free energy associated with the previously calculated density profile from which we subtract the energy that the system would have if the density was uniform. We deduce

$$\gamma = \int_{-\infty}^{+\infty} dx\left(f - \mu\rho + K\left(\frac{d\rho}{dx}\right)^2\right)$$
$$= 2K\int_{-\infty}^{+\infty} dx\left(\frac{d\rho}{dx}\right)^2. \tag{2.6}$$

For a temperature T near the critical temperature T_c, for which $\rho_1 = \rho_2 = \rho^*$, we assume that we can expand the homogeneous free energy in the form

$$f - \mu\rho = f_0 - \alpha(T_c - T)(\rho - \rho^*)^2 + \beta(\rho - \rho^*)^4 \tag{2.7}$$

where α and β are positive coefficients, such that when T is smaller than T_c

$$\rho_{1,2} - \rho^* = \pm\left(\frac{\alpha}{2\beta}(T_c - T)\right)^{1/2} \tag{2.8}$$

the density profile can then be determined by

$$\rho - \rho^* = \left(\frac{\alpha}{2\beta}(T_c - T)\right)^{1/2} \tanh\left(\left(\frac{\alpha}{2K}(T_c - T)\right)^{1/2} x\right) \tag{2.9}$$

and the interfacial tension

$$\gamma \approx \frac{\alpha^{3/2}}{\beta} K^{1/2}(T_c - T)^{3/2}. \tag{2.10}$$

Herpin and Meunier (1974) used light scattering experiments to determine the exponent of interfacial tension, $\mu \approx 1.29$, a value rather far from the one given by equation (2.10). This anomaly must be correlated with the anomaly in the exponent of the correlation length of the homogeneous phase at equilibrium. As shown in paragraph 2.5 the Orstein–Zernike theory concerning fluctuations of such a phase actually specifies that the correlation function of the density $\langle \delta\rho(0)\delta\rho(r) \rangle$ precisely satisfies equation (2.3). The lengthscale over which this function decays, which is by definition the correlation length, must be proportional to the thickness of the interface. Thus the development of the free energy described by equation (2.9) leads to a correlation length $\xi \approx (K/(T_c - T))^{1/2}$, with exponent $\nu = -1/2$, different from the experimental value $\nu \approx -0.63$. Therefore the amplitude of the density fluctuations, which we can similarly deduce from equation (2.9), has an exponent $\beta = 1/2$, different from the experimental value $\beta \approx 0.34$. To a first approximation, if the inaccuracy in the expansion of the free energy near the critical point given by equation (2.7) alone accounts for the observed discrepancies, the expression of the surface tension determined by equation (2.6) is still valid and $\mu = 2\beta - \nu \approx 1.31$, a value now very close to the experimental value. As showed by Fisk and Widon (1969) this small difference must find its origin in the fact that K itself may depends on $T - T_c$. Note that near the critical point $f - \mu\rho \approx (\rho - \rho^*)^2/\rho^*\kappa_T$ where $\kappa_T = (\partial\rho/\partial p)_T$ is the isothermal compressibility of the fluid of measured exponent $\sigma \approx -1.21$. Thus we obtain the surface tension of the same order of magnitude as the correlation length after integrating the free energy over the thickness of the interface, that is $\gamma \approx \xi(\rho - \rho^*)^2/\rho^*\kappa_T$ of exponent $\mu = \nu + 2\beta - \sigma \approx 1.28$. It is compatible with equation (2.6) if $K \approx |T - T_c|^{2\nu - \sigma}$.

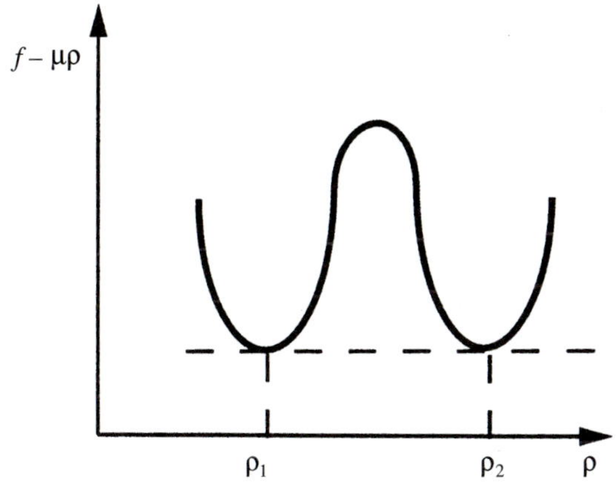

Figure 2.6 Qualitative dependence of the homogeneous part of the thermodynamic potential as a function of the density.

The liquid–solid interface has more specific characteristics because of the feature of the liquid–solid transition itself. The solid phase is different from the liquid phase not only by parameters such as density but also by the symmetries. A liquid phase is isotropic whereas a solid phase is anisotropic. For this reason the liquid–solid transition has to be discontinuous, contrary to the liquid–gas transition which can be continuous for specific values of temperature and pressure.

The anisotropy of the crystalline phase necessarily has an impact on the anisotropy of the liquid-solid surface tension since the density of the atoms in a given crystallographic plane of the surface, as well as the interaction energy of the atoms of that plane with the liquid, depend on the orientation of that plane. This means that the surface tension γ must be a function of the vector normal to the interface $\vec{n}(n_x = \cos\theta, n_y = \sin\theta)$ for a bi-dimensional crystal. If we neglect the effects of thermal fluctuations this dependence has a very peculiar feature. Let us consider a bi-dimensional solid phase of square symmetry. A facet of given orientation is built from a succession of periodically spaced steps as shown in figure 2.7. Let us study the variations of surface tension near the horizontal face of the normal vector $(1, 0)$, where θ is the angle between a face of orientation close to the horizontal face and $\theta \approx \pm 1/n$ where n is the number of atoms per step. Let γ_0 be the surface tension of the horizontal face. If β is the extra energy of a step per unit length we simply have $\gamma = \gamma_0 + \beta/na \approx (\beta/a)|\theta|$. Thus the surface tension linearly increases with the angle of the face, and its derivative as a function of this angle has a discontinuity of gap $2\beta/a$ in $\theta = 0$ (Landau 1950). By symmetry analogous discontinuities appear in the other principal directions of the crystal, that is $\theta = \pi/2, \pi, 3\pi/2$.

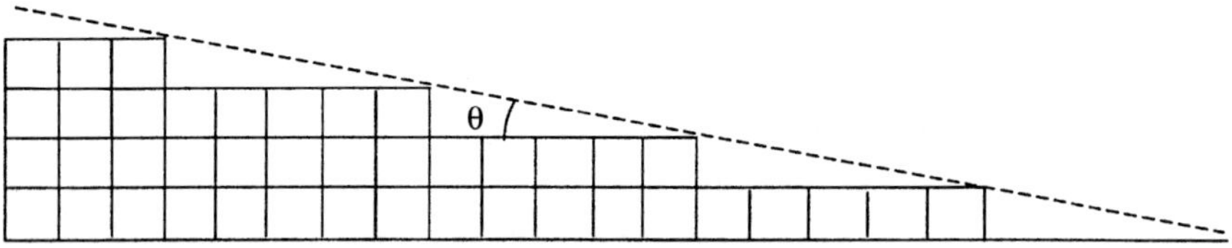

Figure 2.7 Packing of the atoms at the liquid–solid interface at $T = 0$.

At higher temperature the thermal fluctuations control the interaction energy of neighbouring atoms at the surface and annihilate the principal singularities. We then pass the temperature known as the roughening transition. Since the fluctuations of the interface shape become larger than the atomic scale the surface tension becomes regular for all angles but still depends on the orientation of the normal to the interface because of the anisotropy of the crystalline phase.

To understand the nature of the roughening transition let us first consider the following simple example developed by Jackson (1958). The liquid–solid interface is assimilated to a mixture of xN solid blocks and $(1-x)N$ fluid blocks placed onto a single layer, where N is the total amount of blocks and x the fraction of solid blocks. The classical law of statistical physics (Landau and Lifshitz 1969) gives the entropy of this mixture $\Delta S = k(xN\text{Log}(1/x) + (1-x)N\text{Log}(1/(1-x)))$. Hence if E_0 is the energy of a liquid block–solid block bond the free energy of mixing is given by

$$\Delta F = NkT(\alpha x(1-x) + x\text{Log}x + (1-x)\text{Log}(1-x)) \tag{2.11}$$

where $\alpha = zE_0/kT$ with $z = 4$, since we consider a bi-dimensional interface where each block is surrounded with four neighbours. The first term corresponds to the energy of interaction between the liquid and solid blocks whereas the other two terms are associated with the entropy of mixing. In figure 2.8, we plot the dimen-

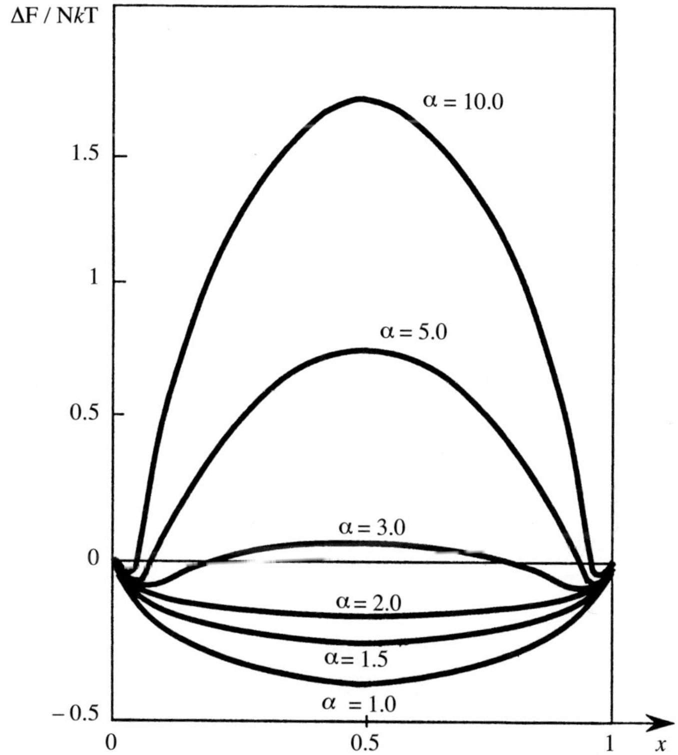

Figure 2.8 Free energy of interface as a function of x for different values of α.

sionless free energy $\Delta F/NkT$ as a function of x. We obtain two kinds of behaviour according to the value of α. For values of α larger than 2 the free energy is minimal at 0 and 1, which corresponds to formed interfaces between either liquid blocks or solid blocks. For values of α smaller than 2 the free energy is minimal for $x = 0.5$, a value for which the solid and liquid blocks coexist in the same proportions. We associate the first behaviour with smooth faces for which the interface is planar on the atomic scale. The second is associated to rough faces for which the fluctuations of the interface position are large in comparison to the atomic scale. A horizontal section of the interface on average then contains a large proportion of liquid and solid phases.

Let z_{3D} be the number of nearest neighbours of an atom included in the solid phase. Then the latent heat of fusion per atom is approximately $L \approx z_{3D} E_0$ so that the roughening transition can be characterised by the Jackson criterion (1958)

$$\frac{z}{z_{3D}} \frac{L}{kT} \approx 2 . \tag{2.12}$$

This criterion is well established experimentally. Metals have an α factor close to 1 and form rough liquid–solid interfaces. Certain transparent organic materials such as succinonitrile or CBr_4 also have an α factor close to 1. They can be directly visualised, contrary to metals, and they replace metals for the morphological characterisation of rough interfaces. Semiconductors have an α factor between 2 and 3 and the solution–solid interfaces can show both morphologies. Complex molecules and polymers have high latent heats and therefore high α factors (from 20 to 100). Consequently, their liquid–solid interfaces are faceted. The latent heat of vaporisation of metals is naturally far larger than the latent heat of fusion. The α factor, about 10 for that case, leads to gas phase faceted growth.

A more rigorous approach consists of determining the free energy of the interface with $F = -kT \mathrm{Log} Z$ where Z is the partition function

$$Z = \int D\xi(\vec{r}) \exp\left(-\frac{E(\xi)}{kT}\right) . \tag{2.13}$$

Here the notation $D\xi(\vec{r})$ means that integral (2.13) is calculated over all the possible interface shapes. The energy associated with a given interface shape is written $E(\xi) = E_c(\xi) + E_r(\xi)$ where

$$E_c(\xi) = \gamma \int_S dS\left(\sqrt{1 + \left(\vec{\nabla}\xi\right)^2} - 1\right) \approx \frac{\gamma}{2} \int_S dS\left(\vec{\nabla}\xi\right)^2 \tag{2.14}$$

is the variation of the capillary energy with the planar interface taken as a reference and

$$E_r(\xi) = \int_S V(\xi) dx dy \tag{2.15}$$

is the lattice energy. The periodic packing of atoms in the crystalline lattice creates an extra periodic energy of the same period a as the lattice, linked to the periodic potential $V(\xi)$ and proportional to $\cos(2\pi\, z/a)$ to a first approximation. Let us begin with the simplest case $V = 0$ which describes an interface between two isotropic phases such as two liquids. Noting that

$$\int_S dS\xi^2 = \int_S dS \sum_q |\xi_q|^2 \tag{2.16}$$

where ξ_q is the amplitude of the Fourier mode wavenumber q, we deduce that $E(\xi)$ is the sum of the energies $E(\xi_q) = \gamma q^2 |\xi_q|^2/2$ of the different modes. As for an ideal gas the partition function is written as the product of the partition functions of each mode

$$Z = \prod_q \int d\xi_q \exp - \frac{\gamma q^2 \xi_q^2 S}{2kT} = \prod_q \sqrt{\frac{2\pi kT}{\gamma q^2 S}}. \tag{2.17}$$

Because of the dependency on cos $(2\pi\xi/a)$ the effect of the periodic potential is to break the independence of the different Fourier modes and therefore to make them interact. Keeping the analogy with the molecular gas means introducing an interaction between molecules which may lead to a phase transition into a liquid phase if the interaction is strong enough. Again, we show that if the lattice potential is strong enough the interface undergoes a phase transition: the roughening transition. The exact calculation of the partition function defined by equation (2.13) has not been achieved yet.

An approximation method called renormalisation and introduced by Wilson in statistical physics of critical phenomena, applied to the case of a liquid–solid interface by Chui and Weeks (1976), can describe the existence and the nature of the transition. The principle of the method is also very well described in the book by Toulouse and Pfeuty (1975). The idea is to introduce a hierarchy of scales $\Lambda_n = ab^n$ where b is a scaling factor larger than 1 and of wavenumber $q_n = (2\pi/a)/b^n$. The calculation of the integral (2.13) is done scale by scale by decomposing $D\xi = \Pi d\xi_n$, where ξ_n is an interface with wavenumbers between q_n and q_{n-1}. We begin with integrating over the interfaces with wavenumbers between $2\pi/a$ and $(2\pi/a)/b$ $(n = 0)$. The remaining integral must be calculated over interfaces with wavenumbers smaller than $(2\pi/a)/b = 2\pi/ab$, as if the unit cell of dimension ab of the underlying lattice were larger. That is where the term renormalisation comes from. The integral performed for a given scale, here $\xi_>$, gives a term analogous to the previous one, but on a larger scale $\xi_<$

$$\begin{aligned} Z &= \int D\xi_< \int D\xi_> \exp - \frac{E_c(\xi_< + \xi_>) + E_r(\xi_< + \xi_>)}{kT} \\ &= \int D\xi_< \exp - \frac{E_c'(\xi_<) + E_c'(\xi_<)}{kT}. \end{aligned} \tag{2.18}$$

The approximation consists of reproducing an energy $E'(\xi) = E_c'(\xi) + E_r'(\xi)$ having the same expression as the one given by equations (2.14) and (2.15) but with different parameters $\gamma(l)$ and $V(l)$, depending on the scaling parameter $l = \mathrm{Log}(\Lambda/a)$. This approximation will be all the more valid if the different lengthscales are similar or if b is close to one. The difference between two neighbouring lengthscales is then $d\Lambda = \Lambda(b-1)$. We derive the equations which determine the dependence of the parameters $\gamma(l)$ and $\mathrm{V}(l)$ in the appendix. These equations were first derived by Kosterliz and Thouless (1973) for the case of bidimensional spin dynamics and are written according to Nozières' notation (1992)

$$\begin{aligned} \frac{dU}{dl} &= U(2-n) \\ \frac{d\gamma}{dl} &= \frac{2\pi^4}{\gamma a^4}U^2 A(n) \end{aligned} \tag{2.19}$$

where $n = \pi T/\gamma a^2$, $U = V\Lambda^2$, and $A(n)$ is a function described in the appendix. We can iterate the integration process and obtain, for each scaling parameter, l and $Z = \exp{-F_l(\xi)/kT}$ where $F_l(\xi)$ is the free energy

$$F_l(\xi) = \frac{\gamma(l)}{2}\int_S \left(\vec{\nabla}\xi\right)^2 dS + \int_S V(l)(\xi)dS\,. \tag{2.20}$$

It is more convenient to introduce the reduced variables $x = 2/n$ and $y = 4\pi\ \ U/T$ and system (2.19) becomes

$$\begin{aligned} \frac{dy}{dl} &= 2y\left(1-\frac{1}{x}\right) \\ \frac{dx}{dl} &= \frac{y^2}{2x}A\left(\frac{2}{x}\right). \end{aligned} \tag{2.21}$$

Since the function A does not cancel out, system (2.21) shows a line of fixed points at $y = 0$. According to the first equation of (2.21) the stability of these points changes at $x = 1$. For $x < 1$ the trajectories determined by system (2.21) in the (x, y) plane are attracted by the line $y = 0$. For $x > 1$ they are repelled. The equation for the trajectories in that space is

$$\frac{dy}{dx} = \frac{x-1}{y}\,\frac{4}{A\left(\frac{2}{x}\right)} \tag{2.22}$$

that is,

$$y^2 = 8\int dx\frac{(x-1)}{A\left(\frac{2}{x}\right)} - C$$

where the integration constant C is fixed by the initial condition at $l = 0$. Near the fixed point $(x = 1, y = 0)$, $y^2 = (4/A)(x-1)^2 + C$, where $A = A(2) \approx 398$, which are hyperbolas sketched in figure 2.9. The initial condition at $l = 0$ corresponds to well-defined values of γ and V, which are γ_0 and V_0. For different temperatures, x_0 and y_0 change but the ratio x_0/y_0 remains constant as shown by the straight line in figure 2.9. According to the initial condition on this line, the trajectory is either attracted by the line of fixed points at high temperature or repelled to infinity along the line $y = (2/\sqrt{A})x$. We then deduce the two following behaviours: at high temperature y decreases as l increases such that the interface behaves as an interface between two fluids; at low temperature y tends to infinity as l increases and the lattice energy controls the capillary term. The partition function is then mainly controlled by the contribution of the planar interface. It is also the origin of the concept of the roughening transition. At high temperature the fluctuations have a large amplitude and the interface has a rough appearance. At low temperature the interface is smooth.

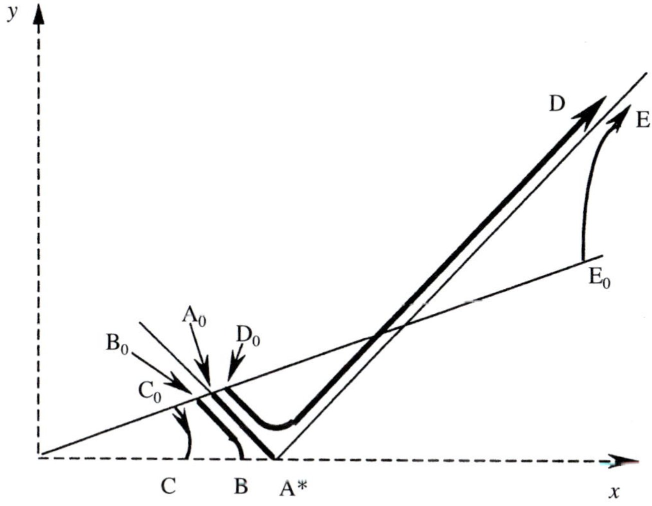

Figure 2.9 Renormalisation trajectories in the *xy* plane.

We obtain the transition temperature T_r by determining the initial condition at the intersection of the two behaviours described above. This temperature is determined by the assumption that $C = 0$ in the equation of the trajectories so that $(\sqrt{A}/2)y_0 = 1 - x_0$, that is

$$T_r = \frac{2\gamma_0 a^2}{\pi}\left(1 + \frac{t_c}{2}\right) \; ; \; t_c = \sqrt{A}\frac{2\pi^2 V_0}{\gamma_0 a^2} = \sqrt{A}\frac{y_0}{x_0} . \tag{2.23}$$

We find the same expression of T_r as in the previous calculation when the lattice energy is weak.

When $T > T_r$ the limiting point x_∞, obtained when l tends to infinity, starting from the point of the line of initial conditions at temperature T, determines the surface tension by $\gamma(T) = (\pi T/2a^2)x_\infty$.

More precisely, we obtain the constant C associated with the trajectory corresponding to an initial condition close to the initial condition of the critical trajectory by introducing the coordinates of the initial point (x_0, y_0) into the equation of the trajectory: $C = -(4/A)t(t+t_c)$, where the reduced temperature, $t = (T - T_r)/T$, is assumed to be small. We then deduce

$$x_\infty = 1 - \frac{1}{2}\sqrt{AC} = 1 - \sqrt{t(t+t_c)} \tag{2.24}$$

and the required surface tension is

$$\gamma(t) = \gamma_0\left(1 + t + \frac{1}{2}t_c - \sqrt{t(t+t_c)}\right). \tag{2.25}$$

Appendix: Determination of the renormalisation equations (2.7).

The important property of the partition function of the free energy given by equation (2.5) is that it can be decomposed into independent scales, that is

$$Z = \prod_{0<q<1/(ab)} \int d\xi_q \exp-\frac{\gamma q^2 \xi_q^2 S}{2kT} \cdot \prod_{1/ab<q<1/a} \int d\xi_q \exp-\frac{\gamma q^2 \xi_q^2 S}{2kT}$$
$$= \int D\xi_< \exp-\frac{E_c(\xi_<)}{kT} \cdot \int D\xi_> \exp-\frac{E_c(\xi_>)}{kT}.$$

so that $E_c(\xi) = E_c(\xi_<) + E_c(\xi_>)$ where $\xi = \xi_< + \xi_>$, $\xi_<$ and $\xi_>$ only contain small and large wavenumbers, respectively. When the lattice energy E_r is taken into account,

$$Z = \int D\xi_< \exp-\frac{E_c(\xi_<)}{kT} \cdot \int D\xi_> \exp-\frac{E_c(\xi_>)}{kT} \exp-\frac{E_r(\xi_< + \xi_>)}{kT}$$
$$= \int D\xi_< \exp-\frac{E_c(\xi_<) + E_r(\xi_<)}{kT}$$

where .

$$E_r(\xi_<) = -kT\mathrm{Log}\left(\int D\xi_> \exp-\frac{E_c(\xi_>)}{kT} \exp-\frac{E_r(\xi_< + \xi_>)}{kT}\right)$$

When E_r has a small amplitude the cumulants expansion gives

$$-kT\mathrm{Log}\left(\left\langle \exp-\frac{E_r}{kT}\right\rangle\right) \approx \langle E_r \rangle - \frac{1}{2kT}(\langle E_r^2 \rangle - \langle E_r \rangle^2) + \ldots$$

where $$\langle E_r \rangle = \int D\xi_> E_r(\xi_< + \xi_>) \exp-\frac{E_c(\xi_>)}{kT}.$$

is the mean value of the lattice energy over a small scale.

By using the specific form of the potential at the first order of the expansion,

$$E_{r1}(\xi_<) = -V_0\langle \int d^2r\left(\cos\left(2\pi\frac{\xi_<}{a}\right)\cos\left(2\pi\frac{\xi_>}{a}\right) - \sin\left(2\pi\frac{\xi_<}{a}\right)\sin\left(2\pi\frac{\xi_>}{a}\right)\right)\rangle$$

$$= -V_0\exp\left(\left(-\frac{2\pi^2}{a^2}\right)\langle\xi_>^2\rangle\right)\int d^2\vec{r}\cos\left(2\pi\frac{\xi_<}{a}\right)$$

since the free distribution is Gaussian. Here

$$\langle\xi_>^2\rangle = \frac{kT}{2\pi\gamma}\int_{\frac{1}{ab}}^{\frac{1}{a}}\frac{qdq}{2} = \frac{kT}{2\pi\gamma}\mathrm{Log}(b)$$

is the mean of the square of the surface fluctuations with scales between a and ab. If the renormalisation is either done over very similar scales or b is close to one, $\langle\xi_>^2\rangle \approx (kT/2\pi\gamma)(b-1)$ and

$$V_0' - V_0 \approx -\frac{\pi kT}{\gamma a^2}V_0(b-1)$$

or by using the definition of U and $dl = b-1$,

$$\frac{dU}{dl} = U(2-n)$$

At the next order of the expansion,

$$E_{r2}(\xi_<) = -\frac{V_0^2}{2kT}\frac{4\pi^2}{a^2}\int d^2\vec{r}d^2\vec{r}'\sin\left(2\pi\frac{\xi_<}{a}\right)\sin\left(2\pi\frac{\xi'_<}{a}\right)\langle\xi_>\xi'_>\rangle\,.$$

But,

$$\langle\xi_>\xi'_>\rangle = \int_{2\pi/ab}^{2\pi/a} d^2\vec{q}\exp(i\vec{q}\cdot(\vec{r}-\vec{r}'))\frac{kT}{\gamma q^2}$$

$$= \frac{kT}{2\pi\gamma}(b-1)J_0\left(\frac{|\vec{r}-\vec{r}'|}{a}\right)$$

so that

$$E_{r2}(\xi_<) = \frac{\pi V_0^2}{2\gamma a^2}(b-1)\int d^2\vec{r}d^2\vec{r}'J_0\left(\frac{|\vec{r}-\vec{r}'|}{a}\right)\left(\cos\left(2\pi\frac{\xi_< + \xi_<'}{a}\right) - \cos\left(2\pi\frac{\xi_< - \xi_<'}{a}\right)\right)\,.$$

Replacing $\cos\,((2\pi/a)(\xi_< - \xi_<'))$ with

$$\left(1 - \frac{2\pi^2}{a^2}\left((\vec{r}-\vec{r}')\cdot\vec{\nabla}\xi_<\right)^2\right)\langle\cos\left(2\pi\frac{\xi_< - \xi_<'}{a}\right)\rangle$$

we isolate in E_{r2} a term proportional to $(\nabla\xi_<)^2$ which corrects the surface tension of

$$\delta\gamma = \frac{2\pi^4 V_0^2}{\gamma a^4}(b-1)\int_0^{+\infty} d\rho\rho^3 J_0(\Lambda\rho)\exp(-2nh(\rho))$$

where .

$$h(u) = \int_0^u \frac{dk}{k}(1 - J_0(k))$$

We then deduce the renormalisation equation of the surface tension

$$\frac{d\gamma}{dl} = \frac{2\pi^4}{\gamma a^4} U^2 A(n)$$

where .

$$A(n) = \int_0^{+\infty} duu^3 J_0(u)\exp(-2nh(u))$$

2.4 THE GIBBS–THOMSON RELATION

As in paragraph 2.1, let us consider one phase (phase 1) contained within another (phase 2), under a constant external pressure p_2. The minimisation of the thermodynamic potential is always $G = F_1 + F_2 + p_2 V + \gamma S$, where F is the thermodynamic potential at a constant temperature of each phase, V is the total volume, S is the separation surface and γ is the interfacial tension between the two phases. The equilibrium condition can now be written as $dG = (p_2 - p_1)\delta V_1 + (\mu_1 - \mu_2)\delta N_1 + \gamma\delta S = 0$, where N_1 is the number of molecules in phase 1 and μ_1 and μ_2 the chemical potentials of the two phases.

The case where the medium 1 is a solid phase deserves special attention. The deformation of the droplet border generates elastic stresses which must be taken into account in the thermodynamic potential (Herring 1952). Since these stresses are generally very weak in comparison to the effects of surface tension they will be neglected in the following study. More importantly, the dependence of the surface tension on the local orientation of the surface should be taken into account

As in paragraph 2.2, the minimisation of the thermodynamic potential with respect to the surface deformation leads to the Laplace formula

$$p_1 - p_2 = \gamma\left(\frac{1}{R_1} + \frac{1}{R_2}\right) \tag{2.1}$$

but the minimisation of the same potential with respect to the number of molecules leads to

$$\mu_1(p_1, T) = \mu_2(p_2, T). \tag{2.2}$$

In the absence of surface tension the pressures are equal in both sides of the interface, and equal to p_0, assuming that the external pressure is constant. The equilibrium temperature T_0 is deduced from equation (2.2). Under the effect of surface tension both the pressure p_1 and the equilibrium temperature are changed. As long as the variations are small, equation (2.2) can be expanded around the point (p_0, T_0) to obtain

$$(p_1 - p_0)v_1 - (T - T_0)s_1 = -(T - T_0)s_2 \tag{2.3}$$

where v_1 is the specific volume of phase 1, and s_1 and s_2 the specific entropies of the two phases.

We can therefore deduce the temperature variation as a function of the interface curvature from the Laplace formula (2.1), or in other words the Gibbs–Thomson relation

$$T = T_0\left(1 - \frac{\gamma}{Q}\left(\frac{1}{R_1} + \frac{1}{R_2}\right)\right) \tag{2.4}$$

where $Q = T_0(s_2 - s_1)/v_l$ is the latent heat per unit volume of phase 1. Table 2.2 shows the numerical values of the solidification temperature and the latent heats of several materials.

	SCN	PVA	Al	Cu
T_0 K	331.2	309.1	933.6	1 358
Q J/m^3	4.7 10^7	2.1 10^7	9.5 10^8	1.6 10^9

Table 2.2 Numerical values of solidification temperatures and latent heats per unit volume of solid.

Thus, due to the presence of surface tension the equilibrium temperature of the drop is decreased with an inverse proportionality to its radius. More precisely, if T_∞ is the uniform temperature of the medium, the equilibrium radius is $R_e = 2d_{0T}/\Delta$, where $\Delta = c_p(T_0 - T_\infty)/Q$ is the amount of supercooling of the medium and $d_{0T} = (\gamma/Q)(c_pT_0/Q)$ the thermal capillary length. Some of the values are given in table 2.3.

	SCN	PVA	Al	Cu
(Q/c_p) K	23.1	11.1	368	409
d_{0T} Å	27.9	37.6	2.50	3.70

Table 2.3 Numerical values of thermal capillary lengths.

Let us consider a solute present in two phases, giving preference to the case where medium 1 is in a solid phase and medium 2 is in a liquid phase. The variation of the thermodynamic potential at a constant external pressure p_l is then written as $dG = (p_l - p_s)\delta V_s + (\mu_s - \mu_l)\delta N_s + (\mu_s' - \mu_l')\delta n_s + \gamma\delta S$, where N_s and n_s are the number of molecules of the solvent and the solute of the solid phase, respectively, and μ and μ' are the chemical potentials of the solvents and the solutes. Moreover, from the already obtained Laplace law the minimisation of the thermodynamic potential with respect to the number of molecules of the solvent and the solution then lead to the following relations

$$\mu_s(p_s, c_s, T) = \mu_l(p_l, c_l, T) \tag{2.5}$$

$$\mu_s'(p_s, c_s, T) = \mu_l'(p_l, c_l, T) \tag{2.6}$$

where c_l and c_s are the solute concentrations in the liquid and the solid. As long as the solutions are weak, or in other words of low concentrations of solute, the chemical potentials are written as

$$\begin{aligned} \mu(P, T, c) &= \mu(P, T) - kTc \\ \mu'(P, T, c) &= \mu'(P, T) + kT\text{Log}c. \end{aligned} \tag{2.7}$$

Analogous to obtaining equation (2.3) we expand equation (2.5) around the equilibrium point of the two pure phases separated by a planar wall, that is

$$(p_s - p_0)v_s - (T - T_0)s_s - kTc_s = -(T - T_0)s_l - kTc_l \tag{2.8}$$

from which we deduce, using the Laplace formula,

$$-Q\frac{(T - T_0)}{T_0} = \gamma\left(\frac{1}{R_1} + \frac{1}{R_2}\right) + \frac{kT_0}{v_s}(c_l - c_s). \tag{2.9}$$

Moreover, the equality of the chemical potentials of the solute leads to

$$kT\text{Log}\frac{c_s}{c_1} = \mu_l'(p_l, T) - \mu_s'(p_s, T) \approx \Delta(p_0, T_0) \tag{2.10}$$

that is $c_s = Kc_l$, where $K = \exp(\Delta/kT_0)$ is the partition coefficient. We can therefore rewrite equation (2.3) in a form analogous to equation (2.4), that is

$$T = T_0\left(1 - \frac{\gamma}{Q}\left(\frac{1}{R_1} + \frac{1}{R_2}\right)\right) + mc_l \tag{2.11}$$

where $m = kT_0^2(K-1)/Qv_s$ is the slope of the liquidus in the phase diagram of a binary mixture presented in figure 2.10. In general K is smaller than 1 even though the equilibrium temperature is lowered by the adjacent solute.

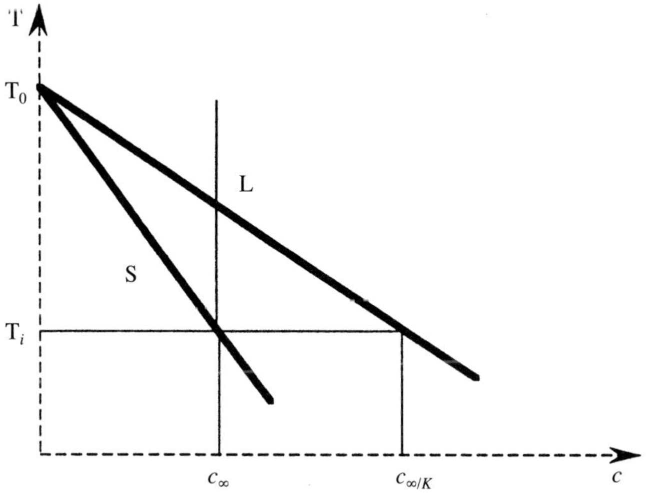

Figure 2.10 Phase diagram of a binary mixture.

In table 2.4 we indicate the values of K and m for different binary mixtures.

	SCN - Ace	Al - Cu	PVA - Eth
K	0.1	0.14	0.16
m °K/mole %	− 2.2	− 6	− 3

Table 2.4 Numerical values of K and m.

In the case of an ideal gas in equilibrium with its solution in a certain solvent,

$$\mu'(P_l, T) + kT\mathrm{Log}c_l = f(T) + kT\mathrm{Log}P_g \tag{2.12}$$

from which we deduce the Henry law

$$c_l = P_g \exp\left(\frac{f-\mu'}{kT}\right) \tag{2.13}$$

according to which the concentration of the dissolved gas is proportional to its partial pressure in the atmosphere.

The effects of anisotropy of the surface tension modify the Gibbs–Thomson relation in the following way. The surface energy included in the thermodynamic potential of paragraph 2.1 is now written in the form $\int \gamma(\vec{n})dS$, and the minimisation of the potential with respect to the deformation of the interface introduces an additional quantity $\int \delta\gamma(\vec{n})dS$ (Landau 1950, Herring 1952). Let R_1 and R_2 be the principal radii of curvature assumed to be positive if they are facing the solid.

Let ϕ_1 and ϕ_2 be the angles normal to the surface in every principal plane with a fixed direction, and du and dv the length elements taken on the surface in the principal planes. We then have:

$$\begin{aligned}\delta\gamma &= \frac{\partial\gamma}{\partial\phi_1}\delta\phi_1 + \frac{\partial\gamma}{\partial\phi_2}\delta\phi_2 \\ &= \frac{\partial\gamma}{\partial\phi_1}\frac{\partial}{\partial u}(\delta\xi) + \frac{\partial\gamma}{\partial\phi_2}\frac{\partial}{\partial v}(\delta\xi)\end{aligned} \tag{2.14}$$

where $\delta\xi$ is the displacement normal to the interface. We therefore deduce, after integration by parts,

$$\int \delta\gamma dS = \int \delta\xi dS\left(\frac{1}{R_1}\frac{\partial^2\gamma}{\partial\phi_1^2} + \frac{1}{R_2}\frac{\partial^2\gamma}{\partial\phi_2^2}\right). \tag{2.15}$$

Thus, the Laplace equation becomes

$$p_s - p_l = \frac{1}{R_1}\left(\gamma + \frac{\partial^2\gamma}{\partial\phi_1^2}\right) + \frac{1}{R_2}\left(\gamma + \frac{\partial^2\gamma}{\partial\phi_2^2}\right) \tag{2.16}$$

and the Gibbs–Thomson law for the pure liquid

$$T = T_0\left(1 - \frac{1}{Q}\left(\frac{1}{R_1}\left(\gamma + \frac{\partial^2\gamma}{\partial\phi_1^2}\right) + \frac{1}{R_2}\left(\gamma + \frac{\partial^2\gamma}{\partial\phi_2^2}\right)\right)\right). \tag{2.17}$$

In particular, if the interface is curved,

$$T = T_0\left(1 - \frac{1}{Q}\frac{1}{R}\left(\gamma + \frac{\partial^2\gamma}{\partial\theta^2}\right)\right) \tag{2.18}$$

where θ is the angle between the normal to the interface and a fixed direction. In the case of a crystal at equilibrium with a liquid phase, $T = \text{cst}$, and the radius of local curvature is proportional to the quantity $\gamma + \partial^2\gamma/\partial\theta^2$. This relation of proportionality has a simple geometric interpretation (Wulff 1901). As shown in figure 2.11, consider a point M on the polar curve $\gamma(\theta)$ at an angle θ, and the point P at an angle Φ on the line perpendicular to OM, O being the origin in the plane of the curve $\gamma(\theta)$.

MP is then the tangent in the equilibrium shape. In other words, the equilibrium shape is the envelope described by the perpendicular to the radius vector of the polar curve $\gamma(\theta)$. Indeed, if r and Φ are the coordinates of the point P, $r = \gamma(\theta)/\cos(\Phi - \theta)$. The straight line MP is a tangent to the equilibrium shape if the conditions

$$\Phi = \theta + \tan^{-1}\left(\frac{1}{\gamma}\frac{d\gamma}{d\theta}\right), \; r = \sqrt{\gamma^2 + \frac{d\gamma}{d\theta}^2} \tag{2.19}$$

are satisfied. Therefore, as expected, the radius of curvature of the equilibrium shape is written as

$$R = \sqrt{\left(\frac{dr}{d\theta}\right)^2 + r^2\left(\frac{d\Phi}{d\theta}\right)^2} = \gamma + \frac{d^2\gamma}{d\theta^2}. \tag{2.20}$$

In the singular direction, the second derivative of the surface tension is infinite and the interface is planar. Thus, the equilibrium shape is generally composed of facets in the singular directions connected by the curved parts.

In the case where the crystalline surface is only made up of facets (figure 2.12), let h_i be the distance from the centre to the facet i, of surface tension γ_i and of surface A_i. From the expression for the crystal volume, $V = (1/3)\Sigma A_i h_i$, and the change in volume, $dV = \Sigma A_i dh_i$, we deduce $dV = (1/2)\Sigma h_i dA_i$. The equilibrium shape which minimises the capillary energy at constant volume $E_c = \Sigma\gamma_i dA_i$ leads to $d(E_c - \lambda dV) = 0$, where λ is a Lagrange multiplier. Hence, from the last expression of dV, h_i/γ_i = const. This is the simplest expression of the Wulff theorem; that the distance of a facet at the crystal centre is proportional to the surface tension.

Appendix 1: Expression of the Gibbs–Thomson term for a crystal of cubic symmetry in the rough phase.

In the rough phase the surface tension of a principal direction is at a maximum because the fluctuations notably increase the surface of contact between the solid and the liquid, contrary to the case in figure 2.7.

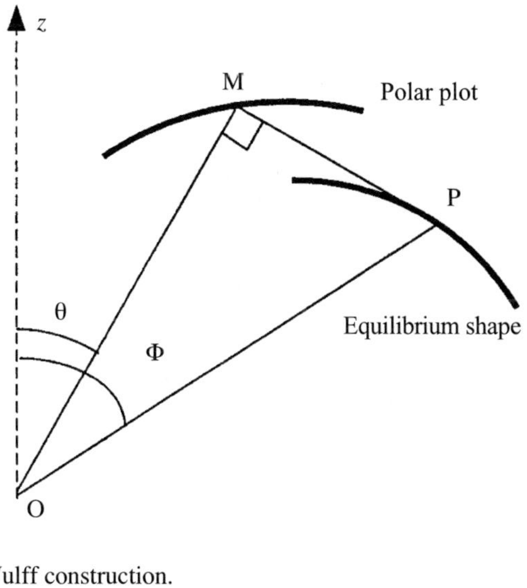

Figure 2.11 Wulff construction.

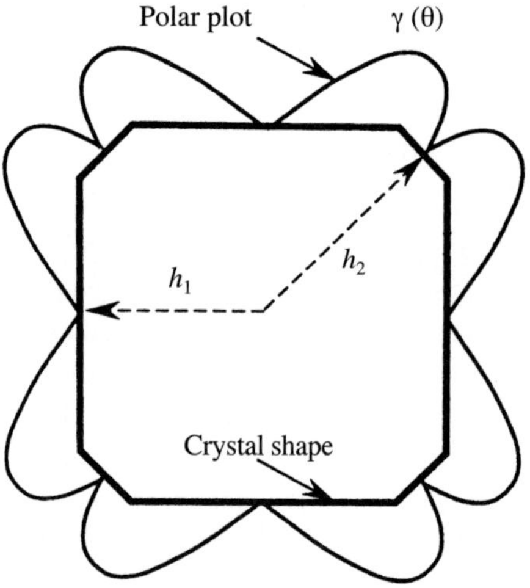

Figure 2.12 Shape of a faceted crystal

The simplest function describing the dependence of the surface energy on the crystalline direction $\vec{n}$ of a crystal of cubic symmetry is

$$\gamma(\vec{n}) = \gamma_0'(1 + \varepsilon'(n_x^4 + n_y^4 + n_z^4))$$
$$= \gamma_0'(1 + \varepsilon'(\cos^4\theta + \sin^4\theta(1 - 2\sin^2\Phi\cos^2\Phi)))$$

where ε' is the coefficient of crystal anisotropy, θ and Φ are the polar and azimuthal angles. In the simplest case of a bidimensional crystal, assumed to be in the plane $\Phi = 0$,

$$\gamma = \gamma_0(1 + \varepsilon\cos(4\theta))$$

where $\gamma_0 = \gamma_0'(1 + 3\varepsilon'/4)$ and $\varepsilon = \varepsilon'/(4 + 3\varepsilon')$. The corresponding Gibbs–Thomson factor is written as

$$\gamma + \frac{\partial^2\gamma}{\partial\theta^2} = \gamma_0(1 - \beta\cos(4\theta))$$

where $\beta = 15\varepsilon$. It is at its minimum at the axis of symmetry.

The measure of *ε* becomes effective once the equilibrium shape is observed (see for example Muschol *et al.*1992). Table 2.5 shows the values of ε and β for some materials.

	NH_4Br	SCN	PVA
ε	0.016	0.005	0.05
β	0.24	0.075	0.75

Table 2.5 Numerical values of anisotropy coefficients.

Appendix: Calculation of the force exerted by the liquid–solid interface on a sphere of radius R bathing in the liquid (Chernov et al. 1976).

If a solid sphere is placed into a liquid in the proximity of a moving liquid–solid interface, it tends to be pushed back by the interface due to a repulsive disjoining pressure. This relative motion with respect to the fluid causes a frictional force which tends to bring it closer to the interface. If the velocity of the interface exceeds a certain critical velocity which we calculate here, the frictional force prevails and the sphere is captured by the solid phase. During the capture, the interface is deformed and its curved surface takes a shape which we associate with a paraboloid

$$z(r) = R + h_0 - \frac{\nu r^2}{2R}$$

where R is the sphere radius, h_0 is the minimum distance between the sphere and the interface, R/ν is the radius of curvature of the interface and where the origin of z is taken to be at the centre of the sphere (figure 2.13). In fact, as long as the sphere is approaching the interface the generated disjoining pressure modifies the chemical potential of the liquid, manifested by a change in the local curvature of the interface. The change in free energy of the system associated with the change in the shape of the interface is written in the same way as in paragraph 2.4, except that we must include the contribution from the disjoining pressure of the liquid film between the sphere at the interface $\delta \int U(h) ds$. Minimising this change with respect to the displacement at the interface we get

$$p_s - p_l = \gamma\left(\frac{1}{R_1} + \frac{1}{R_2}\right) + \Pi(h).$$

Equation (2.3) of paragraph 2.4 enables the derivation of the modified Gibbs–Thomson relation

$$T = T_0\left(1 - \frac{1}{Q}\left(\gamma\left(\frac{1}{R_1} + \frac{1}{R_2}\right) + \Pi(h)\right)\right).$$

If the movement at the interface is sufficiently slow, the medium becomes isothermic and the disjoining pressure must equilibrate the surface pressure in the modified Gibbs–Thomson relation. Following from paragraph 2.1 the disjoining pressure is of the form A_n/h_0^n, with $n = 3$ or 4 so that

$$\frac{\gamma\nu}{R} \approx \frac{A_n}{h_0^n}$$

which determines the distance of approach of the sphere as a function of its radius. The force which the interface exerts on the sphere is then written as

$$F = 2\pi A_n \int_0^R dr \frac{r}{h(r)^n} = \frac{2\pi A_n R}{(1-\nu)(n-1)h_0^{n-1}} \approx \frac{2\pi\gamma\nu}{(1-\nu)(n-1)}\left(\frac{A_n R}{\gamma\nu}\right)^{1/n}.$$

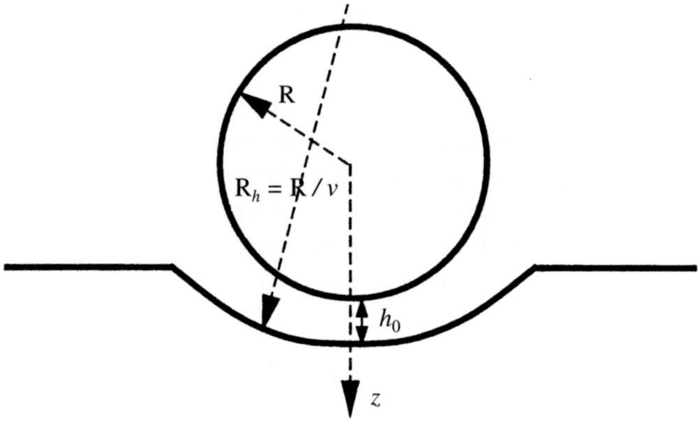

Figure 2.13 Deformation of a liquid–solid interface by the approach of a small sphere.

2.5 INTERFACE FLUCTUATIONS

Let us first determine the characteristics of the fluctuations of a medium in thermodynamic equilibrium at constant pressure and temperature, P and T. The probability of a fluctuation with a given free energy ΔF is written as

$$f_0 \approx \exp\left(-\frac{\Delta F}{kT}\right) \tag{2.1}$$

where $\Delta F = \Delta E - T\Delta S + P\Delta V$, ΔE, ΔS and ΔV are the corresponding fluctuations of energy, entropy and volume. Expanding ΔE with respect to S and V, we get

$$\begin{aligned}\Delta E - T\Delta S + P\Delta V &= \frac{1}{2}\left(\frac{\partial^2 E}{\partial S^2}(\Delta S)^2 + 2\frac{\partial^2 E}{\partial S\,\partial V}\Delta S\Delta V + \frac{\partial^2 E}{\partial V^2}(\Delta V)^2\right)\\ &= \frac{1}{2}(\Delta T\Delta S - \Delta P\Delta V)\end{aligned} \tag{2.2}$$

so that

$$f_0 \approx \exp\left(\frac{\Delta P\Delta V - \Delta T\Delta S}{2kT}\right). \tag{2.3}$$

Taking V and T as independent variables, we can write

$$\begin{aligned}\Delta S &= \left(\frac{\partial S}{\partial T}\right)_V \Delta T + \left(\frac{\partial S}{\partial V}\right)_V \Delta V = \frac{C_V}{T}\Delta T + \left(\frac{\partial P}{\partial T}\right)_V \Delta V\\ \Delta P &= \left(\frac{\partial P}{\partial T}\right)_V \Delta T + \left(\frac{\partial P}{\partial V}\right)_T \Delta V\end{aligned} \tag{2.4}$$

so that

$$f_0 \approx \exp\left(-\frac{C_V}{2kT^2}(\Delta T)^2 + \frac{1}{2kT}\left(\frac{\partial P}{\partial V}\right)_T (\Delta V)^2\right) \quad . \tag{2.5}$$

The fluctuations of temperature and volume are therefore statistically independent, giving $\langle \Delta T \Delta V \rangle = 0$, and their mean squares are determined by Gaussian integrals

$$\langle (\Delta T)^2 \rangle = \frac{kT^2}{C_V}$$
$$\langle (\Delta V)^2 \rangle = -kT\left(\frac{\partial V}{\partial P}\right)_T . \tag{2.6}$$

When the fluctuations are spatially inhomogeneous, Orstein and Zernike (1917) consider the local free energy per unit volume f and rewrite equation (2.1) with

$$\Delta F = \int (f - \langle f \rangle) dV \tag{2.7}$$

where $\langle f \rangle$ is the mean value of f, constant throughout the medium. Let $\langle \rho \rangle$ be the mean density of the medium. If we expand $f - \langle f \rangle$ as a power of $(\rho - \langle \rho \rangle)$ the first term of the expansion must be cancelled out because the total number of molecules $N = \int \rho dV = \int \langle \rho \rangle dV$ is constant. The second term is written as $(\rho - \langle \rho \rangle)^2 / 2\langle \rho \rangle \kappa_T$ where $\kappa_T = (\partial \rho / \partial p)_T / \rho$ is the isothermic thermal compressibility. In fact, $\mu = (\partial f / \partial \rho)_T$ and the thermodynamic laws lead to $(\partial \rho / \partial p)_T / \rho = -1/\rho^2 (\partial \mu / \partial \rho)_T$. In an inhomogeneous medium, the free energy must also depend on the density gradient. Thus, since the free energy is scalar, keeping only the lowest order derivative terms gives

$$f - \langle f \rangle = \frac{(\rho - \langle \rho \rangle)^2}{2\langle \rho \rangle \kappa_T} + \frac{a}{2}\left(\vec{\nabla} \rho\right)^2 \tag{2.8}$$

where a is positive. To determine the amplitude of the inhomogeneous fluctuations we expand $\rho - \langle \rho \rangle$ as a Fourier series

$$\rho - \langle \rho \rangle = \sum_{\vec{k}} \rho_k \exp(i\vec{k} \cdot \vec{r}) . \tag{2.9}$$

Using equations (2.7) – (2.2) we thus obtain

$$\Delta F = \frac{V}{2} \sum_{\vec{k}} \left(\frac{1}{\langle \rho \rangle \kappa_T} + ak^2\right) |\rho_{\vec{k}}|^2 . \tag{2.10}$$

Therefore, the Fourier coefficients of the density are statistically independent and their mean squares can be obtained from equation (2.1),

$$\langle|\rho_{\vec{k}}|^2\rangle = \frac{kT}{V\left(\frac{1}{\langle\rho\rangle\kappa_T} + ak^2\right)}. \tag{2.11}$$

The density correlation function $\theta(\vec{r}) = \langle\rho(0)\rho(\vec{r})\rangle$ is deduced from the inverse Fourier transform of equation (2.4), in other words :

$$\langle\rho(0)\rho(\vec{r})\rangle = \int dV\langle|\rho_{\vec{k}}|^2\rangle \exp(-i\vec{k}\cdot\vec{r}) = \frac{kT}{4\pi\langle\rho\rangle a}\frac{1}{r}\exp\left(-\left(\frac{1}{a\langle\rho\rangle\kappa_T}\right)^{\frac{1}{2}} r\right) \tag{2.12}$$

of corresponding correlation length $\xi = (a\langle\rho\rangle\kappa_T)^{1/2}$. Thus, the density autocorrelation function satisfies the equation

$$\frac{1}{\langle\rho\rangle\kappa_T}\theta - a\Delta\theta = 0 \tag{2.13}$$

in the same form as equation (2.3) in paragraph 2.3.

Although the Laplace law allows us to determine the mean shape of the equilibrium interface, because of the thermal fluctuations of the surrounding atoms or molecules the interface itself will fluctuate, leading to fluctuations of the capillary energy. Let us start with a planar mean shape, and let $\xi(\vec{r}, t)$ describe its fluctuation. The Boltzman–Gibbs law then gives us the probability of such a fluctuation taking place:

$$f_0 \approx \exp\left(-\frac{E(\xi)}{kT}\right) \tag{2.14}$$

where the energy associated with a given shape of interface is written as $E(\xi) = E_c(\xi) + E_r(\xi)$,

$$E_c(\xi) = \gamma\int_S dS\left(\sqrt{1 + \left(\vec{\nabla}\xi\right)^2} - 1\right) \approx \frac{\gamma}{2}\int_S dS\left(\vec{\nabla}\xi\right)^2, \tag{2.15}$$

being the change in the capillary energy with respect to a planar interface and

$$E_r(\xi) = \int_S V(\xi)dS \tag{2.16}$$

being the energy of the network. For an interface between two fluids, $V = 0$. Using

$$\int_S \xi^2 dS = \int_S \sum_q |\xi_q|^2 dS \tag{2.17}$$

where ξ_q is the amplitude of the Fourier mode of wavenumber q,

$$E(\xi) = \int_S \sum_q \frac{\gamma}{2} q^2 |\xi_q|^2 dS. \tag{2.18}$$

Since the different modes are statistically independent, $\langle|\xi_q|^2\rangle$ is deduced from equations (2.14) and (2.18):

$$\langle|\xi_q|^2\rangle = \frac{kT}{\gamma q^2}. \tag{2.19}$$

Integrating over the wavenumber gives the mean displacement of the interface

$$\sqrt{\langle\xi^2\rangle} = \sqrt{\frac{kT}{\gamma}}\sqrt{\mathrm{Log}\left(\frac{L}{a}\right)} \tag{2.20}$$

where L is the linear dimension of the interface. The dimension of the mean displacement is of the order of the interface thickness, i.e. several angströms. Indeed this quantity diverges with L, but logarithmically. Thus $\log(L/a)$ does not usually exceed 10 and the order of magnitude of the fluctuations is given by $(kT/\gamma)^{1/2}$ in table 2.6.

	SCN	PVA	Al	Cu
$\sqrt{kT/\gamma}$ Å	7.2	12.3	3.7	3.3

Table 2.6 Numerical values of $(kT/\gamma)^{1/2}$.

Let us compare this displacement with that of a virtual interface subject to thermal fluctuations of the surrounding medium. Using equation (2.6),

$$\langle(\Delta V)^2\rangle = kT\ V\kappa_T \tag{2.21}$$

where $\kappa_T = -(\partial V/\partial P)_T/V$ is the coefficient of isothermic compressibility of the medium. Thus the fluctuations of the linear dimension of a volume of the order $1/q^3$ is

$$\langle\xi^2(q)\rangle = kT\kappa_T q \tag{2.22}$$

which can be attributed to the deformation of an interface in contact with that volume. For example, if the medium is an ideal gas, $\kappa_T = 1/P$ and $\xi(\lambda) \approx (1/\lambda\rho)^{1/2}$, if the linear dimension of the volume under consideration is λ and ρ the density of the gas.

For a liquid–solid interface the two cases $T > T_r$ and $T < T_r$ must be distinguished. In the first case the energy of the network decreases with the scaling parameter, so that the interface can be considered free but of surface tension $\gamma(l)$. We can therefore write

$$\langle|\xi|^2\rangle = kT\int_1^{l_{max}} \frac{dl}{\gamma(l)} \tag{2.23}$$

where $l_{max} = \mathrm{Log}(L/a)$ is the scaling parameter related to the size L of the interface.

From equation (2.20) in paragraph 2.3, when $T < T_r$ the network energy y, made dimensionless, increases as the scaling parameter l increases. Let us define a correlation length $\xi_{//}$ as a lengthscale at which $y \approx 1$. On smaller scales than $\xi_{//}$, the energy of the network is weak and the interface is free. On larger scales though, the energy of the network is strong, confining the motion of the interface to within a distance of the order of the network unit cell a. To determine that distance, first of all we must evaluate the scaling parameter l_c for which $y \approx 1$. For a temperature slightly lower than T_r the trajectory in the (x, y) plane, of small positive constant C, first goes along the separating line of the fixed point, attains a minimum value of $y_m = \sqrt{C}$ and then distances itself again from the fixed point along the line $y = (2/\sqrt{A})(x-1)$ (figure 2.9). Thus, along the ascending portion of the trajectory

$$\frac{dy}{dl} = 2y(x-1) = \sqrt{A}y\sqrt{y^2 - C} \tag{2.24}$$

which leads to

$$l_c = \frac{2}{\sqrt{A}}\int_{\sqrt{C}}^{1}\frac{dy}{y\sqrt{y^2 - C}} \approx \frac{\pi}{2}\sqrt{t_c\frac{T_r}{T_r - T}}. \tag{2.25}$$

Since the lengthscales are related to the scaling parameter l by $\Lambda \approx a\exp l$, we get

$$\xi_{//} = a\exp l_c \approx a\exp\left(\frac{\pi}{2}\sqrt{t_c\frac{T_r}{T_r - T}}\right) \tag{2.26}$$

which indicates the unusual character of an exponential divergence of the correlation length near the transition temperature.

The mean value of the square of the fluctuation amplitude is determined by equation (2.2), but using $l_{\max} \approx \mathrm{Log}(\xi_{//}/a)$ for which $y \approx 1$:

$$\langle|\xi|^2\rangle = \frac{k\mathrm{T}}{\gamma}\mathrm{Log}(\xi_{//}/a). \tag{2.27}$$

Except near $\mathrm{T_r}$, $\xi_{//}$ is of the order of a, such that $\langle|\xi|^2\rangle$ is of the order of $k\mathrm{T}/\gamma$ and therefore of a^2 because in the usual experimental conditions, T is of the same order of magnitude as T_r. Thus, when $T < T_r$ the interface is nearly planar, that is, it is made of islands of arbitrary size, characteristic of faceted crystalline surfaces. $\xi_{//}$ appears as the characteristic thickness of the region at the edges of the islands, on which the interface altitude gains a network unit cell a. Indeed, the transverse structure of the border region is obtained in a way similar to that of Cahn and Hilliard for a liquid–vapour interface in paragraph 2.3. It minimises the free energy determined by equation (2.3) in paragraph 2.3

$$\gamma(l)\frac{d^2\xi}{dx^2} - \frac{dV(l)}{d\xi}(\xi) = 0 \tag{2.28}$$

between $\xi = 0$ and $\xi \approx a$, where the scaling parameter l is sufficiently large to contain the scale of the border thickness L. From equation (2.28) we deduce $\gamma a/L^2 \approx V/a$, where $L \approx a(\gamma/V)^{1/2}$. On the other hand, $\xi_{//}$ is defined by $y \approx 1$, that is $U = (V/\xi_{//}^2) \approx kT \approx kT_r \approx \gamma_0 a^2$ so that $\xi_{//} \approx L$ as $\gamma \approx \gamma_0$. Since the capillary and the network energies are of the same order of magnitude within the border thickness, the energy of the step β per unit length of the border of the island is of the order $\gamma_0 a^2/\xi_{//}$, which decreases exponentially as a function of T_r as the roughening transition point is approached.

2.6 LIGHT SCATTERING FROM THE INTERFACE FLUCTUATIONS

Let us now consider a planar interface (O, x, y), separating two transparent media with relative dielectric constants ε_1 and ε_2 and relative magnetic permeability equal to one. Maxwell's equations for a monochromatic field for each medium, are, in Gaussian units (Landau and Lifshitz, 1969)

$$i\omega\vec{H} = c \text{ curl } \vec{E}\,; \quad i\omega\varepsilon\vec{E} = -c \text{ curl } \vec{H}. \tag{2.1}$$

For a plane wave the field has an exponential dependence on the spatial coordinates exp $(i\vec{k}\cdot\vec{r})$ where $\vec{k}$ is the wave vector. From equations (2.1) we deduce

$$\omega\vec{H} = c\vec{k}\wedge\vec{E}\,; \quad \omega\varepsilon\vec{E} = -c\vec{k}\wedge\vec{H}. \tag{2.2}$$

Eliminating $\vec{E}$ and $\vec{H}$ we get $k^2 = n^2(\omega/c)^2$ where $n = \sqrt{\varepsilon}$ is the medium index. The energy per unit volume of the wave is

$$I = \frac{1}{8\pi}(\varepsilon E^2 + H^2) \tag{2.3}$$

thus $I = (1/4\pi)\varepsilon E^2$ for a transverse wave.

As shown in figure 2.14, an incident light wave in medium 1, $E_I \exp(i(\vec{k}_i\cdot\vec{r} - \omega_i t))$, with wavenumber $k_i = \sqrt{\varepsilon_1}\omega/c$ forming an angle ψ_0 with the vertical axis Oz, generates a reflected wave $E_r \exp(i(\vec{k}_r\cdot\vec{r} - \omega_f t))$ in medium 1 and a transmitted wave $E_t \exp(i(\vec{k}_t\cdot\vec{r} - \omega_f t))$ in medium 2. Since in the xy plane the surface is stationary and homogeneous, the t, x and y dependences of the waves in the different media must be similar. The waves thus propagate in the same plane, which we choose to be the xz plane. We therefore deduce,

$$\omega_i = \omega_f = \omega$$
$$k_{ix} = k_{rx} = k_{tx}$$
$$k_{rz} = -k_{iz} = -\sqrt{\varepsilon_1}\frac{\omega}{c}\cos\psi_0 \tag{2.4}$$
$$k_{tz} = \sqrt{\varepsilon_2}\frac{\omega}{c}\cos\psi_0' = \frac{\omega}{c}(\varepsilon_2 - \varepsilon_1\sin^2\psi_0)^{1/2}.$$

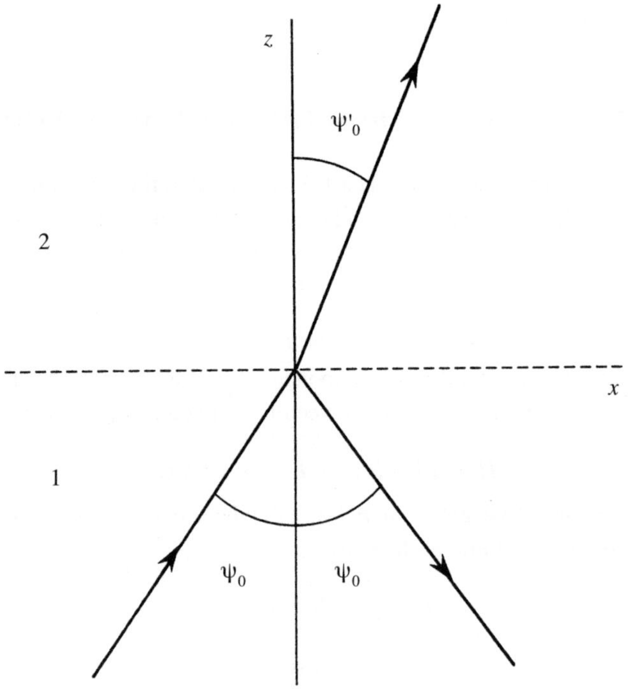

Figure 2.14 Incident, reflected and transmitted waves in a plane diopter.

As long as $n_2 < n_1$, if $\psi_0 < \psi_0^*$, the total internal reflection angle is defined by $\sin\psi_0^* = n_2/n_1$. The wavenumber of the transmitted wave is purely imaginary. The transmitted wave is evanescent and the reflection is total.

Let us first of all suppose that the direction of the incident field is in the plane perpendicular to the incident plane. The boundary conditions imposed by equation (2.1) demand that E_y and H_x are continuous at the interface. As $H_x = -(c/\omega)k_zE_y$,

$$E_i + E_r = E_t,\ k_{iz}(E_i - E_r) = k_{tz}E_{tz} \tag{2.5}$$

from which we can deduce,

$$E_r = \frac{k_{iz} - k_{tz}}{k_{iz} + k_{tz}} E_i = \frac{\sin(\psi_0' - \psi_0)}{\sin(\psi_0' + \psi_0)} E_i. \tag{2.6}$$

The amplitude of the transmitted wave is derived by the relation $E_t^2 + E_r^2 = E_i^2$. The ratio of the reflected and incident intensity is the reflection coefficient $R_+ = \sin^2(\psi_0' - \psi_0)/\sin^2(\psi_0' + \psi_0)$.

The case of parallel polarisation can be treated by changing the roles of the vectors $\vec{E}$ and $\vec{H}$, because $\vec{H}$ is perpendicular to the incident plane in this case. The continuity of H_y and of $E_x = (c/\omega\varepsilon)H_y$ then leads to

$$H_i + H_r = H_t\,;\ \frac{c}{\omega\varepsilon_1}(k_{iz}H_i + k_{rz}H_t) = \frac{c}{\omega\varepsilon_2}k_{tz}H_{tz} \tag{2.7}$$

from which we deduce

$$H_r = \frac{\varepsilon_2 k_{iz} - \varepsilon_1 k_{tz}}{\varepsilon_2 k_{iz} + \varepsilon_1 k_{tz}} H_i = \frac{\tan(\psi_0' - \psi_0)}{\tan(\psi_0' + \psi_0)} H_i. \tag{2.8}$$

The amplitude of the transmitted wave is derived from the relation $H_t^2 + H_r^2 = H_i^2$. The ratio of the incident and reflected intensities is the reflection coefficient $R_{//} = \tan^2(\psi_0' - \psi_0)/\tan^2(\psi_0' + \psi_0)$.

Let us now suppose that the surface separating the two media folds with a weak amplitude which can be written as the superposition of the Fourier coefficients $\xi(q, \Omega)\exp(i(qx - \Omega t))$. The same conditions as in an electromagnetic field but applied to the deformed surface lead to a generation of scattered waves of frequency ω_d and of wavenumber k_d (figure 2.15) in the two media, in addition to the reflected and transmitted waves determined above.

Indeed, for example, when the polarisation of the wave is in the y direction, the continuity of the coefficients E_y at the interface is written as

$$\begin{aligned} &E_{d1} + \xi(ik_{iz}E_i\exp(i(k_{iz}z + k_{ix}x - \omega_i t)) + ik_{rz}E_r\exp(i(k_{rz}z + k_{rx}x - \omega_i t))) \\ &\quad = E_{d2} + \xi ik_{tz}E_t\exp(i(k_{tz}z + k_{tx}x - \omega_i t)) \end{aligned} \tag{2.9}$$

where $E_{d1} = E_{d2}$ from the second of equations (2.4). The continuity of the tangential component of H is written

$$\begin{aligned} &-\frac{c}{\omega_d}k_{d1z}E_{d1} + \frac{c}{\omega_i}\left(-\xi ik_{iz}^2 + \frac{d\xi}{dx}k_{ix}\right)E_i\exp(i(k_{iz}z + k_{ix}x - \omega_i t)) \\ &\quad + \frac{c}{\omega_i}\left(-\xi ik_{rz}^2 + \frac{d\xi}{dx}k_{rx}\right)E_r\exp(i(k_{rz}z + k_{rx}x - \omega_i t)) \\ &= -\frac{c}{\omega_d}k_{d2z}E_{d2} + \frac{c}{\omega_i}\left(-\xi ik_{tz}^2 + \frac{d\xi}{dx}k_{tx}\right)E_t\exp(i(k_{tz}z + k_{tx}x - \omega_i t)). \end{aligned} \tag{2.10}$$

For the different quantities to vary in the same way as time and space, it is necessary that

$$\omega_d = \omega_i + \Omega$$
$$k_{d1x} = h_{d2x} = k_{ix} + q$$
$$k_{d1z} = \sqrt{\varepsilon_1}\frac{\omega_d}{c}\cos\psi \qquad (2.11)$$
$$k_{s2z} = \sqrt{\varepsilon_2}\frac{\omega_d}{c}\cos\psi' = \frac{\omega_d}{c}(\varepsilon_2 - \varepsilon_1\sin^2\psi)^{1/2}.$$

The amplitude of the scattered wave in medium 1 is derived from equations (2.9) and ()

$$E_d = -\frac{\omega_d}{\omega_i}\frac{2i\,k_{iz}(k_{tz} - k_{iz})}{k_{d2z} - k_{d1z}}\xi E_i. \qquad (2.12)$$

It is proportional to the amplitude of the surface deformation. The time scale of the movement of the interface is generally much longer than $1/\omega_i$, $\omega_d \approx \omega_i$. Thus, when the direction of observation of the scattered wave is in the direction of regular reflection,

$$E_d = -2i\,k_{iz}\frac{\sin(\psi_0 - \psi_0')}{\sin(\psi_0 + \psi_0')}\xi E_i \qquad (2.13)$$

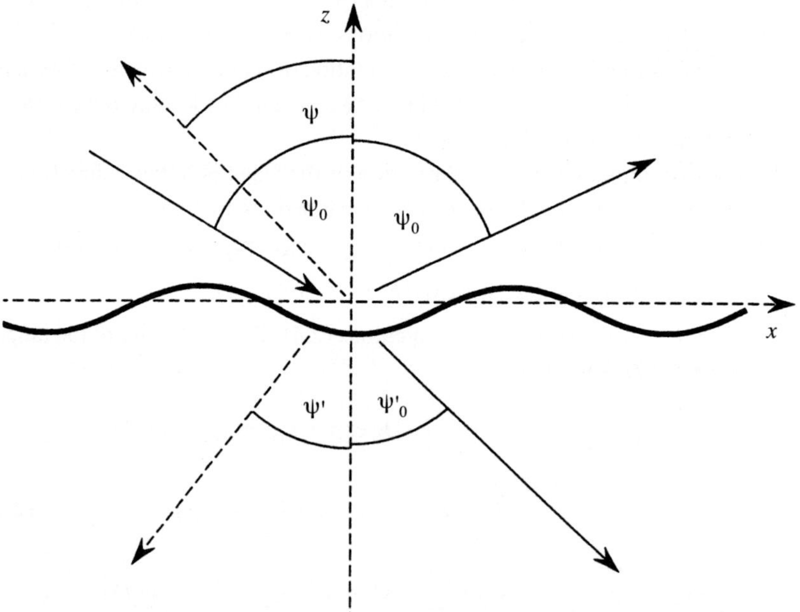

Figure 2.15 Sketch of the scattered waves in both media. The incident wave (plain line) at the angle ψ_0 is reflected with the same angle and is transmitted at ψ_0'. The undulation of the surface generates two scattered waves (dotted lines) at the angles ψ and ψ'.

and therefore proportional to the square root of the factor R_+ in regular reflection.

The same calculation, although more complicated, can be reproduced for polarisations and wave vectors of arbitrary interfaces (Agarwal 1977). We can therefore write

$$E_d = 2i\, k_i \frac{\omega_d}{\omega_i} F(\psi_0, \psi_0', \psi, \psi', \Theta) \xi E_i \tag{2.14}$$

where F is a function of the angles of transmission and reflection of regular and scattered waves, different for each polarisation configuration and Θ is the azimuthal angle formed by projections of wave vectors $\vec{K}_i$ and $\vec{K}_d = \vec{K}_i + \vec{q}$ on the planar interface.

The differential scattering cross-section $d\sigma$ is the ratio of the scattered intensity by the wave vectors q of the interface in the solid angle $d\Omega$ and the incident flux, so

$$d\sigma = \frac{dS' \sum_{\vec{q}} |E_d|^2}{|E_i|^2} = S \frac{\cos\psi}{\cos\psi_0} q dq \frac{|E_d|^2}{|E_i|^2} = S k_i^2 \frac{\cos\psi^2}{\cos\psi_0} 4k_i^2 F^2 |\xi_{\vec{q}}|^2 d\Omega \tag{2.15}$$

where S is the cross-section of the incident beam. If the interface is at thermodynamic equilibrium $\langle |\xi_q|^2 \rangle \approx kT/\gamma q^2$ from equation (2.9) in paragraph 2.5. Since the order of magnitude of q is $k_i \approx 10^7\ \mathrm{m}^{-1}$, that of σ is $(kT/\gamma)k_i^2 S \approx 10^{-4} S$, using table 2.6 in paragraph 2.5.

The intensity of light scattered by the interface fluctuations must be compared to the intensity scattered by the thermodynamic fluctuations of the density and the temperature of the surrounding media. The fluctuations of these quantities lead to a fluctuation of the dielectric constant tensor, $\delta\varepsilon(\vec{r}, t)$ of the medium and therefore of its polarisation $\delta\vec{P}(\vec{r}, t) = \delta\varepsilon(\vec{r}, t)\ \vec{E}(\vec{r}, t) \approx \delta\varepsilon(\vec{r}, t)\ \vec{E}_i(\vec{r}, t)$, where $\vec{E}_i(\vec{r}, t) = E_0 \vec{n}_i \exp(i\,\vec{k}_i \cdot \vec{r} - i\omega_i t)$ is the incident electric field. As shown in figure 2.16, this fluctuation generates a scattered wave of amplitude determined by Maxwell's equations. Indeed, at a given point the electric field is the sum of the incident and scattered fields. Since the total and incident fields satisfy the Maxwell equations, it is the same for the scattered field.

Thus, if E_s, $\vec{D}_s = \varepsilon_0 \vec{E}_s + \delta\vec{P}$ and $\vec{H}_s$ are the scattered electric field, scattered electric and magnetic displacements, respectively, they satisfy the Maxwell equations

$$\vec{\nabla} \wedge \vec{E}_s = -\frac{1}{c}\frac{\partial \vec{H}_s}{\partial t}$$

$$\vec{\nabla} \wedge \vec{H}_s = \frac{1}{c}\frac{\partial \vec{D}_s}{\partial t} \tag{2.16}$$

$$\vec{\nabla} \cdot \vec{H}_s = 0$$

$$\vec{\nabla} \cdot \vec{D}_s = 0.$$

Eliminating $\vec{E}$ and $\vec{H}$, we deduce the equation for the scattered electric shift,

$$\frac{\varepsilon_0}{c^2}\frac{\partial^2 \vec{D}_s}{\partial t^2} - \Delta\vec{D}_s = \vec{\nabla} \wedge \vec{\nabla} \wedge \delta\vec{P}. \tag{2.17}$$

Introducing the Hertz vector defined as $\vec{D}_s = \vec{\nabla} \wedge \vec{\nabla} \wedge \vec{\pi}$, we get

$$\frac{\varepsilon_0}{c^2}\frac{\partial^2 \vec{\pi}}{\partial t^2} - \Delta\vec{\pi} = \delta\vec{P} \tag{2.18}$$

the integration of which leads to

$$\vec{\pi}(\vec{r}, t) = \frac{1}{4\pi}\int_V d\vec{r}' \frac{1}{|r - r'|}\delta\vec{P}(\vec{r}', t') \tag{2.19}$$

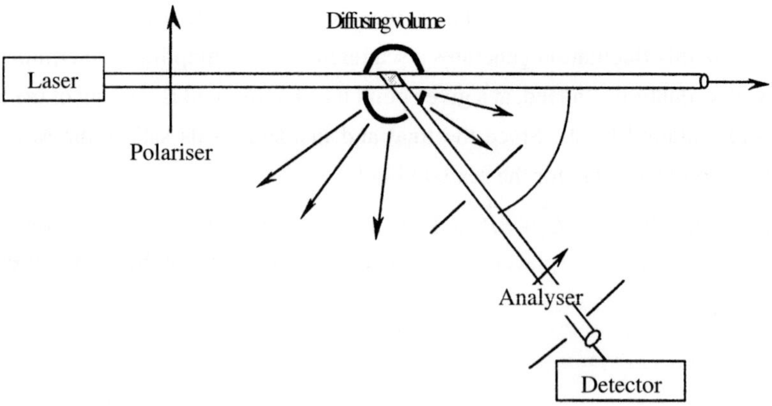

Figure 2.16 Sketch of a light scattering experiment.

where $t' = t - (\sqrt{\varepsilon_0}/c)|r - r'|$ is the retarded time. Using the expressions of $\delta\vec{P}$ and $\vec{D}_s$ as a function of the incident and scattered electric fields, we derive from equation (2.7) the expression for the scattered electric field

$$\vec{E}_s(\vec{r}, t) = \vec{\nabla} \wedge \vec{\nabla}\left(\frac{E_0}{4\pi\varepsilon_0}\int_V d\vec{r}'\frac{1}{|r - r'|}(\delta\varepsilon(\vec{r}', t')\cdot\vec{n}_i)\exp(i(\vec{k}_i\cdot\vec{r} - \omega_i t'))\right). \tag{2.20}$$

If the detection point is at a large distance from the scattering volume, $|\vec{r} - \vec{r}'| \approx r - \vec{r}' \cdot \vec{k}_f/|k_f|$ where $\vec{k}_f$ is the wave vector of the scattered wave and therefore $t' = t - (\sqrt{\varepsilon_0}/c)(r - \vec{r}' \cdot \vec{k}_f/|k_f|)$. The time scale of the evolution of the dielectric constant is much longer than $1/\omega_i$ since it is the same as that of the degrees of freedom of the translation and rotation of molecules. We can therefore approximate $\delta\varepsilon\ (\vec{r}', t')$ by $\delta\varepsilon\ (\vec{r}', t)$ in the integral of equation (2.20) and write $k_f \approx k_i$. These approximations lead to a simplified expression for the coefficients of the scattered electric field parallel to the direction $\vec{n}_f$:

$$E_s(\vec{r}, t) = -\frac{k_f^2 E_0}{4\pi\varepsilon_0 r}\exp(i(k_f r - \omega_i t))\delta\varepsilon_{if}(\vec{q}, t) \tag{2.21}$$

where $\vec{q} = \vec{k}_f - \vec{k}_i$,

$$\delta\varepsilon_{if}(\vec{q}, t) = \vec{n}_f \cdot \delta\varepsilon(\vec{q}, t)\cdot\vec{n}_i \tag{2.22}$$

and

$$\delta\varepsilon\ (\vec{q}, t) = \int_V d^3r'\exp(i\vec{q}\cdot\vec{r}')\delta\varepsilon\ (\vec{r}', t). \tag{2.23}$$

Using the fact that the space scale of the change in the dielectric constant, of the order of the intermolecular distance, is much smaller than $1/q$, we can put $q \approx 0$ into equation (2.23). We therefore derive the intensity of scattered light in the solid angle $d\Omega$,

$$dI_s = \frac{cE_s^2}{4\pi}r^2 d\Omega = \frac{cE_0^2}{4\pi}\frac{1}{16\pi^2}\left(\frac{\omega}{c}\right)^4 V^2\sin^2\Phi\langle\delta\varepsilon^2\rangle d\Omega \tag{2.24}$$

where Φ is the angle between the incident electric field and the direction of observation. The dielectric constant of a pure fluid is a function of the density ρ and the temperature T. Thermal fluctuations of these quantities then lead to a fluctuation of the dielectric constant

$$\langle\delta\varepsilon^2\rangle = \left(\frac{\partial\varepsilon}{\partial\rho}\right)_T^2\langle\delta\rho^2\rangle + \left(\frac{\partial\varepsilon}{\partial T}\right)_\rho^2\langle\delta T^2\rangle . \tag{2.25}$$

Equation (2.6) in paragraph 2.5 gives $\langle(\delta\rho)^2\rangle = (kT/V)\rho^2\kappa_T$ and $\langle(\delta T)^2\rangle = kT^2/Vc_V$ where c_V is the specific heat capacity per unit volume. From equation (2.12) we therefore deduce

$$dI_s = \frac{cE_0^2}{4\pi}\frac{1}{16\pi^2}\left(\frac{\omega}{c}\right)^4 V\sin^2\Phi kT\left(\left(\frac{\partial\varepsilon}{\partial\rho}\right)_T^2\rho^2\kappa_T + \left(\frac{\partial\varepsilon}{\partial T}\right)^2\frac{T}{\rho c_V}\right)d\Omega . \tag{2.26}$$

Dividing by the incident flux $cE_0^2/4\pi$, we deduce the scattering cross-section

$$\frac{d\sigma}{d\Omega} = \frac{1}{16\pi^2}\left(\frac{\omega}{c}\right)^4 V\sin^2\Phi kT\left(\left(\frac{\partial\varepsilon}{\partial\rho}\right)_T^2\rho^2\kappa_T + \left(\frac{\partial\varepsilon}{\partial T}\right)^2\frac{T}{\rho c_V}\right) . \tag{2.27}$$

In general, the term proportional to κ_T in equation (2.27) prevails and

$$\sigma = \frac{1}{6\pi}\left(\frac{\omega}{c}\right)^4 VkT\left(\frac{\partial\varepsilon}{\partial\rho}\right)_T^2\rho^2\kappa_T . \tag{2.28}$$

For an ideal gas, ρ is proportional to $\varepsilon - 1$, ε is independent of temperature and $\kappa_T = \dfrac{1}{\rho kT}$. The scattering cross-section is then $\sigma = \left(\dfrac{2}{3\pi}\right)\left(\dfrac{\omega}{c}\right)^4\left(\dfrac{V}{\rho}\right)(n-1)^2$, where $n = \sqrt{\varepsilon}$ is the refractive index $\varepsilon - 1 = n^2 - 1 \approx 2(n-1)$ (Rayleigh 1881). For water, $n \approx 1 + \alpha\rho\beta$ at $T = 0$ °C where $\alpha \approx 0.34$ and $\beta \approx 0.97$ (ρ in g/cm^3), so that ρ can be considered proportional to $\varepsilon - 1$, as for an ideal gas.

Thus, $\sigma = \dfrac{1}{6\pi}\left(\dfrac{\omega}{c}\right)^4 kTV\kappa_T(n^2-1)^2$. By writing $V = SL$ where $L \approx 1$ cm is the thickness of the scattering volume, $\kappa_T \approx 5 \cdot 10^{-10}\,\mathrm{m}^2/\mathrm{N}$ and $k_i = \dfrac{\omega}{c} = 10^7\,\mathrm{m}^{-1}$, we deduce $\sigma = 10^{-6}S$, two orders of magnitude weaker than the scattering cross-section of an undulating surface. As we will discuss in paragraph 3.7, the growth of a crystal could be generating a mesophase with a compressibility factor one hundred times weaker than that of an ordinary liquid phase. The scattering of the volume becomes comparable to that of the undulating surface.

When the dependence of the dielectric constant on q is not negligible, which is the case for a fluid in conditions close to a second-order critical point, the scattered intensity is porportional to $\langle|\delta\varepsilon_{if}(\vec{q})|^2\rangle \approx (\partial\varepsilon/\partial\rho)^2(\vec{n}_i\cdot\vec{n}_f)^2\langle|\delta\rho(\vec{q})|^2\rangle$ by equations (2.20) and (2.23). However,

$$\langle|\delta\rho(\vec{q})|^2\rangle = \langle\int_V d^3\vec{r}\delta\rho(\vec{r})^*\exp(-i\vec{q}\cdot\vec{r})\int_V d^3\vec{r}'\delta\rho(\vec{r}')\exp(i\vec{q}\cdot\vec{r}')\rangle$$

$$= \int_V d^3\vec{r}\int_V d^3\vec{r}'\langle\delta\rho(\vec{r})^*\delta\rho(\vec{r}')\rangle\exp(i\vec{q}\cdot(\vec{r}'-\vec{r})) \qquad (2.29)$$

$$= V\int_V d^3\vec{R}\langle\delta\rho(0)^*\delta\rho(\vec{R})\rangle\exp(i\vec{q}\cdot\vec{R})$$

such that if we consider the Orstein–Zernike autocorrelation function in paragraph 2.5, the scattered intensity is proportional to $1/(1+q^2\xi^2)$, where ξ is the correlation length of the fluctuations.

Appendix: Correlation functions of the light scattered by an assembly of moving macromolecules.

The electric field of the incident light induces a dipole moment $\vec{\mu} = \alpha\vec{E} \approx \alpha\vec{E}_i$ in each macromolecule which is much larger than for the solvent molecules due to their relative large size. Thus, the total dipole moment of the fluid is that of the macromolecules

$$\vec{P} = \delta\alpha(\vec{r},t)\vec{E}_i = \alpha\vec{E}_i\sum_i a_j(t)\delta(\vec{r}-\vec{r}_j(t))$$

where $a_j(t) = 1$ when the macromolecule is in the scattering volume, and $a_j(t) = 0$ if it is not. This dipole moment generates a scattered field, which is given by equation (2.9) as

$$E_s(\vec{r},t) = -\frac{k_f^2E_0}{4\pi\,\varepsilon_0 r}\exp(i(k_f r-\omega_i\,t))\delta\alpha_{if}(\vec{q},t)$$

where $\vec{q} = \vec{k}_f - \vec{k}_i$ and

$$\delta\alpha_{if}(\vec{q},t) = \vec{n}_f\cdot\delta\alpha(\vec{q},t)\cdot\vec{n}_i\ ;\ \delta\alpha(\vec{q},t) = \alpha\sum_i a_j(t)\exp(i\vec{q}\cdot\vec{r}_j(t))\,.$$

Let us define, at the point of measuring, the so-called heterodyne and homodyne correlation functions $I_1(t) = \langle E_s^*(0)E_s(t)\rangle$ and $I_2(t) = \langle|E_s(0)|^2|E_s(t)|^2\rangle$ respectively proportional to $F_1(\vec{q},t) = \langle\psi^*(\vec{q},0)\psi(\vec{q},t)\rangle$ and $F_2(\vec{q},t) = \langle|\psi(\vec{q},0)|^2|\psi(\vec{q},t)|^2\rangle$, where

$$\psi(\vec{q},t) = \sum_j a_j(t)\exp(i\vec{q}\cdot\vec{r}(t))\,.$$

This terminology comes from the detection methods involving photon beats: heterodyne when the excitation beam interferes with the scattered beam; homodyne when it interferes with itself.

So,

$$F_1(\vec{q},t) = \left\langle\sum_j a_j(0)a_j(t)\exp(i\vec{q}\cdot(\vec{r}_j(t)-\vec{r}_j(0)))\right\rangle\,.$$

The time scale of $a_j(0)a_j(t)$ is the time it takes for a macromolecule to cross the scattering volume, that is $\tau = L^2/D$ where D is the diffusion coefficient of the macromolecule and L the linear dimension of the scattering volume. The time scale of the exponential factor is the time it takes a macromolecule to cover a distance of order $1/q$, that is $\tau_q \approx 1/Dq^2$, much shorter than the previous time scale, since $\tau/\tau_q \approx (q\mathrm{L})^2 \approx 10^6$ under usual conditions. We can therefore approximate $a_j(0)a_j(t)$ as $a_j(0)^2 = a_j(0)$. Thus,

$$F_I(\vec{q}, t) = \left\langle \sum_j a_j(0) \exp(i\vec{q} \cdot (\vec{r}_j(t) - \vec{r}_j(0))) \right\rangle$$

$$\approx \sum_j \langle a_j(0) \rangle \langle \exp(i\vec{q} \cdot (\vec{r}_j(t) - \vec{r}_j(0))) \rangle$$

$$\approx \langle N \rangle F_s(\vec{q}, t)$$

since the movement of the macromolecule is not influenced by whether it is in the scattering volume or not. $\langle N \rangle$ is the average number of molecules in the scattering volume, and

$$F_s(\vec{q}, t) = \langle \exp(i\vec{q} \cdot (\vec{r}_j(t) - \vec{r}_j(0))) \rangle \ .$$

This presence is linked to the probability $G_s(\vec{R}, t) = \langle \delta(\vec{R} - (\vec{r}_j(t) - \vec{r}_j(0))) \rangle$ of moving according to $\vec{R}$ in time t since

$$F_s(\vec{q}, t) = \int d^3\vec{R} \exp(i\vec{q} \cdot \vec{R}) G_s(\vec{R}, t) \,.$$

A macromolecule is subject to small frequent movements due to the collisions with solvent molecules and it therefore diffuses through the medium with a diffusion coefficient D. Thus, from the theory of Brownian motion we can show that G_s satisfies the diffusion equation

$$\frac{\partial G_s(\vec{R}, t)}{\partial t} = D \Delta G_s(\vec{R}, t)$$

so that

$$\frac{\partial F_s(\vec{q}, t)}{\partial t} = -q^2 D F_s(\vec{q}, t)$$

that is,

$$F_s(\vec{q}, t) = \exp(-q^2 Dt)$$

since by definition, $F_s(\vec{q}, 0) = 1$. We can therefore deduce the heterodyne correlation function $F_I(\vec{q}, t) = \langle N \rangle \exp - q^2 Dt$, and the homodyne correlation function $F_2(\vec{q}, t) = \langle N \rangle^2 + \langle N \rangle^2 \exp - 2q^2 Dt$ within the framework of the Gaussian approximation for which $F_2(t) = F_I(0)^2 + F_I(t)^2$ (Berne and Pecora 1976).

BIBLIOGRAPHY

[1] D'Arcy Wentworth Thompson, *On growth and form*, abridged edition1961, Cambridge University Press, translated into French by D. Teyssié, *Forme et croissance,* Seuil, 1994.

[2] D.S. Agarwal, 'interaction of electromagnetic waves at rough dielectric surfaces', *Phys. Rev.,* **B15**, p. 2371, 1977.

[3] B.J. Berne and R. Pecora, *Dynamic Light Scattering with Applications to Chemistry, Biology and Physics*, John Wiley & Sons, inc., 1976.

[4] J.W. Cahn and J.E. Hilliard, 'Free Energy of a Nonuniform System. I. Interfacial free energy', *J. Chem. Phys.* **28**, p. 258, 1958.

[5] A.A. Chernov, D.E. Temkin and A.M. Melnikova, 'Theory of the Capture of Solid Inclusions during the Growth of Crystals from the Melt', *Sov. Phys. Crystallogr.,* **21**, p. 369, 1976.

[6] S.T. Chui, J.D. Weeks, 'Phase transition in the two-dimensional Coulomb gas, and the interfacial routhening transition', *Phys. Rev.,* **B14**, p. 4978, 1976.

[7] B. Deryagin, *Zh. Fiz. Khim.*, **14**, p. 137, 1940.

[8] S. Fisk, B. Widom, 'Structure and free energy of the interface between fluid phases in equilibrium near the critical point', *J. Chem. Phys.* **50**, p. 3219, 1969.

[9] C. Herring, *Structure and Properties of Solid Surfaces*, R. Gomer, C.S. Smith *ed.*, Univ. Chicago. Press, 1952.

[10] J.C. Herpin, J. Meunier, 'Étude spectrale de la lumière diffusée par les fluctuations thermiques de l'interface liquide vapeur de CO_2 près de son point critique. Mesure de la tension superficielle et de la viscosité', *Journal de Physique*, **35**, p. 847, 1974.

[11] J.N. Israelachvili, *Intermolecular and Surface Forces*, Academic Press, 1985.

[12] K.A. Jackson, *Liquid Metals and Solidification*, ASM, Cleveland, Ohio, p.174, 1958.

[13] J.M. Kosterlitz, D.J. Thouless, 'Ordering, Metastability and Phase Transistions in Two-Dimensional Systems', *J. Phys.,* **C6**, p. 1181, 1973.

[14] L. Landau, The Equilibrium Form of Crystals, collected papers Ter Haar, Gordon and Breach ed., p. 540, 1950.

[15] L.D. Landau, E.M. Lifshitz, *Course of theoretical physics*, volume 5, *Statistical physics (2nd revised and enlarged edition)*, Oxford, Pergamon, 1969.

[16] L.D. Landau, E.M. Lifshitz, *Course of theoretical physics*, volume8, *Electrodynamics of continuous media*, Oxford, Pergamon, 1960.

[17] R. Loudon, J.R. Sandercock, 'Analysis of the Light-Scattering Cross Section for Surface Ripples on Solids', *J. Phys. C : Solid St. Phys.*, **13**, p. 2609, 1980.

[18] P. Nozières, 'Shape and Growth of Crystals', in *Solids far From Equilibrium*, C.U.P., p. 1, 1992.

[19] C.W. Park, G.M. Homsy, 'Two-Phase Displacement in Hele-Shaw Cells: Theory', *J. Fluid. Mech.*, **139**, p. 291, 1984.

[20] J.A.F. Plateau, *Statique expérimentale et théorique des liquides soumis aux seules forces moléculaires,* Gauthier-Vilars, Trubner et Cie, F. Clemm, 1873.

[21] Lord Rayleigh, 'Scientific Papers', Cambridge University Press, 1903.

[22] G. Toulouse, P. Pfeuty, *Introduction au groupe de renormalisation et à ses applications*, Presses Universitaires de Grenoble, 1975.

[23] J.D. Van der Waals, 1893, *Verhandel. Konink. Akad. Weten.*, 1, 8, p. 56, French and German translations with new passages have been published in *Z. Physik. Chem.,* **13**, p. 657, 1894, an English translation with an historical introduction in *J. Stat. Phys.*, **20**, p. 197, 1979.

[24] K.G. Wilson, J. Koguth, 'Renormalization Group and the ε Expansion', *Phys. Rep,* **12C**, p. 75, 1974.

[25] G. Wulff, 'Zur Frage der Geschwindigkert des Wachsthums und des Aufläsung der Krystallfläschen', *Z. Krist*, **34**, p. 449, 1901.

CHAPTER 3

First-order phase transition kinetics

3.1 METASTABLE PHASE AND PHASE TRANSITION

Let us consider a planar interface between two pure liquid and solid phases at thermodynamic equilibrium. The change in thermodynamic potential is written as $dG = (\mu_s - \mu_l)\delta N_s$, where μ_s and μ_l are the chemical potentials of the solid and liquid phases. For a given pressure p_0, $\mu_s < \mu_l$ for $T < T_0$. A decrease in the thermodynamic potential requires the number of molecules of the solid phase δN_s to increase, which can be achieved by moving the interface from the solid to the liquid phase. Let us now consider a stable liquid phase while progressively decreasing the temperature. As soon as $T < T_0$ the solid phase becomes more stable, but since this phase transition is a first-order one, liquid–solid interfaces must be present for the solid phase to spread. This process can only be triggered by thermodynamic fluctuations which generate solid phase germs in the liquid phase. If these fluctuations are infrequent the liquid phase can subsist at temperatures lower than T_0 and becomes metastable. The probability of a fluctuation occurring in a medium at constant temperature and pressure is given by the following relation

$$f_0 \approx \exp\left(-\frac{\Delta G}{kT}\right) \tag{3.1}$$

where ΔG is the thermodynamic potential of the fluctuation. So, if we consider a spherical germ of radius *R* coming from the solid phase,

$$\Delta G = \frac{4\pi}{3v_s}(\mu_s(p_0, T) - \mu_l(p_0, T))R^3 + 4\pi\,\gamma\,R^2 \tag{3.2}$$

where v_s is the specific volume of the solid phase and γ the interfacial tension of the liquid-solid interface. When the liquid is supercooled, that is when $T < T_0$, the free energy increases for small radii because of the surface tension, then reaches a maximum and subsequently decreases at large radii since the solid phase is favoured at these temperatures (figure 3.1). The maximum $\Delta G_c = (4\pi/3)\gamma\,R_c^2$ is reached for a critical germ radius

$$R_c = \frac{2\gamma\,v_s}{\mu_l(p_0, T) - \mu_s(p_0, T)} \approx \frac{2d_{0\mathrm{T}}}{\Delta} \tag{3.3}$$

where $\Delta = c_p(T_0 - T)/Q$ is the supercooling of the liquid and $d_{0\mathrm{T}} = \gamma c_p T_0/Q^2$ is the thermal capillary length defined in paragraph 2.4. When $R < R_c$ it is energet-

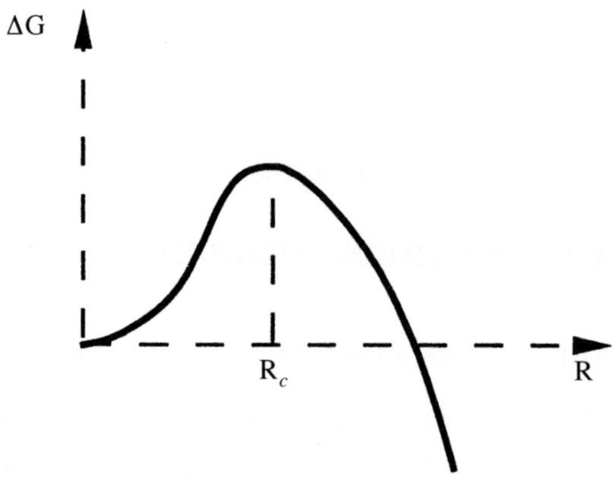

Figure 3.1 Free energy of a germ as a function of its radius.

ically favourable to decrease R in order to break the germ up. For $R > R_c$ it is favourable to increase R for the germ to grow. So at large radii the probability of a germ occurring is progressively higher, which necessarily leads to a quick transformation of the super-cooled liquid phase. Nevertheless it is common to see such a phase existing for a very long time, hence it is called a metastable phase. The fundamental reason is that such a phase is not at thermodynamic equilibrium since its fluctuations cannot be characterised by equation (3.1). To determine the new probability distribution $f(R, t)$ we suppose that the radius of a germ can change by frequent but small-amplitude elementary processes. The function f satisfies the kinetic equation known as the Fokker–Planck equation

$$\frac{\partial f}{\partial t} = -\frac{\partial J}{\partial R} \tag{3.4}$$

where

$$J = \frac{dR}{dt}f - B(R)\frac{\partial f}{\partial R} \tag{3.5}$$

is the flux in the radii space and dR/dt the growth rate of a germ as determined in paragraph 5.4. The function f is normalised so that $f(R, t)dR$ is the number of germs per unit volume between R and $R + dR$. The equilibrium distribution function f_0 must be a solution of this equation when $J = 0$. We obtain the relation $dR/dt = -B(d(\Delta G)/dR)kT$ which determines $B(R)$. The new sought after solution corresponds to phase transition dynamics where germs which are initially very small grow bigger. Thus we are naturally looking for a solution with a constant flux J_0 (Becker and Döring 1936, Zeldovich 1942). Using the previous relation between $\frac{dR}{dt}$ and B we obtain by integration of equation (3.5)

$$\frac{f}{f_0} = -J_0 \int \frac{dR}{Bf_0} + \text{Cst.} \tag{3.6}$$

The constant of integration and the flux J_0 have to be determined by the boundary conditions for the function f at small and large radii. The germs formed at small radii must cross the potential gap $\Delta G(R_c)$. If this quantity is large compared to kT, the germs seldom overcome this gap and they can be considered to be at thermodynamic equilibrium for small R. We then set $f = f_0$ when R tends to 0. At large radii the equilibrium solution diverges whereas the sought after solution must remain finite. We then impose that f/f_0 tends to 0 at large radii. We then obtain

$$\frac{f}{f_0} = J_0 \int_R^{+\infty} \frac{dR}{Bf_0} \tag{3.7}$$

and so

$$\frac{1}{J_0} = \int_0^{+\infty} \frac{dR}{Bf_0} . \tag{3.8}$$

f_0 is defined within an arbitrary constant defined by the normalisation of f. If we assume again that $\Delta G(R_c) > kT$, the function $1/f_0$ has a very sharp maximum around R_c and we can expand ΔG around R_c to calculate integral (3.8). We then obtain the number of nucleating germs per unit volume and unit time

$$J_0 = 2\sqrt{\frac{\gamma}{kT}}\, B(R_c) f_0(R_c) \exp\left(-\frac{4\pi\,\gamma\, R_c^2}{3kT}\right) . \tag{3.9}$$

In order to evaluate the order of magnitude of the exponential in equation (3.9) let us consider for instance succinonotrile, a transparent organic material (Kurz and Fisher 1989). We determine $R_c = 5.6\times10^{-9} m/\Delta$ from table 2.4 of paragraph 2.4 and from equation (3.4), and the exponential factor of equation (3.9) is written as $\exp(-253/\Delta^2)$. Thus, in common experimental conditions where $\Delta \approx 10^{-2}$, the factor under consideration is exponentially small and the liquid phase is metastable. The transition occurs in the presence of an initial germ larger than the critical radius, created by a cold tip for example, and growing in the metastable phase. As is described in the following chapters, the kinetics of the phase transition are governed by the dynamics of the liquid–solid interface.

Appendix: Nucleation of gas bubbles at the liquid–solid interface.

When the concentration of dissolved gas in a liquid exceeds a critical value dependent on the pressure p_0 and temperature T_0 of the medium, $\mu_g(p_0, T_0, c) < \mu_l(p_0, T_0, c)$ and gas

For a curved interface,

$$\Delta\mu = \mu_s(p_l + \delta p, T_0 - \Delta T) - \mu_l(p_l, T_0 - \Delta T)$$
$$\approx v_s \delta p - \frac{Q v_s}{T_0}\Delta T \tag{3.5}$$

and then by using the Laplace law,

$$\vec{v} \cdot \vec{n} = \lambda \Delta T - \alpha v_s \frac{\gamma}{R}. \tag{3.6}$$

In particular assuming that the interface is almost planar, $h(\vec{r}, t) \approx v_0 t + \xi(\vec{r}, t)$ and also using equation (3.3), we obtain

$$\frac{\partial \xi}{\partial t} = D\Delta\xi\,;\ D = \frac{\lambda\gamma T_0}{Q} \tag{3.7}$$

which describes the relaxation of the curved interface towards the flat plane, by kinetics. For a water-ice interface (Güttinger *et al.* 1978), $\lambda = 10^{-4}$ m/sK, $\gamma = 3 \cdot 10^{-2}$ J/m^2 and $Q/T_0 = 1.12\times10^6$ J/m^{3}K, so that $D = 2.7\times10^{-8}$ cm^2/s.

3.3 NUCLEATION OF ISLANDS ON A FACET

When the interface is in the smooth phase the interface is planar on a molecular scale. It cannot grow layer by layer. The transformation of the liquid layer next to the solid surface occurs in the same way as in paragraph 3.1. As mentioned in paragraph 2.5, the surface of a facet at equilibrium is made of small islands of arbitrary size and of thickness a confined by borders of thickness ξ. When the temperature of the liquid decreases or when its pressure increases the free energy of the solid phase decreases and results in a negative free energy proportional to the surface inside the island. The total energy of an island is given by

$$\Delta G = \frac{\pi R^2 a}{v_s}(\mu_s(p_0, T) - \mu_l(p_0, T)) + 2\pi\beta R \tag{3.1}$$

where $\beta \approx (\gamma a^2)/\xi_{//}$ is the step energy per unit length of the island and R its circumference. The free energy first increases with the radius and then goes through the maximum $\Delta G_c = \pi\,\beta\,R_c$ for the critical radius

$$R_c = \frac{\beta\, v_s}{a(\mu_l(p_0, T) - \mu_s(p_0, T))} = \frac{\beta\, v_s}{a\,\Delta\mu}. \tag{3.2}$$

Using a similar analysis to that in paragraph 3.1, we can determine the number of nucleating islands per unit surface area and unit time

$$J_0 = \sqrt{\frac{\beta}{kTR_c}}B(R_c)f_0(R_c)\exp\left(-\frac{\pi\beta R_c}{kT}\right) . \tag{3.3}$$

When $R > R_c$ the island expands on the surface with a velocity v_M, which we determine the same as in paragraph 5.4, as

$$v_M = \frac{\gamma}{\eta\beta}\frac{\Delta\mu}{v_s}a . \tag{3.4}$$

After a time t_0, J_0St_0 islands have nucleated on the surface S. To complete the layer each island just has to grow over a distance of $r_0 \approx (1/J_0t_0)^{1/2}$. This distance is covered over a time of about $t_0 = r_0/v_M$. We then deduce the growth rate of the facet $v = a/t_0$, that is $v = a(J_0v_M^2)^{1/3}$. If the distance covered by each island is larger than $\sqrt{S}$, which is the case when $\Delta\mu$ is small, a single island covers the whole facet and

$$v = a\ S\ J_0 . \tag{3.5}$$

The measurement of the growth rate of a superfluid solid interface of ^{4}He has been achieved by Wolf and coworkers (1985). A horizontal faceted crystal grows upwards between two parallel plates. On the outside, the crystal is at equilibrium with the fluid above. The supercooling of the interface is achieved by an extra-pressure $\delta p_L = \rho_L gh$ where h is the difference between the level of the liquid–solid interfaces outside and inside the plates. By using the growth law $v = a(J_0v_M^2)^{1/3}$ and the expression of R_c given by equation (3.2), we deduce that v is proportional to the exponential factor $\exp(-\pi\beta^2/3FkT)$, where $F = (\rho_s - \rho_L)\ agh$. The linear relation measured between $\text{Log}(v/h)$ and $\text{Log}(1/h)$ shows that the growth regime corresponds to the one described in this paragraph. The slope allows the measurement of the energy of a step $\beta(T)$. As expected from the theory in paragraph 2.5, $\beta(T)$ exponentially decays to 0 near the roughening transition temperature.

3.4 THE FRANK AND READ SOURCES

Experiments show that a smooth face generally grows more rapidly than expected from the previous theory. The presence of defects, such as a screw dislocation sketched in figure 3.2, accounts for this discrepancy. The border of the dislocation is a step which is a preferred place for the deposition of atoms. The crystal can then grow by rotation of the step around the centre of the dislocation, and hence generates an extra layer at each turn without any critical aggregate nucleation.

The motion of the step cannot be linear since the points which are the furthest away from the centre move too fast. According to paragraph 5.6 the velocity of propagation of a curved step is written as

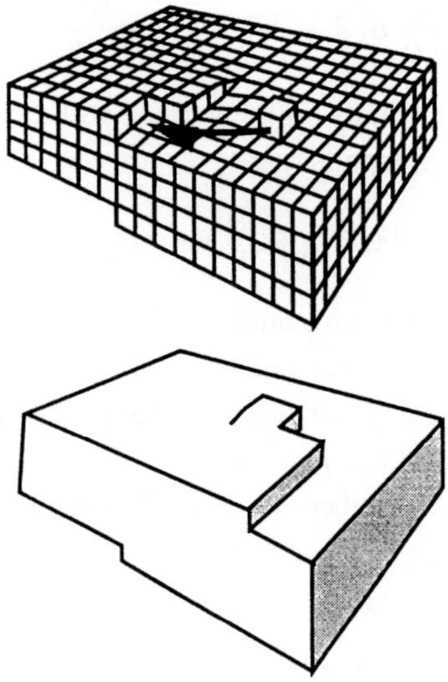

Figure 3.2 Screw dislocation.

$$v = v_M - \frac{\gamma}{\eta R} = v_M\left(1 - \frac{R_c}{R}\right) \tag{3.1}$$

where R_c is the critical radius of nucleation of islands determined by equation (3.3) in paragraph 3.3. Again, we find that island growth is only possible for $R > R_c$.

If the angular rotation velocity ω of the step is constant its shape in polar coordinates $\theta(r)$ satisfies the equation

$$\frac{\omega r}{\sqrt{1+\Psi^2}} = v_M\left(1 - R_c\left(\frac{\frac{d\psi}{dr}}{\sqrt{1+\Psi^2}^3} + \frac{\psi}{r\sqrt{1+\Psi^2}}\right)\right) \tag{3.2}$$

where $\psi = r\, d\theta/dr$. Far from the centre, the curvature of the step is progressively smaller since it winds around circles of progressively larger radii. The step winds at normal and constant velocity and $\theta = (\omega/v_M)r$ is the solution. It is an Archimedean spiral of pitch $\lambda = 2\pi\, v_M/\omega$. When we get closer to the centre the curvature effects become more important and the shape of the step deviates from the Archimedean spiral. When we neglect the size of the centre, the boundary condition near the centre is $\psi(0) = 0$. This constraint is large enough to set the rotation frequency to $\omega \approx 0.32 v_M/R_c$ (Burton, Cabrera and Frank 1951). The growth rate of the crystal is then $v_0 = a(\omega/2\pi)$, that is

$$v_0 = a\frac{v_M}{R_c} \approx \frac{a\eta}{\gamma}v_M^2 \approx \frac{\gamma a^3}{\eta\beta^2 v_s^2}\Delta\mu^2, \tag{3.3}$$

proportional to the square of the amount of supercooling. Contrary to the regime of island nucleation, the increase in the number of dislocations – and of the spirals – does not increase the facet growth. In fact, each spiral appears as a pile with its top, and in this case its centre, rising by a height a over time $T = 2\pi/\omega$. Each dislocation generates one of these piles and their tops rise at the same speed. On the other hand the heights of these piles depend on the number of dislocations. A step covers the distance d between two dislocations over the time d/v_M. Meanwhile, the top rises by $h = v_0 d/v_M \approx a\ d/R_c$

A similar spiral form grows in a reactive, so-called excitable medium (Winfree 1972). The equivalent of the crystalline step is then a reaction-diffusion wave similar to the flame structure described in paragraph 5.5 with an important difference: the reactive medium comes back to its original state. This form is not a growth form which separates two media of different properties. If a circular wavefront splits, it generates two free ends which wind at a well defined rotation frequency, the velocity of propagation of an end being slowed down by a curvature effect similar to the one in equation (3.1) (Zykov 1987). The problem of determination of the rotation frequency is more complicated than that developed above, in that v_M depends on the concentration of a diffusion field different to the one which structures the wave (Pelcé and Sun 1992, Karma 1993).

3.5 ELECTRODEPOSITION

When a metal electrode M is plunged into a solution of ions M^+ the electrode can grow or be dissolved by the electrode reaction

$$M \leftrightarrow M^+ + e^- . \tag{3.1}$$

This process is similar to the crystallisation since the metal atoms can move from a disordered phase – the solution – to a metallic ordered phase, in which they settle on reticular planes. Thus, it is not surprising to see shapes of electrodes similar to dendritic or faceted crystalline forms with growth modes similar to the ones previously described (Bockris and Reddy 1970). Two phenomena make the growth dynamics more complicated. If the electrolyte far from the electrode is locally electrically neutral the same is not true near the electrode, over a distance of the same magnitude as the Debye length which we determine in paragraph 5.7. A fraction of the metal atoms dissociates into ions and electrons, generating a double layer of positive charges in the solution and negative charges in the electrode. These charges generate a difference of potential on both sides of the interface which affects the transfer kinetics of metal atoms near the electrode. Besides, ions in solution can take part in chemical reactions, which will also modify the kinetics of electrodeposition. A few examples of those will be presented in the next chapter. The growth dynamics then become quite analogous to the dynamics of the propagation of a flame, the main difference being that the former takes place in a condensed phase whereas the latter generally takes place in the gaseous phase. It can be concluded from the previous remarks that, in the absence of an electric potential difference across the interface, the kinetics of growth are directly analogous to equation (3.1) where the indices c and f must be replaced by r (reduction) and o (oxidation), respectively:

$$v = \alpha F\left(k_r c_{M+} \exp -\frac{E_r}{RT} - k_o \exp -\frac{E_o}{RT}\right). \tag{3.2}$$

The activation energies of both processes (here per mole) appear in figure 3.3 which represents the free energy of an ion as a function of its distance from the electrode. In the presence of an electric field, the free energy of an ion is modified by a correction term with a linear dependence on the distance to the electrode.

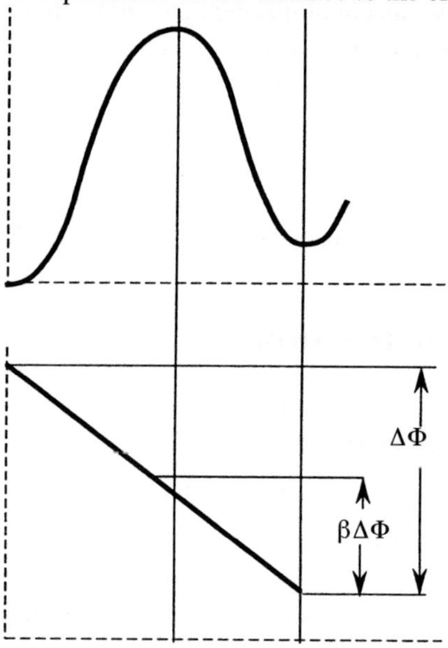

Figure 3.3 Free energy of ions as a function of their distance to the interface.

Let $\Delta\Phi$ be the potential difference felt by an ion as it goes from the solution to the electrode and $\beta\Delta\Phi$ the corresponding change of the activation energy of the reduction process. Then $(\beta - 1)\Delta\Phi$ is the change in the activation energy of the oxidation process and the kinetics of growth satisfies the Butler-Volmer law

$$v = \alpha F\left(k_r c_{M+} \exp -\left(\frac{E_r + \beta F\Delta\Phi}{RT}\right) - k_o \exp -\left(\frac{E_o + (\beta - 1)\, F\Delta\Phi}{RT}\right)\right). \tag{3.3}$$

At thermodynamic equilibrium, $v = 0$, that is

$$k_r c_{M^+} \exp -\left(\frac{E_r + \beta F \Delta\Phi_e}{RT}\right) = k_o \exp -\left(\frac{E_o + (\beta - 1)\, F\Delta\Phi_e}{RT}\right). \tag{3.4}$$

We deduce the Nernst equation

$$\Delta\Phi_e = \frac{(E_o - E_r)}{F} + RT\ \mathrm{Log}\left(\frac{k_r}{k_o}\right) + \frac{RT}{F}\ \mathrm{Log} c_{M^+}. \tag{3.5}$$

At thermodynamic equilibrium we can also write

$$\mu_M = \mu_{M^+} + \mu_e \tag{3.6}$$

that is,

$$\mu^0_M = \mu^0_{M^+} + kT\ \mathrm{Log} c_{M^+} + e\ \Phi_s + \mu^0_e - e\ \Phi_M. \tag{3.7}$$

We find equation (3.5) by identifying the constants of the two equations. Using the equilibrium point as a reference point we write $\Delta\Phi = \Delta\Phi_e + \eta$ and the Butler–Volmer law is simplified to

$$v = v_0\left(\exp -\frac{\beta\eta}{RT} - \exp\frac{(1-\beta)F\eta}{RT}\right). \tag{3.8}$$

3.6 FLUCTUATIONS OF A MOVING INTERFACE

We have already discussed the fluctuations of an interface at equilibrium in the previous chapter. The problem is more difficult when the interface is in motion, as shown by the following model of interface dynamics introduced by Kardar, Parisi and Zhang (1986)

$$\frac{\partial\xi}{\partial t} = \nu\Delta\xi + \frac{\lambda}{2}\left(\vec{\nabla}\xi\right)^2 + \eta(\vec{r}, t). \tag{3.1}$$

Here $\xi(\vec{r}, t)$ is the shape of the interface, $\eta(\vec{r}, t)$ is a random and external perturbation of mean value 0 and of correlation

$$\langle \eta(\vec{r}, t)\eta(\vec{r}', t')\rangle = 2\mu\delta^d(\vec{r} - \vec{r}')\delta(t - t') \tag{3.2}$$

where ν, λ and μ are the characteristic coefficients of the dynamics and d the dimension of the surface in motion. Even for a crystalline surface $d = 2$, we treat the case of an arbitrary dimension to show its determining effect on the amplitude of fluctuations. The deterministic part of equation (3.1) can be obtained by the motion of an interface with a normal velocity $\vec{v} \cdot \vec{n} = v_0 - \nu/R$, the same as for the progression of a crystalline step on a facet or a reaction-diffusion wave according to paragraph 5.6. The coefficient ν has the dimension of a diffusion coefficient. In the first case ν is defined by equation (3.7) in paragraph 3.2, in the second case ν is the diffusion coefficient of the diffusing species. Assuming that the amplitudes of the external perturbations are weak the shape of the interface is close to the intrinsically stable planar shape. We write $\xi = h - v_0 t$, where $h(\vec{r}, t)$ is the position of the interface and we expand the dynamics of the interface of amplitude ξ, simply keeping the first non-linear term. So

$$\begin{aligned} \vec{v} \cdot \vec{n} &\approx v_0\left(1 - \frac{1}{2}(\overrightarrow{\nabla\xi})^2\right) + \frac{\partial \xi}{\partial t} \\ \frac{1}{R} &\approx -\Delta\xi \end{aligned} \tag{3.3}$$

and equation (3.1) is deduced with $\lambda = v_0/2$.

When $\lambda = 0$ this is the equation of Edwards and Wilkinson (1981). If linear, this equation can easily be solved by Fourier transform

$$\tilde{\xi}(\vec{k}, \omega) = \frac{\tilde{\eta}(\vec{k}, \omega)}{\nu k^2 - i\omega} \tag{3.4}$$

where $$\tilde{\xi}(\vec{k}, \omega) = \int d^d r \, dt \, \exp i(\omega t - \vec{k} \cdot \vec{r}) \xi(\vec{r}, t). \tag{3.5}$$

From equation (3.2) we deduce

$$\langle \eta(\vec{k}, \omega) \eta(\vec{k}', \omega') \rangle = 2\mu \, \delta^d(\vec{k} + \vec{k}')\delta(\omega + \omega') \tag{3.6}$$

so that

$$\langle \tilde{\xi}(\vec{k}, \omega) \, \tilde{\xi}(\vec{k}', \omega') \rangle = \frac{\langle \tilde{\eta}(\vec{k}, \omega) \, \tilde{\eta}(\vec{k}', \omega') \rangle}{(\nu k^2 - i\omega)(\nu k'^2 - i\omega')}. \tag{3.7}$$

After integrating over ω,

$$\langle \tilde{\xi}(\vec{k}, t) \, \tilde{\xi}(\vec{k}', t) \rangle = \frac{\mu}{\nu k^2}(1 \quad \exp - 2\nu k^2 t)(2\pi)^d \delta^d(\vec{k} + \vec{k}'). \tag{3.8}$$

In particular, we can deduce the value of the mean square of the amplitude of the deformation of a surface of linear dimension L (Nattermann and Tang 1992):

$$\begin{aligned} \langle \xi^2 \rangle &= \int \frac{d^d\vec{k}}{(2\pi)^d} \frac{d^d\vec{k}'}{(2\pi)^d} \langle \xi(\vec{k}, t) \xi(\vec{k}', t) \rangle \\ &= K_d \frac{\mu}{\nu} \int_{2\pi/L}^{\pi/a} \frac{dk}{k^{3-d}} (1 - \exp - 2\nu k^2 t) \end{aligned} \tag{3.9}$$

where $K_d = 2^{d-1}\pi^{d/2}\Gamma(d/2)$, and after integrating by parts,

$$\begin{aligned} \langle \xi^2 \rangle &= \frac{\mathrm{K}_d}{d-2} \frac{\mu}{\nu} \left(\frac{\pi}{a}\right)^{d-2} \\ &+ \frac{\mu}{\nu} L^{2-d} f_d\left(\frac{\nu t}{L^2}\right) + \mathrm{O}\left(\exp\left(-2\pi^2 \frac{\nu t}{a^2}\right)\right) \end{aligned} \tag{3.10}$$

where

$$f_d(x) = \frac{K_d}{(2-d)(2\pi)^{2-d}} \times \left(1 - \exp(-8\pi^2 x) + (8\pi^2 x)^{1-\frac{d}{2}} \int_{8\pi^2 x}^{+\infty} y^{\frac{d}{2}-1} \exp(-y)dy\right). \tag{3.11}$$

Thus, when $d \neq 2$, $\langle \xi^2 \rangle \approx L^{2\alpha} f(t/L^z)$ where $\alpha = (2-d)/2$ and $z = 2$. In the realistic case, $d = 2$,

$$\langle \xi^2 \rangle = \frac{\mu}{2\pi\nu} \operatorname{Log}\left(\frac{L}{a} g\left(\frac{\nu t}{L^2}\right)\right) + O\left(\exp\left(-2\pi^2 \frac{\nu t}{a^2}\right)\right) \tag{3.12}$$

where $\operatorname{Log}(g(x)) = -\operatorname{Log}2 + (1/2)\mathrm{E}i(-8\pi^2 x)$. In particular, when $\nu t >> L^2$, $\langle \xi^2 \rangle \approx (\mu/2\pi\nu)\operatorname{Log}(L/a)$ and we find the amplitude of the equilibrium fluctuations by identifying $\mu/2\pi\nu$ to kT/γ.

When $\lambda \neq 0$ the amplitude of the fluctuations still satisfies a scaling law of the form $\langle \xi^2 \rangle \approx L^{2\alpha} f(t/L^z)$, but in general with exponents ξ and z different from the ones found for $\lambda = 0$. We must first note that the equation for the vector field $\vec{v} = -\vec{\nabla}\xi$ is written from equation (3.1) as

$$\frac{\partial \vec{v}}{\partial t} + \lambda\left(\vec{v} \cdot \vec{\nabla}\right)\vec{v} = \nu \Delta \vec{v} - \vec{\nabla}\eta(\vec{r}, t). \tag{3.13}$$

When $\lambda = 1$, this becomes the Burger equation with a noise term for the velocity field $\vec{v}$ of a fluid, an equation well established in mechanics of turbulent fluids (Burgers 1974). In the previous equation the scaling transformation $\vec{v} \to b^{\alpha}\vec{v}$, $t \to b^z t$, leads to

$$b^{\alpha - z} \frac{\partial \vec{v}}{\partial t} + b^{2\alpha - 2}\left(\vec{v} \cdot \vec{\nabla}\right)\vec{v} = \nu\, b^{\alpha - 2} \Delta \vec{v} - b^{-\frac{d}{2} - \frac{z}{2} - 1} \vec{\nabla}\eta(\vec{r}, t). \tag{3.14}$$

Since the first two terms have to change with the same scaling factor because they form the total acceleration $D\vec{v}/Dt = \partial\vec{v}/\partial t + \vec{v} \cdot \vec{\nabla}\, \vec{v}$ of the fluid, $\alpha + z = 2$. The same constraint on the exponents applies to equation (3.1) for all dimensions d.

The exponent α can be obtained at $d = 1$ by determining the probability distributions of the solutions of equation (3.1), $\mathrm{P}(\xi, t)$. By analogy with equation (3.4) in paragraph 3.1, it satisfies the Fokker–Planck equation in the amplitude space $\xi(\vec{r}, t)$

$$\frac{\partial P}{\partial t} = -\int d^d r \frac{\delta}{\partial \xi}\left(\frac{\partial \xi}{\partial t} P\right) + \mu \int d^d r \frac{\delta^2 P}{\partial \xi^2}$$
$$= -\int d^d \vec{r} \frac{\partial}{\partial \xi}\left(\left(\nu \Delta \xi + \frac{\lambda}{2}\left(\vec{\nabla} \xi\right)^2\right) P\right) + \mu \int d^d \vec{r} \frac{\partial^2 P}{\partial \xi^2}. \tag{3.15}$$

Here, the dynamic variable is no longer a scalar R but a function $\xi(\vec{r}, t)$. The derivative towards $\xi(\vec{r}, t)$ is a functional derivative. It is worth mentioning that this equation only has an exact solution in the case $d = 1$,

$$P(\xi, t) = \exp\left(-\int dx \frac{\mu}{2\nu}\left(\frac{\partial \xi}{\partial x}\right)^2\right). \tag{3.16}$$

We then find the Gaussian distribution law of the interface at equilibrium in paragraph 2.5, which leads to $\alpha = 1/2$ for $d = 1$. The constraint on the exponents implies that $z = 3/2$; an exponent different to that of the interface at equilibrium, since this constraint involves the non-linear term.

The determination of the exponent α for arbitrary dimensions has only been achieved by numerical and perturbation methods so far. The numerical simulations of Grossman, Guo and Grant (1991) give $\alpha \approx 1/2$ for $d = 2$, $\alpha \approx 1/4$ for $d = 3$ for values of λ between 1 and 100, showing that the exponent α must decrease with the dimension. A perturbation method of renormalisation for small λ has been developed by Kardar, Parisi and Zhang (1986). Since this method only gives correct values of α for $d = 1$ we will not introduce it here, but will simply mention that it consists of determining the variation laws for the coefficients λ, μ and ν as a function of the considered scaling factor l, as in paragraph 2.3. The structure of the fixed points associated with these evolution laws determines the exponents. Using other perturbation methods Schwartz and Edwards (1992) determine $\alpha = 0.29$ for $d = 2$, Bouchaud and Cates (1993), $\alpha = 0.36$ for $d = 2$.

The main interest of equation (3.1) is to show that the motion of an interface can notably amplify the fluctuations for $d = 2$, ranging from an exponent $\alpha \approx 0$ at $\lambda = 0$ to an exponent $\alpha \approx 1/2$ at $\lambda = 1$, and then in the latter case can lead to a high dependency of the amplitude of the fluctuations on the size of the system. For a quantitative application let us evaluate the value of λ for which the equilibrium fluctuations are modified. The non-linear term of equation (3.1) becomes of the same order of magnitude as the diffusion term when $\nu\xi/l^2 \approx \lambda\xi^2/l^2$, where l is an arbitrary lengthscale, that is $\xi \approx \nu/\lambda$. Using the order of magnitude of $\xi^2 \approx (\mu/\nu)\mathrm{Log}(L/a)$ given by the equilibrium fluctuations we obtain $(\mu\lambda^2/\nu^3)\mathrm{Log}(L/a) \approx 1$. Using the renormalisation method introduced above Tang, Natterman and Forrest (1990) have evaluated the characteristics of the fluctuations in this transition regime.

3.7 ANOMALOUS LIGHT SCATTERING BY A MOVING INTERFACE

As will be described in chapter 9, the fluctuations of the dynamic quantities characterising a liquid near a moving liquid–solid interface play a key role in the morphology of crystalline dendrites since they appear as initial conditions for the amplifying dynamics of the secondary branching. The simplest way to analyse them consists of doing a light scattering experiment near a moving liquid–solid interface (Güttinger, Bilgram et Känzig 1979).

A liquid–solid interface is set into motion at constant speed v, in a classic configuration of directional growth described in paragraph 5.4. As shown in figure 3.4 a laser beam is shone onto the interface at an angle Ψ_0 with the normal to the interface. The scattered light is collected by a photomultiplier in the directions described by the polar angle Ψ and the azimuthal angle Θ. The total intensity and the static and dynamic spectra of the scattered light are analysed for different wave vectors $q = \left|\vec{k}_f - \vec{k}_i\right|$.

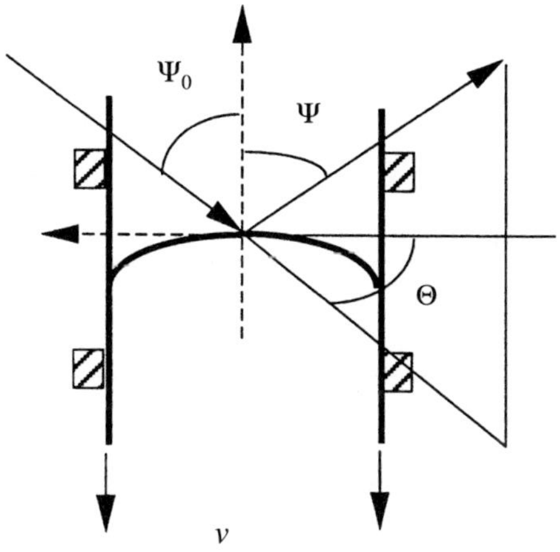

Figure 3.4 Experimental set-up. The interface is at rest in the laboratory reference frame. The tube containing this reference frame moves downwards at speed v and the solid grows at the same speed towards the liquid. The light scattered by the incident beam (at the angle ψ_0) is observed in the direction (ψ, Θ).

Qualitatively similar results are obtained for different materials, rough or faceted such as succinonitrile, cyclohexane, ice or salol:

- The scattered light intensity becomes abnormally high (two to three orders of magnitude higher than for an ordinary fluid) as soon as the speed of the crystal exceeds a critical velocity of about 1 μm/s. This velocity is far smaller than the critical velocity for morphological instabilities of rough interfaces. For faceted materials the critical velocity is far smaller, even non-existent.

- When the growth velocity exceeds the critical velocity, scattered light appears after a few hours – the smaller the growth speed, the longer it takes for scattered light to appear.
- As shown in figure 3.5 the intensity of anomalous scattered light increases with growth velocity. It exhibits a hysteresis, that is, the anomalous scattered light remains when the velocity decreases at smaller growth velocities than the critical velocity from which it appears.
- The light scatters from an area of the liquid phase near the interface, which is a few micrometers thick:

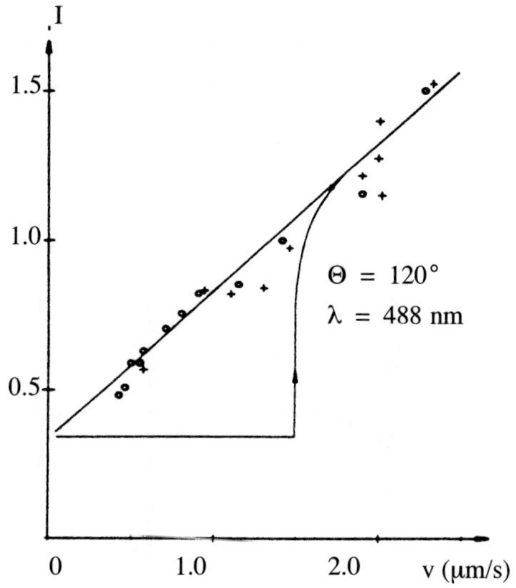

Figure 3.5 Scattered intensity (in arbitrary units) as a function of the growth rate (Güttinger *et al.* 1979).

- As shown in figure 3.6 for a fixed azimuthal angle Θ the intensity of the scattered light increases with the incidence angle – especially when the angle of incidence gets close to the angle of total internal reflection – and then saturates at higher angles. The solid curve is the theoretical variation for a rough interface, as calculated in paragraph 2.6.
- Figure 3.7 shows two static spectra, the first one by Steininger and Bilgram (1989) for cyclohexane (figure 3.7a) and the second by Cummins *et al.*, (1986) for salol (figure 3.7b). The solid curves are the theoretical variations; one calculated in paragraph 2.5 using an Ornstein–Zernike-type autocorrelation function as below, the other by Mie theory (see H.C. van de Hulst 1962) for light scattering by particles of radii larger than the wavelength of light.

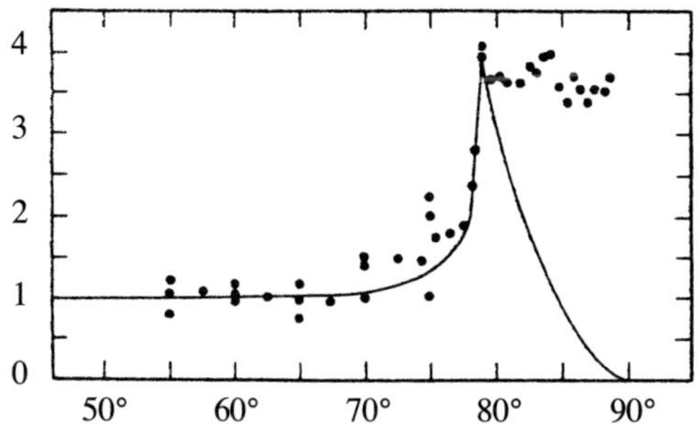

Figure 3.6 Dependence of scattered light (in arbitrary units) with the incident angle (v = 1.3 μm/s). The solid line is the result of the calculation of the rough interface from paragraph 2.6 (Böni *et al.*, 1983).

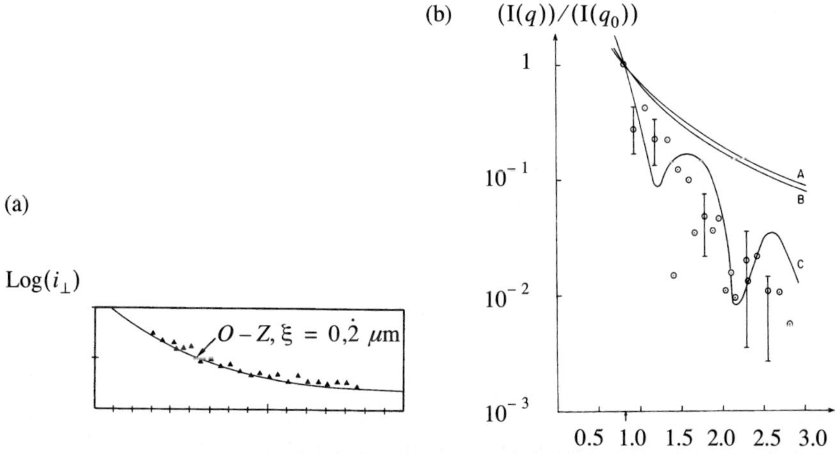

Figure 3.7 (a) Scattered intensity for a polarisation normal to the scattering plane as a function of the scattering angle. Experimental data ($\lambda = 488$ nm). The solid curve is the Orstein–Zernike function for $\xi = 0.2\ \mu$m (Steininger and Bilgram, 1989); (b) scattered intensity normalised to 1 for $q_0 = 0.825 \times 10^5 \text{cm}^{-1}$. The theoretical curves are A: Orstein–Zernike for $\xi = 0.3\mu$m ; B: Orstein–Zernike for $\xi = \infty$; C: Mie scattering for a radius of 0.3 μm. (Cummins *et al.*, 1986).

We obtain the dynamic spectrum of light scattered using the homodyne correlation function calculated in the appendix of paragraph 2.6, $F_2(\vec{q}, t) = \langle I(0)\ I(t)\rangle \approx \langle n(0)\ n(t)\rangle$, I and n being the intensity of light and the number of photons received by the photomultiplier, respectively, and $\vec{q} = \vec{k}_f - \vec{k}_i$ the transfer wave vector, the difference between the incident and scattered wave vectors. As shown in figures 3.8a and 3.8b this function can be written as $F_2(\vec{q}, t) = A + B\,\exp(-2\Gamma(q)t)$, where $\Gamma = Dq^2$. Thus, this function does not depend on the orientation of $\vec{q}$ towards the plane of the interface.

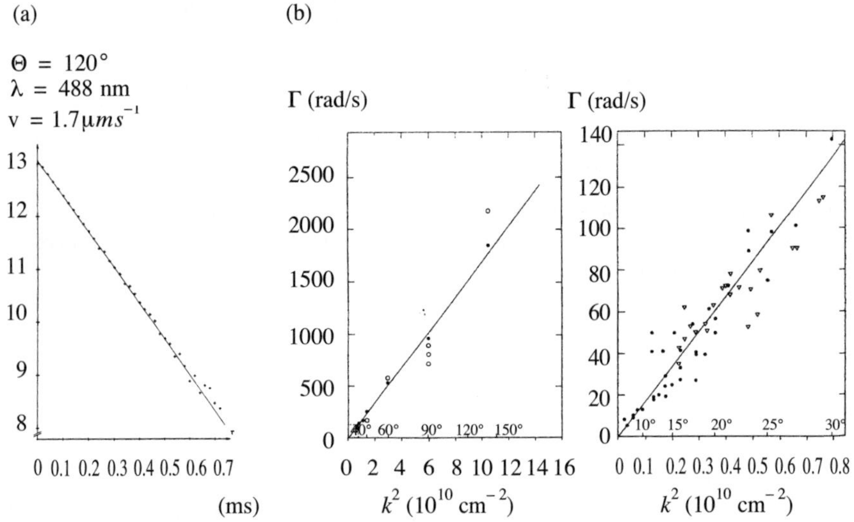

Figure 3.8 (a) $\mathrm{Log}(F_2 - A)$ as a function of *t* (Güttinger *et al.*, 1979), (b) Γ function as a function of the transfer wavenumber when it is parallel or perpendicular to the interface. (Böni *et al.*, 1983).

In all the systems studied $D \approx 10^{-9}$–$10^{-10}\,\mathrm{cm^2/s}$ (except for pure water where $D \approx 10^{-8}\,\mathrm{cm^2/s}$). It changes with the concentration of dissolved gases in the liquid and with the intensity of scattered light.

The experimental data simultaneously show typical features of scattering by interface fluctuations as shown in figure 3.6 and of scattering by volume fluctuations as shown by the isotropy of the factor $\Gamma(\vec{q})$. It is likely that several phenomena occur at the same time, among which are the three following processes: the fluctuations of the liquid–solid interface, the existence of a dynamic mesophase upstream from the interface and the nucleation of microbubbles from dissolved gases present in the liquid.

According to paragraph 2.6 the interface fluctuations scatter the light, the scattered intensity being proportional to the square of the amplitude of the fluctuations as long as this amplitude remains far smaller than the wavelength of light. Equilibrium interface dynamics are well described by the Edwards–Wilkinson equation of paragraph 3.6 and lead to a correlation function which decays with the term $\exp(-Dq^2t)$ where $D \approx 10^{-8}\,\mathrm{cm^2/s}$, in agreement with the measured homodyne autocorrelation function. However, in this case the transfer wave vector is parallel to the interface, as opposed to the observation that transfer wave vectors of all directions lead to the same decay law of the correlation function. Moreover, these dynamics describe the interface at equilibrium as well as the interface in motion and cannot account for the appearance of the phenomenon when the interface is moving. More suitable dynamics would be the Kardar, Parisi and Zhang equation (1986) described in paragraph 3.6 for which the amplitude of the fluctuations is directly linked to the motion of the interface. Nevertheless, the growth rate of 1 μm/s is far too small for the observation of such high-amplitude fluctuations if the cause of the fluctuations is thermodynamical. In fact, identifying μ/ν as in paragraph 3.6, at kT/γ and ν at D we deduce $\lambda^2 \approx (D^2\gamma/kT)/\mathrm{Log}(L/a)$, that is $\lambda \approx (D/a)/(\mathrm{Log}(L/a))^{1/2}$. For usual sample sizes the logarithmic factor L is only able to increase by one order of magnitude at most, so that with $D \approx 10^{-8}\mathrm{cm}^2/\mathrm{s}$ and $a \approx 10\mathrm{\AA}$, we have $\lambda \approx 1$ mm/s. The sample size corresponding to $v \approx 1\ \mu\mathrm{m/s}$ would be huge: $L \approx a\ \exp(10^6)$. Besides, the phenomenon of anomalous light scattering exists for both rough (succinonitrile) and faceted (salol) interfaces. All these observations lead to discarding a mechanism of anomalous light scattering by interface fluctuations.

A second hypothesis deals with the existence of a liquid mesophase on a scale of a few microns, upstream from the interface, which is generated by the slow motion of the interface. Although it is not confirmed by the usual laws of thermodynamics of irreversible processes (Mazur and Keizer 1988), the dependency of the amplitude of the scattered light with the transfer wave vector would be compatible with an Orstein–Zernike type density correlation function described in paragraph 2.5:

$$\langle \delta\rho(0)^* \delta\rho(\vec{r})\rangle \approx \frac{1}{r} \exp\left(-\frac{r}{\xi}\right). \tag{3.1}$$

The intensity of the scattered light by a volume of fluid determined by equation (2.24) in paragraph 2.6 is only valid if the correlation length of the density is small compared to $1/q$, where q is the transfer wave vector, which is true for a common fluid far away from a second order critical point. This approximation is not necessarily valid for the considered mesophase. We must remember the dependency of the dielectric constant on q, and by using the autocorrelation function (1) in equation (2.29) of paragraph 2.6, an intensity of scattered light which is proportional to the factor $1/(1+q^2\xi^2)$. As shown in figure 3.7a, such a dependency of the intensity of the scattered light on q^2 is compatible with the corresponding experiment, if $\xi \approx 0.2\ \mu\mathrm{m}$. In this assumption, the dependency of the scattered intensity on the incident angle sketched in figure 3.6 would be explained by the multiple reflection of the light by the liquid-mesophase and solid-mesophase interfaces (Böni *et al.* 1983). According to the theory of Landau–Placzek (1934) the fluctuations of entropy at constant pressure lead to a dynamic spectrum of scattered light decaying as $\exp(-Dq^2t)$ where D is the thermal diffusion coefficient. The theory of mode coupling (Kadanoff and Swift 1968, Kawasaki 1969) allows us to write a relation between the diffusion coefficient D and the correlation length ξ of the fluctuation with the Stokes–Einstein relation, $D = kT/6\pi\eta\xi$ where η is the viscosity of the fluid. Assuming that the order of magnitude of the viscosity is the same as for a common fluid, a diffusion coefficient of the order of $D = 10^{-8}\text{cm}^2/\text{s}$ corresponds to a correlation length of about $0.1\ \mu\text{m}$, which is in good agreement with the experimental data.

The third assumption deals with the existence of microbubbles of dissolved gas nucleating near the interface (Cummins *et al.,* 1986). In usual experimental conditions the considered liquids contain dissolved gases from the atmosphere such as N_2, O_2, CO_2 and Ar, their concentration being determined as a function of their partial pressure by the Henry law. When the liquid–solid interface is set in motion it rejects the dissolved gases as impurities of increased concentration near the interface. As shown in the appendix of paragraph 3.1, as soon as the concentration of the gas reaches a critical value, gas microbubbbles nucleate at the interface and grow at a constant rate depending on the local over-saturation. We obtain a typical length-scale of the size of the bubbles with the Henry law given by equation (2.13) of paragraph 2.4. Using the Laplace law to determine the pressure P_g of the gas in the microbubble as a function of the partial pressure P of dissolved gases, we obtain

$$R = \frac{2\gamma c_0}{P(c_l - c_0)} \tag{3.2}$$

where c_0 is the concentration of saturation. Each microbubble has a diffusion coefficient determined by the Stokes-Einstein equation, $D = kT/6\pi\eta R$, which gets smaller when the radius of the bubble increases. A diffusion coefficient of the order of $D = 10^{-9}\ \text{cm}^2/\text{s}$ corresponds to a size of a microbubble of about 1 μm. The spectrum of static diffusion of the figure is then determined by Mie theory (see H.C. van de Hulst 1962). As shown by the appendix of paragraph 2.6 a set of macromolecules diffusing in a medium generates a homodyne correlation function of scattered light decaying as $\exp(-2Dq^2t)$, in good agreement with the observed functions in figures 3.8. Two comments: the size of the microbubble is of the same order as the wavelength of light whereas the calculation in the appendix of paragraph 2.6 is only valid for far smaller sizes; since the observed correlation function has a purely exponential decay, it must be associated with a monodisperse distribution of bubbles whereas the growth dynamics described above necessarily leads to a polydisperse distribution. A favourable factor: the pumping of the dissolved gases leads to a decrease in the intensity of scattered light and in the diffusion coefficient, in agreement with the increase in the average radius of the bubbles described by equation (3.2).

As soon as their radius reaches a critical value R_{cap} the bubbles are trapped by the interface. In fact, according to the second appendix in paragraph 2.4, the interface repels a sphere of radius R with a force $F = ((2\pi\gamma\nu)/(1-\nu)(n-1))(A_nR/\gamma\nu)^{1/n}$. As soon as the Stokes force, $F = 6\pi\eta R v$, exerted by the fluid on the sphere owing to the motion of the interface at speed v, exceeds the repelling force, that is, from the critical radius

$$R_{cap} = \left(\frac{\gamma \nu}{3\eta v(1-\nu)(n-1)}\right)^{\frac{n-1}{n}} \left(\frac{A_n}{\gamma \nu}\right)^{\frac{1}{n-1}} \tag{3.3}$$

the sphere is trapped by the interface. At small distances between the sphere and interface $n = 3$ and $R_{cap} \approx v^{-2/3}$. More comprehensive models take into account electrostatic effects, impurities and gravity. As shown in figure 3.9 the trapping law can always be written as $R_{cap} \approx v^{-m}$ and it links growth rates of about 1 µm/s to trapping sizes of the order of 1 µm, where, depending on the models, m changes from 2/3 to 2 (Körber 1988). Thus, nucleation and trapping of bubbles create perturbations of the interface, the amplitude of which can exceed the amplitude of the thermodynamic fluctuations by several orders of magnitude.

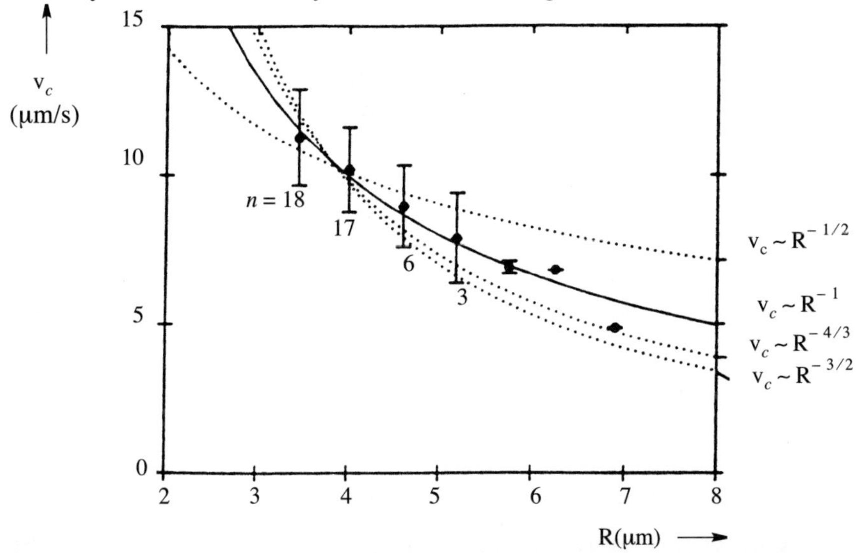

Figure 3.9 Critical speed v_c as a function of the radius R of the particle. Comparison with several power laws discussed in the text (Körber, 1988).

BIBLIOGRAPHY

[1] E. Becker, W. Döring, 'Kinetische Behandlung der Keimbildung in übersättigten Dämpfen', *Ann. Physik,* **24**, p. 719, 1935.

[2] J. O'M Bockris, A.K.N. Reddy, *Modern electrochemistry*, Plenum Press, 1970.

[3] P. Böni, J.H. Bilgram, W. Känzig, 'Properties of the Solid–Liquid Interface Layer of Growing Ice Crystals: A Dynamic Light Scattering Study', *Phys. Rev.,* **A 28**, p. 2953, 1983.

[4] J.P. Bouchaud, M.E. Cates, 'Self-Consistent Approach to the KPZ Equation', *Phys. Rev.,* **E 47**, p. 1455, 1993.

[5] W.K. Burton, N. Cabrera, F.C. Frank, 'The growth of crystals and the equilibrium structure of their surfaces', *Phil. Trans. Roy. Soc.*, **243**, p. 299, 1951.

[6] J.M. Burgers, *The Nonlinear Diffusion Equation*, D. Reidel publishing company, 1974.

[7] A.V. Butler, *Proc. Roy. Soc.,* **A 157**, p. 423, 1936.

[8] H.Z. Cummins, G. Livescu, H. Chou, M.R. Srinivasan, 'Origin of Diffusional Light Scattering at the Non-Equilibrium Crystal-Melt Interface', *Solid State Communications*, **60**, p. 857, 1986.

[9] S.F. Edwards, D.R. Wilkinson, 'The Surface Statistics of a Granular Aggregate', *Proc. Roy. Soc.,* **A 381**, p. 17, 1982.

[10] T. Erdey-Gruz, M. Volmer, *Z. Physik. Chem.*, **150A**, p. 203, 1930.

[11] J.C. Géminard, P. Oswald, D. Temkin, J. Malthete, 'Rapid Solidification of a Columnar Hexagonal Liquid Crystal', *Europhys. Lett.*, **22**, p. 69, 1993.

[12] B. Grossmann, H. Guo, M. Grant, 'Kinetic Roughening of Interfaces in Driven Systems', *Phys. Rev.,* **A 43**, p. 1727, 1991.

[13] H. Güttinger, J.H. Bilgram, W. Känzig, 'Dynamic Light Scattering at the Ice Water Interface during Freezing', *J. Phys. Chem. Solids*, **40**, p. 55, 1979.

[14] K.A. Jackson, B. Chalmers, 'Kinetics of Solidification', *Can. J. Phys.,* **34**, p. 473, 1956.

[15] L.P. Kadanoff and J. Swift, 'Transport Coefficients near the liquid–gas critical point', *Phys. Rev.,* **166**, p. 89, 1968.

[16] M. Kardar, G. Parisi, Y.C. Zhang, 'Dynamic Scaling of Growing Interfaces', *Phys. Rev. Lett.*, **56**, p. 889, 1986.

[17] A. Karma, 'Scaling Regime of Spiral Wave Propagation in Single-Diffusive Media', *Phys. Rev. Lett.*, **68**, p. 397, 1992.

[18] K. Kawasaki, 'Decay rate of concentration fluctuation of a binary mixture in the non-hydrodynamical regime', *Phys. Lett.*, **A 30**, p. 325, 1969.

[19] C. Körber, 'Phenomena at the Advancing Ice–Liquid Interface: Solutes, Particles and Biological Cells', *Quarterly Review of Biophysics*, **21**, 2, p. 229, 1988.

[20] W. Kurz and D.J. Fischer, *Fundamentals in Solidification*, third edition, Trans Tech publications, 1989.

[21] E.M. Lifshitz, L.P. Pitaevskii, *Physical kinetics*, Pergamon Press, 1981.

[22] L. Landau, G. Placzek, *Physik. Z. Sowjetunion,* **5**, p. 172, 1934.

[23] P. Mazur and J. Keizer, 'Anomalous Light Scattering at the Surface of Growing Crystals: A Theoretical Analysis Based on Irreversible Thermodynamics', *Phys. Rev. A*, **38**, p. 5267, 1988.

[24] T. Nattermann, L.H. Tang, 'Kinetic Surface Roughening. I. The Kardar–Parisi–Zhang equation in the Weak-Coupling Regime', *Phys. Rev.*, **A 45**, p. 7156, 1992.

[25] P. Pelcé, J. Sun, 'Wave Front Interaction in Steadily Rotating Spirals', *Physica D*, **48**, p. 353, 1991.

[26] M. Schwartz, S.F. Edwards, 'Nonlinear Deposition: a New Approach', *Europhys. Lett.*, **20**, p. 301, 1992.

[27] R. Steininger, J. Bilgram, 'Light Scattering at the Solid–Liquid Interface of Cyclohexane', *Helvetica Physica Acta*, **62**, p. 215, 1989.

[28] L.H. Tang, T. Nattermann, B.M. Forrest, 'Multicritical and Crossover Phenomena in Surface Growth', *Phys. Rev. Lett.*, **65**, p. 2422, 1990.

[29] H.C. Van De Hulst, *Light Scattering by Small Particles*, John Wiley, New York, 1962.

[30] A.T. Winfree, 'Spiral Waves of Chemical Activity', *Science*, **175**, p. 634, 1972.

[31] P.E. Wolf, F. Gallet, S. Balibar, E. Rolley, P. Nozieres, 'Crystal growth and Crystal curvature near roughening transitions in $hcp\,^4He$', *J. Physique*, **46**, p. 1987, 1985.

[32] Ya.B. Zeldovich, *On the Theory of New Phase Formation. Cavitation*, selected works, vol. 1, p. 120, Princeton University Press, 1942.

[33] V.S. Zykov, *Simulation of Wave Processes in Excitable Media*, Manchester University Press, 1987

CHAPTER 4

Chemical reactions

4.1 CHEMICAL EQUILIBRIUM

A chemical reaction is another cause of exchange of matter. Such a reaction can be written as

$$\sum_i \nu_i A_i = \sum_i \nu_i' A_i' \tag{4.1}$$

where A_i and A_i' are the symbols of the chemical reagents, and ν_i and ν_i' are the stoichiometric coefficients of the reaction. The = sign indicates that the reaction is reversible and leads to a state of thermodynamic equilibrium of all the reagents. Let $G = \Sigma n_i \mu_i + \Sigma n_j' \mu_j'$ be the thermodynamic potential at constant temperature and pressure, where n and μ are the concentration and the chemical potential of a given species, respectively. At equilibrium G is at a minimum such that $dG = \Sigma \mu_i dn_i + \Sigma \mu_j' dn_j' = 0$. Since the change in concentration of each species is limited by the reaction: $dn_i = (\nu_i/\nu_1)dn_1$, $dn_j' = -(\nu_j'/\nu_1)dn_1$, we obtain the condition for chemical equilibrium

$$\sum_i \nu_i \mu_i = \sum_j \nu_j' \mu_j'. \tag{4.2}$$

Since the chemical potentials are a function of the concentration, the pressure and the temperature, condition (4.2) constitutes a relation between these different variables. In general, this equation has several possible solutions. Nevertheless, if all the reacting species are ideal gases there is only one solution (Zeldovich 1938). The equilibrium condition takes a more useful form as long as the chemical species are ideal gases or the solutes are at low concentrations (a weak solution). Indeed in this case $\mu = \mu_0(P, T) + kT\mathrm{Log}\,n$, and condition (4.2) can be written in the form

$$K(P, T) = \frac{\Pi n_j'^{\nu_j'}}{\Pi n_i^{\nu_i}}, \tag{4.3}$$

a law commonly known as the 'law of mass action', where the equilibrium constant is defined by $K(P, T) = \exp((\Sigma \nu_i \mu_{0i} - \Sigma \nu_j' \mu_{0j}')/kT)$. This relation can also be applied when the solvent participates in the reaction. Indeed, for the usual concentrations of the solute, the change in the solvent chemical potential due to the solute concentration can be neglected compared to the change in the solute chemical potential. The solvent chemical potential only enters into the equation as a constant which renormalises the equilibrium constant.

For ideal gases the dependence of the equilibrium constant on pressure and temperature can be determined. Indeed, $\mu_0(P, T) = \varepsilon - c_p T \mathrm{Log} T - T\xi + kT \mathrm{Log} P$, where ξ is the chemical gas constant and ε the internal energy originating from the electronic configuration of the gas molecules. We therefore deduce

$$K(P, T) = \exp(\Sigma \nu_j' \xi_j' - \Sigma \nu_i \xi_i) P^{(\Sigma \nu_i - \Sigma \nu_i')} T^{(\Sigma \nu_j' c_{pj}' - \Sigma \nu_i c_{pi})} \exp\left(\frac{\Sigma \nu_i \varepsilon_i - \Sigma \nu_j' \varepsilon_j'}{kT}\right) \quad (4.4)$$

that is, the product of the temperature and pressure to the power of a constant term and a term with an exponential dependence on the temperature. We will see later on that the last term has great importance in explosive gas reactions because the exponential term is very sensitive to small changes in temperature which greatly affect the equilibrium constants of the reactions.

The chemical reaction is usually accompanied by a heat exchange with the surroundings. At constant pressure and temperature, the amount of heat corresponding to the loss of dn_1 molecules of species 1 is written as $dQ = -(\Sigma h_i dn_i + \Sigma h_j' dn_j') = (\Sigma \nu_j' h_j' - \Sigma \nu_i h_i) dn_1 / \nu_1$, where h is the free enthalpy per molecule. We usually define the reaction heat as the heat which corresponds to the consumption of ν_1 molecules of species 1, that is,

$$Q = \sum_i \nu_i h_i - \sum_j \nu_j' h_j'. \quad (4.5)$$

The reaction is said to be exothermic (and respectively endothermic) if Q is positive (respectively negative). As long as the reactants and the products are gases $h = -T^2 \partial / \partial T (\mu / T)_p = \varepsilon + c_p T$, we can relate the equilibrium constant and the reaction heat using equation (4.4) by the relation

$$K(P, T) = \mathrm{cst}\ P^{(\Sigma \nu_i - \Sigma \nu_j')} T^{(\Sigma \nu_j' c_{pj}' - \Sigma \nu_i c_{pi})} \exp\left(\frac{Q}{kT}\right) \quad (4.6)$$

so that the reaction heat determines the exponential dependence of the equilibrium constant. It is the same for weak solutions, except for the value of the exponential prefactor in equation (4.6).

4.2 CHEMICAL KINETICS

If the chemical equilibrium only concerns the properties of the reactants and products the same cannot be said for the reaction kinetics. Starting from arbitrary initial concentrations of the reactants, there are usually many possible pathways to reach chemical equilibrium, each with their own kinetics. It is therefore necessary to introduce the notion of elementary reaction in which the products are directly formed from the reactants. Thus, the general reaction is a chain of elementary reactions. By the stoichiometry of the reaction the derivatives dn/dt are related by the common ratio $W = -(1/\nu_i)dn_i/dt = (1/\nu_j')dn_j'/dt$ called the reaction rate. W is a function of the pressure, the temperature and the reactant concentrations. Before determining its expression let us note that since the concentrations of reactants and products are stable at equilibrium, the reaction rates of forward and backward reactions W and W' must be equal. Since this condition is the same as the law of mass action in paragraph 4.1, it is clear that W must be of the form

$$W = k_0 \exp\left(-\frac{E}{kT}\right)\Pi n_i^{\nu_i} \tag{4.1}$$

where E is the reaction activation energy and k_0 still depends on the pressure and temperature, but not exponentially. The condition of chemical equilibrium allows the determination of $K(P, T) = (k_0/k_0')\exp(-(E-E')/kT)$. Thus, from equation (4.6) in paragraph 4.1, $Q = E'-E$, or even for an exothermic reaction, the reaction rate of the backward reaction is more sensitive to temperature than the forward reaction. Consider for example the case of a bimolecular reaction $(\nu_1 = \nu_2 = 1)$ in which two molecules react to form a third molecule. The reacting molecules must have sufficient thermal energy to overcome the potential barrier associated with the change in their electronic structure. The rate of the reaction is therefore proportional to the product of the concentrations of the two species weighted by the Arrhenius coefficient, corresponding to the passage over the barrier of potential energy E.

For reactions in the gaseous phase the activation energy is often large compared to the thermal energy. Contrary to the liquid–solid phase transition in paragraph 2.2 or even for liquid–gas transitions which essentially use Van der Waals' forces, the chemical reaction breaks and forms covalent bonds. The energy of the covalent bond is a hundred times larger, of the order of 10 kcal/mole or $6.9 \cdot 10^{-20}$ J per molecule. This results in a very strong dependence of the reaction rate on temperature, as is illustrated in the following example: the order of magnitude of the activation energy for a combustion reaction is 50 kcal/mole. Thus at room temperature, the Arrhenius coefficient is about $6 \cdot 10^{-37}$, whereas at 1000 K it is about 10^{-11}. At a given temperature there is therefore a strong dependence of the same coefficient on the elementary reaction under consideration. For example, an activation energy of 10 kcal/mole always leads to an Arrhenius coefficient of $6 \cdot 10^{-8}$ at room temperature. This large difference in the order of magnitude in the reaction rate means that a reactant mixture can stay far from the equilibrium concentrations (frozen mixture) at a given temperature because the forward reaction is for instance much faster than the backward one.

Let us consider a reaction producing hydrochloric acid HCl

$$H_2 + Cl_2 = 2HCl. \tag{4.2}$$

This reaction can occur by binary collisions between molecules of hydrogen and chlorine. Nevertheless, they must overcome an activation energy barrier of 162 kcal / mole corresponding to the dissociation of the hydrogen and chlorine molecules. The chain reaction which we will now describe is much faster. The first elementary reaction consists of the dissociation of a molecule of chlorine with the help of a third molecule which is chemically inert

$$Cl_2 + M = Cl + Cl + M \tag{4.3}$$

of reaction constant k_1, much more favourable than the analogous reaction with a molecule of hydrogen, as it is associated with a moderate dissociation energy (57 kcal/mole compared to 103 kcal/mole). Reaction (4.3) is reversible with a reaction constant k_1'. The chlorine atoms formed in this way can react with a hydrogen molecule to form hydrochloric acid

$$Cl + H_2 = HCl + H \tag{4.4}$$

with a reaction constant k_2 with a weak activation energy ($\approx$ 5 kcal/mole). Hydrogen atoms thus formed can further react with chlorine molecules

$$H + Cl_2 = HCl + Cl \tag{4.5}$$

with a reaction constant k_3. The chlorine atom plays the role of a catalyst because it is reproduced by the reaction. Indeed, by adding together reactions (4.4) and (4.5), we get

$$Cl + H_2 + Cl_2 = 2HCl + Cl. \tag{4.6}$$

As determined by the initial reaction (4.1) the concentration of the chlorine atoms stays approximately constant, being a catalyst and a product at the same time, in reaction (4.6). From the appendix, since H and Cl are produced slowly and consumed rapidly, their concentration is quasi-stationary

$$\frac{d[H]}{dt} = k_2[Cl][H_2] - k_3[H][Cl_2] \approx 0 \tag{4.7}$$

$$\frac{d[Cl]}{dt} = 2k_1[M][Cl_2] - 2k_1'[M][Cl]^2 \approx 0 \tag{4.8}$$

then

$$[H] = \frac{k_2}{k_3}\left(\frac{k_1}{k_1'}\right)^{1/2}[H_2][Cl_2]^{1/2}; \ [Cl] = \left(\frac{k_1}{k_1'}\right)^{1/2}[Cl_2]^{1/2}. \tag{4.9}$$

The rate of production of hydrochloric acid can be determined as

$$W = \frac{d[\mathrm{HCl}]}{dt} = k_2[\mathrm{Cl}][\mathrm{H_2}] + k_3[\mathrm{H}][\mathrm{Cl_2}]$$
$$= 2k_2\left(\frac{k_1}{k_1'}\right)^{1/2}[\mathrm{Cl_2}]^{1/2}[\mathrm{H_2}]. \tag{4.10}$$

As we would expect, this reaction rate is not proportional to the product of the reactant concentrations, as it is in a simple bimolecular reaction. The effective reaction rate constant k_{HCl} is the product of the rate constant of the propagation of a reaction chain with a low activation energy and the square root of the equilibrium constant of the initiation reaction $K_I = k_1/k_1'$. Taking the numerical values $k_2 = 10^{-10}\exp(-6000/RT)\ \mathrm{cm^3/s}$, $K_I = 2.5\times10^{25}\exp(-56800/RT)\ \mathrm{cm^3/s}$ we deduce $k_{\mathrm{HCl}} = 10^3\exp(-34400/RT)\ \mathrm{cm^3/s}$, with a lower activation energy than each of the two reactions involving the dissociation of the initial molecules.

Let us finally consider the reaction at the electrode

$$\mathrm{I_3^-} + 2\mathrm{e} \leftrightarrow 3\mathrm{I^-} \tag{4.11}$$

in three stages:

$$\mathrm{I_3^-} \leftrightarrow \mathrm{I_2} + \mathrm{I^-} \tag{4.12a}$$

$$\mathrm{I_2} \leftrightarrow 2\mathrm{I} \tag{4.12b}$$

$$2(\mathrm{I} + \mathrm{e} \rightarrow \mathrm{I^-}). \tag{4.12c}$$

The Butler–Volmer kinetics in paragraph 3.5 applied to reaction (4.12c) is written as

$$i = 2F\left(k_3 c_{I^-}\exp\left((1-\beta)\frac{F\Delta\Phi}{RT}\right) - k_{-3}c_I\exp\left(-\beta\frac{F\Delta\Phi}{RT}\right)\right). \tag{4.13}$$

Reactions (4.12a) and (4.12b) are at equilibrium with equilibrium constants K_I and K_2,

$$c_I = (K_2 c_{I_2})^{1/2} = \left(K_I K_2 \frac{c_{\mathrm{I_3^-}}}{c_{\mathrm{I^-}}}\right)^{1/2} \tag{4.14}$$

so that

$$i = 2F\left(k_3 c_{\mathrm{I^-}}\exp\left((1-\beta)\frac{F\Delta\Phi}{RT}\right) - k_{-3}K_1K_2^{1/2}\left(\frac{c_{\mathrm{I_3^-}}}{c_{\mathrm{I^-}}}\right)^{1/2}\exp\left(-\beta\frac{F\Delta\Phi}{RT}\right)\right). \tag{4.15}$$

Appendix: Quasi-stationary approximation

Let us consider the consecutive reactions

$$R \rightarrow X \rightarrow P$$

of rate constants k_1 and k_2, where R is the reactant, X an intermediate radical and P the product. The concentrations of different reacting species satisfy the equations

$$\frac{d[R]}{dt} = -k_1[R]; \quad \frac{d[X]}{dt} = -k_1[R] - k_2[X]; \quad \frac{d[P]}{dt} = -k_2[X]$$

Thus,

$$[R](t) = [R](0)\exp(-k_1 t)$$

$$[X](t) = [R](0)\frac{k_1}{k_2 - k_1}(\exp(-k_1 t) - \exp(-k_2 t)).$$

When the radical is consumed much faster than it is produced, $k_2 > k_1$ and, on short timescales $1/k_2$, the term $\exp(-k_2 t)$ becomes negligible. Thus, after a time of that order of magnitude $k_1[R] \approx k_2[X]$ and $dX/dt \approx 0$.

4.3 KINETICS OF AN EXPLOSIVE REACTION

Explosive reactions employed in premixed flame propagations generally have very complex kinetics. The kinetic mechanisms of a large number of those reactions are described for instance in the book by Zeldovich and coworkers (1980) and Linan and Williams (1993). For example, a hydrocarbon–oxygen reaction takes place in some three hundred intermediate reactions involving a large number of molecules and radicals. These reactions have diverse activation energies and each one can play a more or a less important role according to the given temperature and pressure. A given kinetic mechanism can only be valid for a specific region of temperature and pressure. For example, the reaction mechanism of hydrogen–oxygen can be summarised by three important stages. In the region $P \approx 1$ atm., $T \approx 1200$ K, and when the oxygen limits the reaction, first we have the group

$$OH + H_2 \Leftrightarrow H_2O + H \qquad (4.R1)$$

$$H + O_2 \Leftrightarrow OH + O \qquad (4.R2)$$

$$O + H_2 \Leftrightarrow OH + H \qquad (4.R3)$$

where the general reaction $(2(R1) + (R2) + (R3))$

$$H + O_2 + 3H_2 \Rightarrow 2H_2O + 3H \qquad (4.CB)$$

generates active centres, H radicals, capable of propagating the reaction. It is a chain branching reaction (CB), meaning that it creates more radicals than it uses and is weakly exothermic. Reaction (4.R2) produces radicals O and OH slower than it consumes them. According to the appendix in paragraph 4.2 their concentration can be considered constant, such that

$$[O] = \frac{k_{R2}[H][O_2]}{k_{R3}[H_2]} \tag{4.1}$$

and

$$[OH] = \frac{2k_{R2}[H][O_2]}{k_{R1}[H_2]}. \tag{4.2}$$

The rate of production of H radicals by the reaction (4.CB) is therefore $W_{CB} = 2k_{R2}[H][O_2]$. Since the total reaction is very exothermic, the heat of reaction is mainly produced by the recombination of the radicals. The simplest case is a simple recombination of H radicals by chain termination (4.CT)

$$H + H + M \Leftrightarrow H_2 \tag{4.CT}$$

which is strongly exothermic ($Q \approx 103$ kcal/mole) but of a very low activation energy. The reaction rate is therefore $W_{CT} = k_{CT}[H]^2$. Another reaction slowly generates the radical HO_2 by a three body reaction

$$H + O_2 + M \Rightarrow HO_2 + M \tag{4.R4}$$

which can then rapidly be recombined with H by

$$H + HO_2 \Leftrightarrow OH + OH \tag{4.R5}$$

$$H + HO_2 \Leftrightarrow H_2 + O_2. \tag{4.R6}$$

There, the concentration of HO_2 is quasi-stationary as well, so

$$[HO_2] = \frac{k_{R5}}{k_{R6} + k_{R7}}[O_2]. \tag{4.3}$$

The production of radicals OH by reaction (4.R6) changes its given quasi-stationary concentration by equation (4.2)

$$[OH] = \frac{2k_{R2} + \dfrac{2k_{R5}k_{R6}}{k_{R6} + k_{R7}}}{k_{R1}} \frac{[H][O_2]}{[H_2]}. \tag{4.4}$$

The rate production of H radicals by reaction (4.CB) therefore becomes $W_{CB} = (2k_{R2} + 2k_{R5}k_{R6}/(k_{R6} + k_{R7}))[H][O_2]$, and that of the consumption of H by reactions R5 to R7 is $W_C = -2k_{R5}[H][O_2]$. The kinetic mechanism of the reaction can be summarized by the three stages

$$R + H \rightarrow 3H + P_1 \tag{4.BC}$$

$$R + H \rightarrow P_2 \tag{4.C}$$

$$M + H + H \rightarrow P_3 + M. \tag{4.CT}$$

The first is a chain-branching reaction, with a high activation energy E_I and it is weakly exothermic. Its rate constant is $k_1 = (k_{R2} + k_{R5}k_{R6}/(k_{R6} + k_{R7}))$. The second is a competing reaction with a low activation energy and the rate constant is $k_2 = 2k_{R5}$. The third is a termination reaction with a negligible activation energy and it is strongly exothermic with a rate constant $k_3 = k_{TC}$. By regrouping reactions (4.CB) and (4.C), the total production of H radicals is proportional to $(2k_1 - k_2)$ $[H]$. Contrary to k_2, k_1 is very sensitive to temperature and rapidly decreases with it. Thus, below a temperature T^* which satisfies $2k_1(T^*) = k_2$, the H radicals are no longer generated and completely inhibit the reaction. At slightly higher temperatures, $2k_1 - k_2 < k_3$, such that $[H]$ can be considered quasi-stationary, according to the appendix in paragraph 4.2. For $T > T^*$, $[H] = (2k_1 - k_2)/k_3\ [R]$, for $T < T^*$, $[H] = 0$. We are therefore brought back to the case of a single reaction $R \rightarrow P$, of rate constant $k = (k_1 + k_2)(2k_1 - k_2)/k_3\ [R]$ for $T > T^*$, and 0 if not (Clavin 1985).

4.4 THE REACTION OF CHLORITE–IODIDE–MALONIC ACID

The reaction of chlorite (ClO_2^-), iodide (I^-) and malonic acid (MA) is the first example of a reaction which takes place spontaneously at a steady-state and is spatially inhomogeneous (Castet *et al.* 1990). Before studying the origin of these modulations in paragraph 6.8, we describe here a simplified mechanism of the reaction (Lengyel and Epstein 1991). After the fast production of iodine I_2 and ClO_2 from I^- and ClO_2^-, the iodine I_2 reacts with the malonic acid by

$$MA + I_2 \rightarrow IMA + I^- + H^+ \tag{4.1}$$

with reaction rate $W_I = k_{1a}[MA][I_2]/(k_{1b} + [I_2])$. In a second reaction, ClO_2 reacts with I^- to give iodine

$$ClO_2 + I^- \rightarrow ClO_2^- + \frac{1}{2}I_2 \tag{4.2}$$

with the simple reaction rate $W_2 = k_2[ClO_2][I^-]$. Finally, we have the reaction

$$ClO_2^- + 4I^- + 4H^+ \rightarrow Cl^- + 2I_2 + 2H_2O \tag{4.3}$$

with reaction rate

$$W_3 = k_{3a}[ClO_2^-][I^-][H^+] + k_{3b}[ClO_2^-][I_2][I^-]/(k_{3c} + [I^-]^2).$$

Also assuming that the malonic acid and the iodine are in excess and that the pH is high the reaction mechanism can again be reduced to two species $X = I^-$ and $Y = ClO_2^-$ by

$$I_2 \rightarrow X \tag{4.4}$$

with constant reaction rate $r_1 = k_{1a}[MA]_0[I_2]_0/(k_{1b} + [I_2]_0)$, and

$$X \rightarrow Y \tag{4.5}$$

with reaction rate $r_2 = k_2[ClO_2]_0[X]$, and finally

$$4X + Y \rightarrow P \tag{4.6}$$

with reaction rate $r_3 = k_{3b}[I_2]_0[X][Y]/(k_{3c} + [X]^2)$.

4.5 ELECTRODE REACTION

Despic (1961) and Drazic (1962) showed that the reaction kinetics at the electrode

$$Fe^{2+} + 2e \leftrightarrow Fe \tag{4.1}$$

also depends on the pH of the solution, involving a more complex mechanism than that in equation (4.1). One of the proposed mechanisms assumes three stages

$$Fe^{2+} + H_2O \leftrightarrow FeOH^+ + H^+ \tag{4.R1}$$

$$FeOH^+ + e \rightarrow FeOH \tag{4.R2}$$

$$FeOH + H^+ + e \leftrightarrow Fe + H_2O \tag{4.R3}$$

If reactions (4.R1) and (4.R3) are equilibrated the limiting reaction is (4.R2) and applying the Butler–Volmer equation in paragraph 3.5 for this reaction,

$$i = 2F\left(k_{-2}\theta_{FeOH}\exp\left((1-\beta)\frac{F\Delta\Phi}{RT}\right) - k_2 c_{FeOH^+}(1-\theta_{FeOH})\exp\left(-\beta\frac{F\Delta\Phi}{RT}\right)\right) \tag{4.2}$$

where θ_{FeOH} is the fraction of the electrode surface covered by the adsorbed Fe OH. This last quantity is deduced from the equilibrium condition of reaction (4.R3), that is,

$$k_{-3}(1-\theta_{FeOH})\exp\left((1-\beta)\frac{F\Delta\Phi}{RT}\right) = k_3 c_{H^+}\theta_{FeOH}\exp\left(-\beta\frac{F\Delta\Phi}{RT}\right) \tag{4.3}$$

so assuming $\theta_{FeOH} < 1$,

$$\theta_{FeOH} = \frac{1}{K_3 c_{H^+}}\exp\left(\frac{F\Delta\Phi}{RT}\right) = \frac{1}{K_3 K_e c_{OH^-}}\exp\left(\frac{F\Delta\Phi}{RT}\right) \tag{4.4}$$

where K_e is the equilibrium constant of water. The equilibrium condition applied to reaction (4.R1) is written as

$$K_1 = \frac{c_{FeOH^+} c_{H^+}}{c_{Fe^{2+}}} \tag{4.5}$$

We can therefore deduce, by expressing θ_{FeOH} and c_{FeOH^+} determined by equations (4.4) and (4.5) in equation (4.2),

$$i = 2F\left(k_{-2}\frac{1}{K_3 K_2}c_{OH^-}\exp\left((2-\beta)\frac{F\Delta\Phi}{RT}\right) - k_2\frac{K_1}{K_e}c_{Fe^{2+}}c_{OH^-}\exp\left(-\beta\frac{F\Delta\Phi}{RT}\right)\right) \tag{4.6}$$

This mechanism is validated by experimental measurements, i.e. by a linear current dependence with respect to c_{OH^-} and $c_{Fe^{2+}}$, and with $\beta \approx 1/2$ the slopes $d(\Delta\Phi)/d(\mathrm{Log}\,i) = 2.3RT/F(2-\beta) = 0.04\mathrm{V}$ for the positive potentials where the oxidation reaction dominates and $d(\Delta\Phi)/d(\mathrm{Log}i) = -RT/F\beta = 0.12\mathrm{V}$ for the negative potentials.

BIBLIOGRAPHY

[1] J.O'M. Bockris, A.K.N. Reddy, *Modern Electrochemistry*, Vol. 1 and 2, Plenum Press, New York, 1970.

[2] V. Castet, E. Dulos, J. Boissonade, P. de Kepper, 'Experimental Evidence of a Sustained Standing Turing-Type nonequilibrium Chemical Pattern', *Phys. Rev. Lett.,* **64**, p. 2953, 1990.

[3] P. Clavin, 'Dynamical Behaviour of Premixed Flame Fronts in Laminar and Turbulent Flows', *Prog. Energy. Combust. Sci.*, **11**, p. 1, 1985.

[4] J. O'M. Bockris, D. Drazic, A. R. Despic, 'The Electrode Kinetics of the Deposition and Dissolution of Iron', *Electrochim. Acta*, **4**, p. 325, 1961.

[5] J. O'M. Bockris, D. Drazic, 'The Kinetics of Deposition and Dissolution of Iron: Effect of Alloying Impurities', *Electrochim. Acta*, **7**, p. 293, 1962.

[6] I. Lengyel, I. Epstein, 'Modeling of Turing Structures in the Chlorite–Iodide–Malonic Acid–Starch Reaction System', *Science*, **251**, p. 650, 1991.

[7] A. Linan, F.A. Williams, 'Fundamental Aspects of Combustion', Oxford Engineering Science Series, **34**, 1993.

[8] Ya. B. Zeldovich, 'Dokazatel'svo edinstvennosti resheniya uravnenii zakona deistvuy-ushchikh mass' (A Proof of the Uniqueness of the Solution of the Equations for the Law of Mass Action), *Zh. Fiz. Khim*, **11**, p. 635, 1938.

[9] Ya.B. Zeldovich, 'Tsepnye reaktsii v goryachikh plamenakh – priblizhennaya teoriya skorosti plameni' (Chain Reactions in Hot Flames – an Approximate Theory of Flame Velocities), *Kinetika i Kataliz*, **11**, p. 305, 1961.

[10] Ya.B. Zeldovich, G.I. Barenblatt, V.B. Librovich, G.M. Makhviladze, *The Mathematical Theory of Combustion and Explosion*, Nauka, Moscow, 1980; english translation, Consultants Bureau, a division of Plenum Publishing Corporation, New York, 1985.

CHAPTER 5

Simple growth forms

5.1 LOW REYNOLDS NUMBER HYDRODYNAMICS

According to equations (1.2) and (1.12) in paragraph 1.4, and (1.9) in paragraph 1.5 the velocity field $\vec{w}$ of a Newtonian and incompressible viscous fluid satisfies the Navier–Stokes equation

$$\rho\left(\frac{\partial \vec{w}}{\partial t} + \left(\vec{w} \cdot \vec{\nabla}\right)\vec{w}\right) = -\vec{\nabla} p + \eta \Delta \vec{w} \tag{5.1}$$

with the condition of incompressibility

$$\vec{\nabla} \cdot \vec{w} = 0 . \tag{5.2}$$

Here ρ is the volumetric mass of the fluid, p the pressure and η the dynamic viscosity. At a solid wall the molecules of the fluid are held by molecular cohesion forces and we can assume that the velocity of the fluid is zero at the wall, $\vec{w} = 0$. In specific cases such as a polymer flow we can take into account a small slipping velocity. In the case of an interface between two immiscible moving fluids we will assume that the velocities of the fluids on both sides of the interface are equal and equal to the velocity of the interface itself.

Since the fluid is incompressible its momentum density flux tensor, determined by equations (1.12) in paragraph 1.4 and (1.9) in paragraph 1.5, is written more simply as $\Pi_{ik} = p\delta_{ik} + \rho w_i w_k - \eta(\partial w_i / \partial x_k + \partial w_k / \partial x_i)$. To determine the boundary conditions for a velocity field at a free surface dividing two viscous fluids we integrate equation (1.12) in paragraph 1.4 over a small volume element containing the surface element and we make the dimension normal to the interface tend to zero. Transforming the second part of this equation into a surface integral we obtain

$$\Pi_{ik}^{(1)} n_k^{(1)} + \Pi_{ik}^{(2)} n_k^{(2)} = 0 \tag{5.3}$$

where $\vec{n}$ is the vector normal to the surface and is oriented outside the small volume element and $\vec{n}^{(1)} = -\vec{n}^{(2)} = \vec{n}$. This relation also means that the forces with which the fluids interact on each other are equal and of opposite direction. Indeed, the force exerted by a fluid on a surface element dS is just the momentum flux through this element, $\Pi_{ik} n_k dS$. Relation (5.3) is written more simply by using the equality of the

Moreover, if we assume that the interface is slow enough the pressure gap generated by capillary effects, as determined in the appendix of paragraph 2.2, is

$$p_1 - p_2 = -\gamma\left(\frac{2}{b} + \frac{\pi}{4}\frac{1}{R}\right). \tag{5.6}$$

Let us consider now the more general case of a steady-state flow between two moving surfaces, almost parallel, with a large radius of curvature compared to the gap between the two surfaces. Let us assume that the lower surface ($z = 0$) is planar and fixed. Let x be the coordinate along the relative motion and z be the transverse coordinate. The upper surface ($z = h(x)$) is driven with a velocity of component U along the x axis and V along the z axis, where V is far smaller than U. Then v will be small compared to u, with both u and v slowly varying along the x direction compared to their variation in z. Equations (5.6) and (5.2) in paragraph 5.1 are written in the form

$$\frac{dp}{dx} = \eta\frac{d^2u}{dz^2},\ \frac{dp}{dz} = 0 \tag{5.7a}$$

$$\frac{du}{dx} + \frac{dv}{dz} = 0. \tag{5.7b}$$

At the plates ($z = 0$ and $z = h$) we must apply the boundary condition of zero relative velocity, that is

$$\begin{aligned} z = 0,\ u = 0,\ v &= 0 \\ z = h,\ u = U,\ v &= U\frac{dh}{dx} + V. \end{aligned} \tag{5.8}$$

The second of equations (5.7a) gives p independent of z. The component u is deduced from the first of the above equations (5.7a) and from the boundary conditions (5.8)

$$u = \frac{1}{2\eta}\frac{dp}{dx}(z-h)z + U\frac{z}{h}. \tag{5.9}$$

We then determine dv/dz from equation (5.7b). Integrating the previous expression between $z = 0$ and $z = h$ we determine the Reynolds equation of lubrication (1886) relating p and h:

$$\frac{d}{dx}\left(h^3\frac{dp}{dx}\right) = 6\eta\left(U\frac{dh}{dx} + 2V\right). \tag{5.10}$$

The force exerted by the fluid on the unit moving surface has the components: p along the z axis, $-\eta\ \partial u/\partial z$ along the x axis evaluated at $z = h$, or the latter, $-(h/2)dp/dx - \eta U/h$, according to paragraph 5.1. When the surfaces have well-defined shapes, which is the case for the rotation of an axle in a lubricated cylindrical cavity from appendix 1, the profile $z = h(x)$ must be introduced into equation (5.10) and the pressure field is then determined by solving the equation.

When the surface $z = h(x)$ is free, as in the case of the spreading of a liquid droplet on a substrate (appendix 4), or in the case of the formation of a film along a plate initially dipped into a fluid and then pulled vertically upwards (appendix 2), it is the shape of the free surface which must be determined. With this in mind, we introduce the boundary condition (5.5) in paragraph 5.1 which relates the pressure gap at the interface to the viscous friction and to the curvature effects related to surface tension. Since the free surface is almost parallel to the fixed and planar surface, these boundary conditions become

$$p = -\gamma\frac{\partial^2 h}{\partial x^2} \tag{5.11a}$$

$$\eta\frac{du}{dz} = 0\,. \tag{5.11b}$$

The new longitudinal velocity field is therefore written as

$$u = \frac{1}{2\eta}\frac{dp}{dx}(z - 2h)z\,. \tag{5.12}$$

In the same way as previously, we determine dv/dz from equations (5.7) and we integrate the expression between 0 and h. As $v(h) = U\ dh/dx$, where U is the velocity of the displacement of the steady-state profile, we obtain the following equation

$$\gamma\frac{d}{dx}\left(h^3\frac{d^3 h}{dx^3}\right) = 3\eta U\frac{dh}{dx}\,. \tag{5.13}$$

Appendix 1: Friction momentum exerted by a viscous fluid on a moving cylinder of radius R' of mass m, rotating at an angular speed Ω inside a cylinder of a slightly larger radius R.

Due to its own weight the axle is not centred on the cavity during its rotation (figure 5.2). The thickness of the annular space between the two cylinders, far smaller than the radii of the cylinders, therefore changes with the angular coordinate θ. Let r be the distance between the centres of the two cylinders and $d = R - R'$. Then, to a first approximation, the thickness h of the annular space is

$$h = d + r\cos\theta$$

and the origin of the angular coordinate is determined by the direction joining the two centres of the cylinders where the thickness is the largest.

Since the thickness h depends only on the angular coordinate, the Reynolds equation (5.10) in paragraph 5.2 is simply integrated and we obtain

$$\frac{1}{R}\frac{dp}{d\theta} = 6\eta\Omega R\left(\frac{1}{h^2} - \frac{h_0}{h^3}\right)$$

where h_0 is the thickness of a section where $dp/d\theta = 0$. Assuming that the pressure is high enough everywhere so that the phenomenon of cavitation does not occur, the values which between a liquid L and a solid S for instance, the equality of the energy flux projected on the normal direction is written in the moving frame, where the interface is at rest,

$$\vec{j}_L \cdot \vec{n} = \rho_L h_L \vec{v}_L \cdot \vec{n} - \lambda_L \left(\vec{\nabla T}\right)_L \cdot \vec{n} = \rho_S h_S \vec{v}_S \cdot \vec{n} - \lambda_S \left(\vec{\nabla T}\right)_S \cdot \vec{n} = \vec{j}_S \cdot \vec{n}. \quad (5.5)$$

From the equation of continuity through the interface

$$\rho_S \vec{v}_S \cdot \vec{n} = \rho_L \vec{v}_L \cdot \vec{n} = -\rho_S \vec{v} \cdot \vec{n} \quad (5.6)$$

we deduce that the temperature gradient undergoes a change proportional to the normal velocity of the interface

$$(h_L - h_S)\rho_S \vec{v} \cdot \vec{n} = \left(\lambda_S \left(\vec{\nabla T}\right)_S - \lambda_L \left(\vec{\nabla T}\right)_L\right) \cdot \vec{n}. \quad (5.7)$$

For a liquid–solid interface the latent heat per unit solid volume is defined as $Q = (h_L - h_S)\rho_S$ and we obtain the relation

$$Q\vec{v} \cdot \vec{n} = \left(D_S c_{pS} \left(\vec{\nabla T}\right)_S - D_L c_{pL} \left(\vec{\nabla T}\right)_L\right) \cdot \vec{n}. \quad (5.8)$$

A similar relation is also obtained by using the conservation of the number of molecules of solute. If we neglect the diffusion of solute in the solid phase we deduce, from equation (2.10) in paragraph 2.4, the concentration gradient around the interface in the liquid phase as a function of the normal velocity of propagation

$$c_L(1-K)\vec{v} \cdot \vec{n} = -D\vec{\nabla c_L} \cdot \vec{n}. \quad (5.9)$$

We must take into account any potential heat source and chemical reaction in the medium in equations (5.3) and (5.4). If Q is the quantity of heat produced per unit time and per unit fluid volume and W the number of molecules of solutc created per unit time and per unit volume we can write

$$\frac{\partial T}{\partial t} = D_T \Delta T + \frac{Q}{\rho c_p}$$
$$\frac{\partial c}{\partial t} = D\Delta c + W. \quad (5.10)$$

In the case of a point-like heat source $Q(\vec{r}, t) = Q_0 \delta(\vec{r} - \vec{r_0})\delta(t - t_0)$ we have to introduce the Green function $G(p/p_0)$ which satisfies the following equation

$$\frac{\partial G}{\partial t} = D_T \Delta G + \delta(\vec{r} - \vec{r_0})\delta(t - t_0) \quad (5.11)$$

where $p = (\vec{r}, t)$ and $p_0 = (\vec{r_0}, t_0)$, with the condition $G = 0$ if $t < t_0$. We easily determine the solution of this equation using the Fourier transform, that is

$$G(p|p_0) = \int \frac{d^3\vec{k}d\omega}{(2\pi)^4} \hat{G}(\vec{k}, \omega)\exp(-i\vec{k}(\vec{r} - \vec{r_0}))\exp(-i\omega(t - t_0)) \quad (5.12)$$

where

$$\hat{G}(\vec{k}, \omega) = \frac{1}{-i\omega + D_T k^2}. \tag{5.13}$$

We first integrate over the variable ω, that is, using the residue theorem applied to the simple pole $i\omega = D_T k^2$,

$$G(p|p_0) = \int \frac{d^3\vec{k}}{(2\pi)^3} \exp(-i\vec{k}(\vec{r}-\vec{r_0}))\exp(-D_T k^2(t-t_0)) . \tag{5.14}$$

We integrate over each coordinate k_x, k_y, k_z the same way, producing first a square of the component of the considered wavenumber and then calculating the resulting Gaussian integral

$$G(p|p_0) = \frac{\theta(t-t_0)}{(4\pi D_T(t-t_0))^{3/2}} \exp -\frac{|\vec{r}-\vec{r_0}|}{4D_T(t-t_0)} \tag{5.15}$$

where $|\vec{r}-\vec{r_0}| = ((x-x_0)^2 + (y \;\; y_0)^2 + (z-z_0)^2)^{1/2}$ is the distance between the two points considered and $\theta = 1$, if $t < t_0$, and 0 otherwise.

The model of growth of viscous fingering in a Hele–Shaw cell defined in paragraph 5.2 and the models of crystalline growth defined in this paragraph share many common points. In all these cases it comes down to dealing with problems of free boundaries. We must determine the dynamics of an interface between two media; this dynamics is determined by field equations defined in these two media (fluid velocity, temperature and concentration) and by boundary conditions at the interface and at the limits of the domain considered. The value of the velocity, temperature or solute concentration potentials varies proportionally to the curvature of the interface in all cases. The rate of displacement of the interface is proportional to the gradient of the mentioned quantities.

However, we must note several differences. The first one is related to the equation of the considered field. The velocity potential satisfies the Laplace equation whereas temperature and concentration satisfy the diffusion equation. Let us consider the growth of a crystal at a velocity V in a space of dimension L. The order of magnitude of the advective term is $V\delta c/L$, whereas the order of magnitude of the diffusive term is $D\delta c/L^2$. Therefore when the Péclet number $P = VL/D$ is small, the advective term is negligible and we find the Laplace equation again. The second difference deals with crystalline anisotropy. Contrary to the interface between two fluids the liquid–solid interface is anisotropic. This anisotropy intervenes in the boundary condition which determines the value of the considered field at the interface. We can artificially create this anisotropy in the case of viscous fingering if we engrave a lattice of arbitrary symmetry on the plates of the Hele–Shaw cell. We will see later that these two differences lead to clear effects distinguishing the dynamics of these interfaces.

5.4 PLANAR LIQUID-SOLID INTERFACES AND SPHERICAL GERMS

When the amount of supercooling Δ is small, according to the change of variable used in appendix 1, the inertial term $\partial T/\partial t$ becomes negligible at the scale of the radius of the germ ($u \approx u_0$) and the temperature field satisfies the Laplace equation. The temperature field is therefore simply written as

$$T = T_\infty + (T_f - T_\infty)\frac{R}{r} . \tag{5.1}$$

From equation (5.8) in paragraph 5.3, we deduce that

$$\frac{dR}{dt} = \frac{D_T}{R}(T_f - T_\infty)\frac{c_p}{Q} . \tag{5.2}$$

In particular when we take into account the effects of surface tension,

$$\frac{dR}{dt} = \frac{D_T}{R}\left(\Delta - \frac{2d_{0T}}{R}\right) \tag{5.3}$$

where d_{0T} is the capillary length defined in paragraph 2.4 (Langer 1980). Then, the expansion of the germ cancels for a critical radius $R_c = 2d_{0T}/\Delta$, the nucleation radius determined by equation (3.3) in paragraph 3.1. This equilibrium is unstable since it maximises the free energy of the germ. Near the equilibrium point, $dR/dt \approx D_T(\Delta/R_c^2)(R - R_c)$. We can deduce the Zeldovich factor $B(R_c) = D_T\Delta^3 kT/32\pi\,\gamma d_{0T}^2$, introduced in the formula of the nucleation rate determined by equation (3.9) in paragraph 3.1, taking the limit when R tends to R_c of $B(R) = -(dR/dt)kT/(d(\Delta G/dR))$.

A similar calculation can be done for a germ growing from an isothermic oversaturated solution at the concentration c_∞. We then use the model determined by equations (5.4) and (5.9) in paragraph 5.3 and we obtain

$$\frac{dR}{dt} = \frac{D}{R}\left(\Delta - \frac{2d_{0c}}{R}\right) \tag{5.4}$$

where $\Delta = (c_0 - c_\infty)/c_0(1 - K)$ is the oversaturation and $d_{0c} = \gamma T_0/Q|m|c_0(1 - K)$ is the chemical capillary length.

When Δ is larger than 1 solutions featuring a constant velocity can be determined if the effects of kinetic relations are taken into account. In fact for such a solution the temperature field is a function of only the variable $x - Vt$, such that

$$V\frac{dT}{dx} = D_T\frac{d^2T}{dx^2} \tag{5.5}$$

the solution of which is of the form

$$T = T_\infty + (T_f - T_\infty)\exp\left(-x\frac{V}{D_T}\right), x > 0$$
$$T = T_f \qquad , x < 0. \tag{5.6}$$

For convenience we will still use the same variable x in the moving frame where the interface is at rest, at $x = 0$. According to equation (5.8) in paragraph 5.3, $VQ/c_p = V(T_f - T_\infty)$. The temperature of the interface will therefore be set at a value independent of the velocity. Besides, according to the effects of kinetic relations described in paragraph 3.2 $T_f = T_0(1 - \alpha V/Q)$, we deduce

$$V = \xi\frac{Q}{c_p}(\Delta - 1) \tag{5.7}$$

where $\xi = Q/T_0\alpha$.

In the configuration of a directional growth of a liquid–solid interface a sample of liquid is drawn at a constant speed V in a temperature gradient $G = (T_l - T_s)/L$ where T_l and T_s are the temperatures, held constant, of the two points L apart (figure 5.7). A liquid–solid interface is formed, near the liquid–solid transition temperature, and this interface is therefore in relative motion with respect to the liquid with velocity V. Far in front of the interface the concentration of solute is c_∞. Due to its motion the interface pushes the solute back into the liquid and creates a diffusion layer.

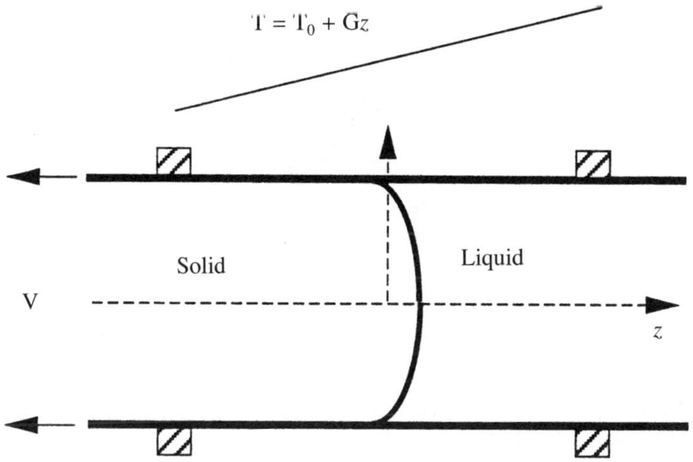

Figure 5.7 Typical configuration of directional growth.

To determine the resulting concentration profile we solve the diffusion equation for the concentration field in the moving frame where the interface is at rest and in the same way as for equation (5.6) we deduce

$$\begin{aligned} c &= c_\infty + (c_i - c_\infty)\exp\left(-x\frac{V}{D}\right), \quad x > 0 \\ c &= c_\infty, \qquad\qquad\qquad\qquad\qquad x < 0 \end{aligned} \tag{5.8}$$

since, owing to the conservation of the molecules of solute in a steady-state regime, the concentration far behind the front must be the same as that far ahead of the front. At the interface, equations (2.11) in paragraph 2.4 and (5.9) in paragraph 5.3 lead to $c_i = c_\infty/K$. Since c_i is determined from the local temperature, by the relation $T = T_0 + mc_i = T_r + Gz_i$, where T_r is a reference temperature which sets the origin of the positions, the position of the front of the temperature gradient is deduced. It is worth noticing that the concentration profile (5.8) can lead to a thermodynamic instability of the liquid upstream from the interface if the growth rate is high enough. As shown in figure 5.8, if the growth velocity V exceeds the velocity of 'constitutional supercooling' $V_c = GD/(|m|c_\infty(1-K))$, the local temperature of the liquid can be lower than the local temperature of solidification, which leads to a different crystallisation mode.

The velocity v_M of a crystalline step such as the one considered in paragraph 3.3 can be determined as below (Nozières 1992). When the chemical potentials of the

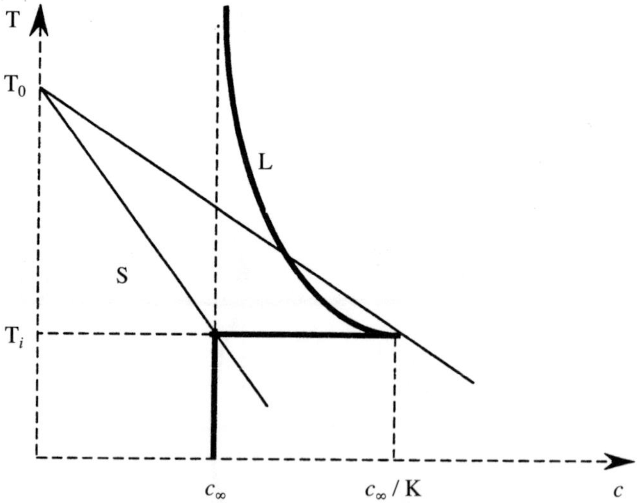

Figure 5.8 Concentration profile (in bold) plotted on the diagram of the binary mixture of figure 2.10 (paragraph 2.4). Here the velocity of the interface is higher than the critical velocity of the 'constitutional super-cooling'.

liquid and solid phases are equal, the transformation of a liquid layer into a solid one does not change the free energy. It implies the periodicity of the lattice potential $V_0(\xi)$ introduced in paragraph 2.3. When the solid phase is more stable the transformation of the same layer decreases the Gibbs free energy by $\Delta G = \Delta\mu\ a\ S/v_s$. It is equivalent to modifying the lattice potential by $V(\xi) = V_0(\xi) - (\Delta\mu/v_s)\xi$. The position of the interface satisfies the dissipative dynamics

$$\eta\frac{\partial\xi}{\partial t} = -\frac{\delta E}{\delta\xi} = \gamma\Delta\xi - \frac{dV_0}{d\xi} + \frac{\Delta\mu}{v_s} \tag{5.9}$$

where we have taken into account the new lattice potential in the interface energy E in paragraph 2.3. Looking for the planar solutions of equation (5.9) propagating at a constant speed v_M in the x direction, consequently of the form $\xi(x - v_M t)$, we deduce

$$v_M\frac{d\xi}{dx} = \gamma\frac{d^2\xi}{dx^2} - \frac{dV_0}{d\xi} + \frac{\Delta\mu}{v_s}. \tag{5.10}$$

When $\Delta\mu = 0$, $v_M = 0$ and we find again equation (2.28) in paragraph 2.5 for the structure of the edge of the island. If we multiply equation (5.10) by $d\xi/dx$ and if we integrate in the range $x = \pm\infty$, we deduce

$$\eta v_M\int_{-\infty}^{+\infty} dx\left(\frac{d\xi}{dx}\right)^2 = \frac{\Delta\mu}{v_s}a. \tag{5.11}$$

At low $\Delta\mu$ the structure of the edge of the island is close to equilibrium and we can approximate the integral of equation (5.11) by β/γ where β is, as in paragraph 2.5, the energy per unit length of the edge of the island. We then deduce

$$v_M = \frac{\gamma}{\eta\beta}\frac{\Delta\mu}{v_s}a. \tag{5.12}$$

Appendix 1: Self-similar growth of a spherical germ

Let us consider a spherical germ of radius $R(t)$ growing from a pure liquid phase supercooled at the temperature T_∞. Let us suppose that this germ reaches a radius such that the effects of surface tension become negligible. In the solid phase the temperature is constant and equal to T_0. In the liquid phase we can determine a simple solution for which the temperature field depends only on the self-similar variable $u = r/\sqrt{D_T t}$. In fact if we consider the new variables as u and t, we can write

$$\frac{\partial}{\partial t} = \frac{\partial}{\partial t} - \frac{1}{2t}u\frac{\partial}{\partial u}$$

$$\frac{\partial}{\partial r} = \frac{1}{\sqrt{Dt}}\frac{\partial}{\partial u}$$

so that, from the equation of diffusion, the self-similar temperature field satisfies

$$\frac{d^2T}{du^2} + \frac{dT}{du}\left(\frac{2}{u} + \frac{u}{2}\right) = 0 .$$

Since the interface is isothermal its position must correspond to a well-defined value of u, $u_0 = R(t)/\sqrt{D_T t}$ and therefore to a growth rate $dr/dt = (1/2)(D/t)^{1/2}$. Using the boundary condition of equation (5.8) in paragraph 5.3, we therefore obtain

$$\frac{dT}{du} = -\frac{Q}{c_p}\frac{u_0}{2}\frac{u_0^2}{u^2}\exp\left(\frac{u_0^2 - u^2}{4}\right).$$

The constant u_0 is defined by integrating the preceding equation between u_0, for which $T = T_0$, and infinity, for which $T = T_\infty$. Thus we obtain

$$\Delta = \frac{u_0}{2}\int_{u_0}^{+\infty} du \frac{u_0}{u^2}\exp\left(\frac{u_0^2 - u^2}{4}\right) \tag{5.13}$$

where $\Delta = c_p(T_0 - T_\infty)/Q$ is the supercooling of the liquid. Then u_0 grows from 0 when Δ is small, according to the relation $u_0 \approx (2\Delta)^{1/2}$ and tends to infinity when Δ tends to 1 (figure 5.9). This type of solution only exists for supercooling smaller than 1.

Appendix 2: Kinetics of coalescence.

Let us consider an assembly of germs growing from an isothermic supersaturated solution featuring an oversaturation Δ. During their growth the germs expel impurities and the oversaturation decreases. The smallest germs become smaller than the critical radius and dissolve to the benefit of the largest ones. The kinetics of this coalescence regime can be defined as follows (Lifshitz and Slyozov 1964). Assuming that the germs do not interact directly their radii grow as predicted by equation (5.4) in paragraph 5.4, which we can rewrite here as

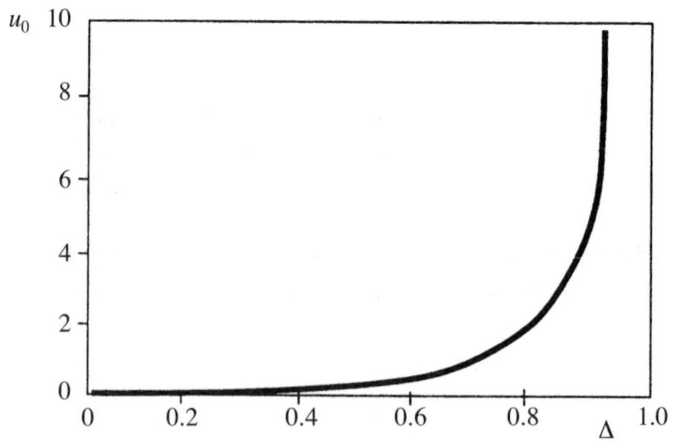

Figure 5.9 u_0 as a function of Δ.

$$\frac{dR}{dt} = \frac{D}{R}\left(\Delta - \frac{2d_{0c}}{R}\right)$$

where the oversaturation Δ now depends on time. We introduce the distribution function in the space of the radii of the grains, $f(R, t)$ such that

$$N(t) = \int_0^{\infty} f(R, t)dR$$

is the number of grains per unit volume. The distribution function f satisfies the continuity equation

$$\frac{\partial f}{\partial t} + \frac{\partial}{\partial R}\left(f\frac{dR}{dt}\right) = 0$$

as well as the relation of conservation of the impurities

$$\Delta(t) + \frac{4\pi}{3}\int_0^{\infty} R^3 f(R, t)dR = \Delta_0 .$$

It is interesting to notice that the problem raised has a self-similar solution for which the mean radius of a germ increases proportionally to $t^{1/3}$, therefore more slowly than the law in $t^{1/2}$ for a single germ, because of the simultaneous decrease in the oversaturation. To determine this solution we introduce the critical radius $R_c(t) = d_{0c}/\Delta(t)$, the variable $x(t) = R_c(t)/R(0)$, the time $\tau = 3\text{Log}x(t)$ and the dimensionless radius $u = R/R_c(t)$. In the new coordinates u and τ, the set of equations is rewritten in the form

$$\frac{du}{d\tau} = \frac{\delta(u-1) - u^3}{3u^2}$$

where $\delta(\tau) = (1/x^2)(dt/dx)$. If we introduce the new distribution function $\phi(u, \tau) = f(R, t)/R_c(t)$, we obtain

$$N(t) = \int_0^{\infty} \phi(u, \tau)du$$

and

$$\frac{\partial \phi}{\partial \tau} + \frac{\partial}{\partial u}\left(\phi\frac{du}{d\tau}\right) = 0 .$$

The total conservation of impurities is now written as

$$A\exp\tau\int_0^{u} u^3\phi(u, \tau)du = 1$$

where $A = 4\pi R_c^3(0)/3\Delta_0$.

From the last equation we deduce that ϕ must be of the form $\phi(u, \tau) = \exp(-\tau\Phi(u))$. Thus, $du/d\tau$ and δ are independent of τ. We deduce that $d(x^3)/dt = 3/\delta$ and therefore that $R_c(t) = R_c(0)(3t/\delta)^{1/3}$. The integral mentioned above has a finite value only for $\delta = 27/4$. We therefore deduce $\Phi(u) = \lambda P(u)$, where

$$P(u) = \frac{3^4}{2^{5/3}} \frac{u^2 \exp\left(-1/\left(1-\frac{2}{3}u\right)\right)}{(u+3)^{7/3}\left(\frac{3}{2}-u\right)^{11/3}}, \quad u < \frac{3}{2}$$

$$P(u) = 0, \qquad u > \frac{3}{2}$$

is normalised to one and $\lambda = 0.9/A$. We can then determine the mean number of grains per unit volume:

$$N(t) = \lambda \exp(-\tau) \int_0^\infty P(u) du = \frac{9\lambda}{4t}$$

and the mean dimensionless size of a grain as well:

$$\bar{u} = \frac{\lambda \exp(-\tau) \int_0^\infty u P(u) du}{N(t)} = 1 .$$

If we go back to the initial variables we deduce that the mean radius of a grain increases as $R_c(t)$ and so as $t^{1/3}$. This growth is slower than the $t^{1/2}$ law for a single germ since the over-saturation decreases as $\Delta(t) = (9\gamma)/(Dt)^{1/3}$.

5.5 FLAME STRUCTURES

A chemical reaction does not take place homogeneously in general. Initiated in a point of space the reaction propagates towards the neighbouring points by the diffusion of the reactants with a well-defined rate. If τ is the order of magnitude of the reaction time and D the coefficient of diffusion of a reactant limiting the reaction, the distance over which the reactant diffuses before being consumed is of the order of $l \approx (D\tau)^{1/2}$. The velocity of propagation of the place of the reaction is therefore of the order of $V \approx l/\tau \approx (D/\tau)^{1/2}$ (Mikhel'son 1889). The whole determination of this velocity of propagation requires the mathematical solution of the reaction diffusion equations associated to the studied system.

Let us first study a template case applied to the growth of a colony of microorganisms (Kolmogorov *et al.* 1927). The reaction is modelled as $A + B \rightarrow 2A$, where A is the microorganism and B the nutritious substance. If A and B have the same diffusion coefficient the net concentration $[A]+[B]$ is constant since a molecule of B is replaced by a microorganism A. The reaction rate is therefore written as $W = k[A][B] = k[A]([B]_0 - [A])$ where $[B]_0$ is the initial concentration in $[B]$.

The equation for the concentration of A is thus

$$\frac{\partial [A]}{\partial t} = D\Delta A + k[A]([B_0]-[A]) \tag{5.1}$$

where D is the diffusion coefficient of the two species, which after making the variable dimensionless leads to

$$\frac{\partial u}{\partial t} = D\Delta u + ku(1-u)\,. \tag{5.2}$$

We are looking for a solution of (5.2) in the form of a propagating wave $u(x+Vt)$ with the boundary conditions $u = 0$ for $x = -\infty$ and $u = 1$ for $x = +\infty$. Although very qualitative the following argument allows us to find the propagation velocity V. Far upstream from the wave u is small and we can neglect the non-linear term in equation (5.2). Thus $u' = u\exp(-kt)$ satisfies the equation of diffusion, the solution with point-source initial condition of which is written as

$$u' = \frac{C}{\sqrt{t}}\exp\left(-\frac{x^2}{4Dt}\right). \tag{5.3}$$

We deduce

$$u = \frac{C}{\sqrt{t}}\exp\left(-\frac{x^2}{4Dt}+kt\right). \tag{5.4}$$

Thus, the place where u is constant satisfies $kt - x^2/4Dt - \mathrm{Log}(t/2)$, that is when t is large, $x = -(4Dk)^{1/2}t$. The propagation velocity of the reactive wave is therefore $V = 2(Dk)^{1/2}$.

Let us consider another typical example of the propagation of premixed flames by thermal effects. The temperature of the gas T satisfies the equation

$$\rho c_p\frac{\partial T}{\partial t} = \lambda\Delta T + QW(Y,T) \tag{5.5}$$

where Q is the heat released per unit volume, λ the thermal conductivity of the gas, ρ its mass density and c_p its specific heat at constant pressure. If we assume a simple reaction $A \rightarrow B$, the concentration Y of the reactant satisfies

$$\rho\frac{\partial Y}{\partial t} = \rho D\Delta Y - W(Y,T) \tag{5.6}$$

where D is the diffusion coefficient of the reactant. According to the law of Arrhenius the reaction rate $W(Y,T)$ is of the form $W(Y,T) \approx (\rho/\tau)Y^n\exp(-E/RT)$ where τ is the collision time of the reactive molecules and n the degree of the reaction. Moreover, assuming that the thermal diffusivity and the diffusion coefficient of the reactant are equal (or the Lewis number is $Le = (\lambda/\rho c_p D) = 1$), $(Q/c_p)Y + T = (Q/c_p)Y_0 + T_0 = T_b$, where T_0 and T_b are the temperatures of the fresh and burnt gases and Y_0 the initial concentration of the reactive. We are therefore back to a single reaction diffusion equation for the temperature

$$\frac{\partial T}{\partial t} = D_T\Delta T + \frac{Q}{c_p\tau}\left(\frac{c_p(T_b-T)}{Q}\right)^n\exp\left(-\frac{E}{RT}\right) \tag{5.7}$$

where $D_T = \lambda/\rho c_p$ with a reaction term with a different structure from the previous term. The first difference is found in the initial state of the gas: the gas is not really in a steady state since the reaction rate does not cancel at $T = T_0$. Nevertheless as we have already mentioned in paragraph 4.2, due to the high activation energy of the explosive reactions, this rate is exponentially small at room temperature and can be neglected. If we linearise the reaction rate around T_0 as in the preceding example we can determine a factor k proportional to $(1/\tau)\exp(-E/RT_0)$ which is therefore also exponentially small. This would lead to too small a propagation velocity to be compared to the experimental results and it constitutes the second difference.

The region where the reaction occurs is not in front of the wave but close to the hottest place where the reaction rate is high. This reason accounts for the differences mentioned above. The flame structure can then be determined as follows (Zeldovich and Frank-Kamenetskii 1938). We are looking for a solution of equation (5.7) propagating at constant velocity V, that is of the form $T(x + Vt)$

$$V\frac{dT}{dx} = D_T\frac{d^2T}{dx^2} + \frac{Q}{c_p\tau}\left(\frac{c_p(T_b - T)}{Q}\right)^n \exp\left(-\frac{E}{RT}\right). \tag{5.8}$$

Upstream from the reaction region the reaction rate is almost negligible and the temperature satisfies

$$V\frac{dT}{dx} = D_T\frac{d^2T}{dx^2} \tag{5.9}$$

that is

$$T = T_0 + A\exp x\frac{V}{D_T}. \tag{5.10}$$

The temperature therefore increases from the fresh gases to the reaction region where the reaction rate cannot be neglected anymore. As shown in figure 5.10 this region is small compared to the diffusion length $l = D_T/V$. In fact because of the high activation energy of the reaction the exponential term of the production rate can be expanded near the temperature where it is high: $\exp(-E/RT) \approx \exp(-E(T_b - T)/RT_b^2)$. Therefore a temperature difference of $\delta T \approx T_b(RT_b/E)$ is sufficient to notably decreases the reaction rate. The coefficient A in equation (5.10) is therefore close to $T_b - T_0$. The heat flux in the reaction zone is of the order of $\lambda\delta T/l_r$, where l_r is the thickness of the reaction region sought. This flux must be high enough to increase the temperature from T_0 to T_b over the diffusion thickness, that is, according to equation (5.10), it must be equal to $\lambda(T_b - T_0)/l$. The thickness of the reaction region is deduced from $l_r \approx l/\beta$ where $\beta = E(T_b - T_0)/RT_b^2$ is the Zeldovich number, usually larger than one.

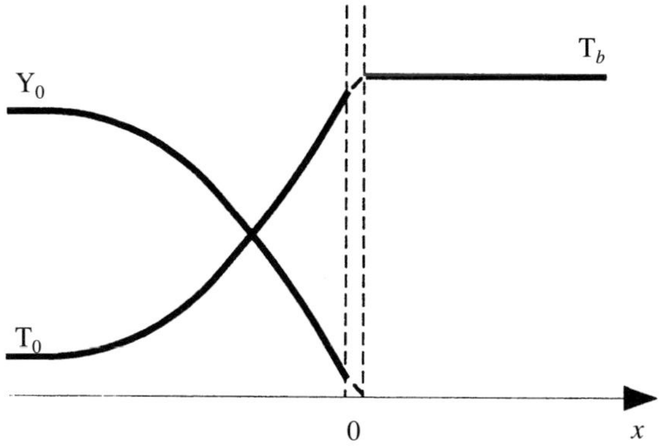

Figure 5.10 Flame structure.

Since the thickness of the reaction region is far smaller than the thickness of the diffusion region the advective term VdT/dx is negligible and the temperature satisfies

$$\lambda\frac{d^2T}{dx^2} = -\frac{\rho Q}{\tau}\exp\left(-\frac{E}{RT_b}\right)\left(\frac{c_p(T_b-T)}{Q}\right)^n\exp\left(-\frac{E(T_b-T)}{RT_b^2}\right) \quad (5.11)$$

the integration of which leads to the heat flux given out

$$\lambda\frac{dT}{dx} = \left(\frac{2\Gamma(n+1)\lambda\rho QY_0^n(T_b-T_0)}{\tau\beta^{n+1}}\right)^{1/2}\exp\left(-\frac{E}{2RT_b}\right). \quad (5.12)$$

We obtain the propagation velocity of the flame by identifying the flux in equation (5.12) and the one determined by equation (5.10) where $A = T_b - T_0$,

$$V = \left(\frac{2\Gamma(n+1)}{\beta^{n+1}}\right)^{1/2} Y_0^{\frac{n-1}{2}}\left(\frac{D_T}{\tau}\exp\left(-\frac{E}{RT_b}\right)^{1/2}\right). \quad (5.13)$$

The propagation velocity is written in the form of the square root of the ratio of the thermal diffusivity and a reaction time but contrary to the preceding case the reaction time is close to that of the final state, in agreement with the experimental results. From the kinetic theory of gases $D_T \approx l_c c \approx c^2\tau$, where l_c and c are the mean free path of the gas molecules and the speed of sound, respectively. Then to within a prefactor $V \approx c\exp(-E/2RT_b)$, since the exponential factor decays by several orders of magnitude of the velocity of the flame compared to the speed of sound. For instance since $c = 300$ m/s, when $E/RT_b = 16$ the flame propagates at very low velocity, $V \approx 10$ cm/s, has a diffusion thickness $l = D_T/V_0 \approx 0.1$ mm and a reaction thickness $l_r \approx 100$ μm, so that the region lit up by the radiating products of the reaction has a very small thickness (figure 5.11).

Appendix 1: Influence of the dependence of the reaction rate with the diffusive variable on the propagation velocity.

The study of these particular cases can be generalised to the study of a reaction-diffusion wave of velocity V for the variable u satisfying

$$V\frac{du}{dx} = D\frac{d^2u}{dx^2} + F(u)$$

where $F(0) = F(1) = 0$, at $x = -\infty$, $u = 0$, $du/dx = 0$ and at $x = +\infty$, $u = 1$, $du/dx = 0$ (Zeldovich *et al.* 1985). It is equivalent to introducing $y = du/dx$ and to looking for a trajectory $y(u)$ solution in the space of phases (u, y) of

$$\frac{dy}{du} = \frac{V}{D}y - \frac{1}{D}F(u)$$

which links the two fixed points $(0, 0)$ and $(1, 0)$.

We first determine the nature of the fixed points, or, the structure of the trajectories passing by in their proximity. From that point of view we are searching for a solution in the proximity of a fixed point of the form $y = s(u-u_0) + \phi(u-u_0)$ where $\phi << u-u_0 << 1$ and we linearise the preceding equation. We then deduce the characteristic equation

$$s^2 - \frac{V}{D}s + \frac{1}{D}F'(u_0) = 0$$

with solutions either of the same sign when $F'(u_0) > 0$, or of opposite sign when $F'(u_0) < 0$. Moreover we obtain

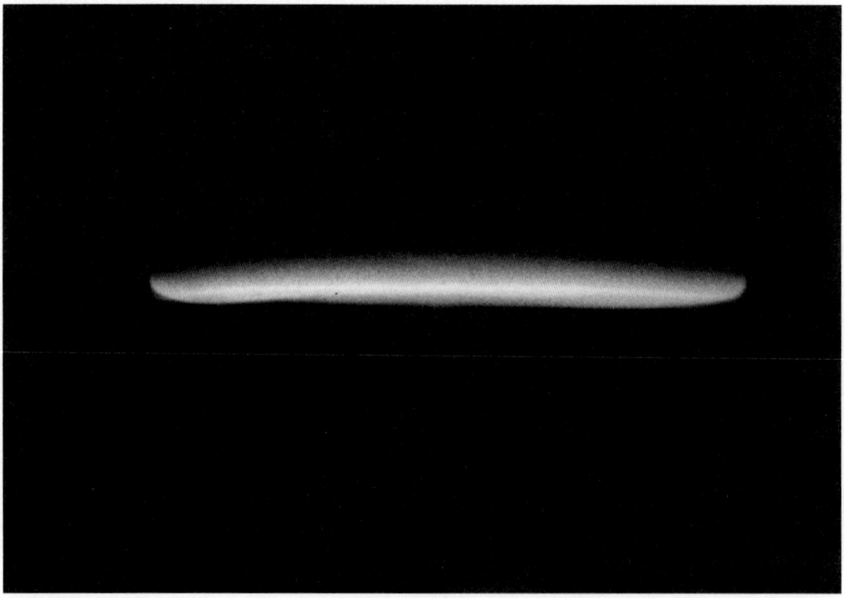

Figure 5.11 Planar flame observed from the edge (Quinard 1984).

$$s(u-u_0)\frac{d\phi}{du} = \left(\frac{V}{D}-s\right)\phi$$

the solution of which is either zero or of the form $w(u-u_0)^\alpha$ where $\alpha = (V/D-s)/s$ when $\alpha > 1$.

Since the fixed point $u_0 = 0$ satisfies $F'(u_0) > 0$ the two roots are real and positive and only one of them is associated to an expansion of the form $y = s(u-u_0) + w(u-u_0)^\alpha$, w being an arbitrary constant. Since the fixed point $u_0 = 1$ satisfies $F'(u_0) < 0$ the two roots are of opposite signs associated with the expansions at $\phi = 0$. A trajectory starting from the fixed point $u_0 = 1$ can only be of the form $y = s_2(u-u_0)$ where s_2 is the positive root. The other trajectory is actually at $y < 0$ for $u < 0$, which implies $dy/du > 0$. The trajectory can only cross the y axis below the origin.

Thus we have two possible trajectories: the first one joins the two fixed points with the directions s_2 for a well-given value of the propagation velocity; the second one joins the fixed point $u_0 = 0$ with the direction s_1 and the fixed point $u_0 = 1$ with the direction s_2 for any value of the propagation velocity higher than $2(DF'(0))^{1/2}$. In fact in the first case we only have one parameter available to link the extremities of the trajectory: the velocity V. In the second case we also have the arbitrary constant w, where the only constraint is to have two real and positive roots. In general a continuum of possible propagation velocities therefore exists.

A study of stability of these solutions (Aronson and Weinberger 1972) shows that, starting from sufficiently localised initial conditions, the wave finally steadily propagates at the lowest possible velocity. According to the form of the reaction term the minimum velocity is either $2(DF'(0))^{1/2}$ as in the first example or proportional to the area of the reaction rate as in the second one (Zeldovich *et al.* 1985).

Appendix 2: Simplified structure of an $H_2 - O_2$ *flame.*

A simplified flame structure can be obtained in the same way as before but starting from the three reaction-diffusion equations for the temperature, the radical H and the reactive O_2 obtained from the reactions in paragraph 4.3 (Zeldovich 1961, Joulin *et al.* 1985). The reaction (CB) favours the propagation of the flame by producing radicals. The reaction (C + CT) at the same time favours and inhibits the propagation by releasing heat and recombining radicals, respectively. Thus the heat released from the back of the flame by the recombination reaction diffuses forwards and triggers the reaction generating radicals. Since the activation energy of the latter reaction is high it takes place, as above, in a very thin region of the space but not close to the point of maximum temperature since the recombination reaction which occurs later on further increases the temperature. The profiles of the different concentrations are sketched in figure 5.12.

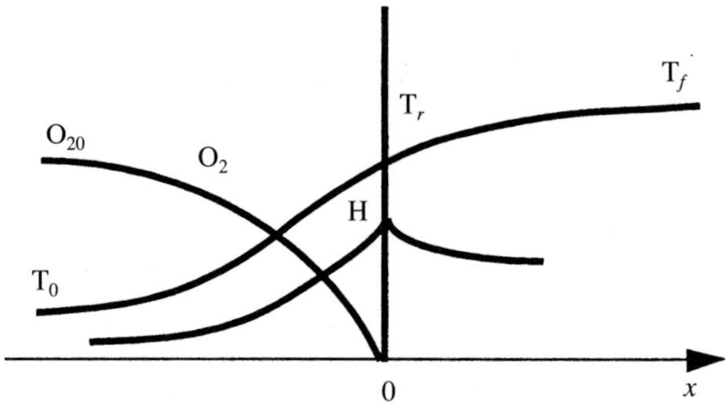

Figure 5.12 Simplified structure of an $H_2 - O_2$ flame (Zeldovich *et al.* 1985).

The concentration profile of the reactive O_2 decreases as in the simplified structure of the lflame. The temperature profile increases differently since there is no discontinuity of the slope on the scale of the reaction region because of the spread of the release of heat. Lastly, the concentration of radicals increases as far as the reaction region where they have been generated because of the diffusion and then decreases because of the recombination reaction.

Several quantities can be calculated at the same time as the flame structure : the relative temperature of the reaction region T_r/T_0, the relative maximum concentration of the radicals H reached in the reaction region $[H]_r/[O_{20}]$ and the velocity of propagation of the flame V. These quantities are determined in figure 5.13a and figure 5.13b, respectively, as a function of the reaction constant k_2 or more exactly as a function of the parameter $y = k_2/k_1 \exp(-E/RT_b)$ which characterises the ratio of the reaction rates of recombination and chain branching. The ratio T_r/T_0 increases as y increases to get close to T_f/T_0 when the recombination reaction is so fast that it occurs simultaneously as the chain branching reaction (figure 5.13a). The heat is released in the reaction region as in the simplified example in the beginning of the paragraph. Besides, as y increases the maximum concentration of radical decreases since the recombination reaction is faster. The velocity curve therefore undergoes a maximum, it increases first by the effect of the increase in temperature in the reaction region and then decreases by the effect of the decrease in the maximum concentration of radical H (figure 5.13b).

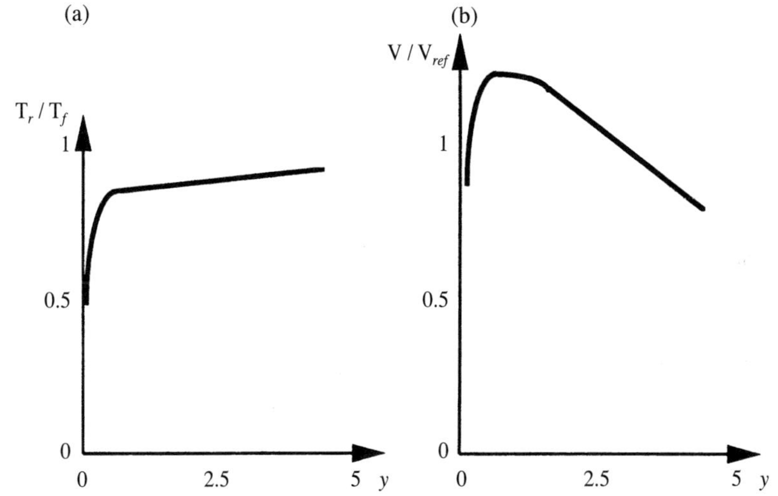

Figure 5.13 (a) Temperature of the reaction region as a function of y (Zeldovich *et al.* 1985); (b) plot of velocity as a function of y (Zeldovich *et al.* 1985).

5.6 CURVATURE EFFECTS

As it gets curved the local structure of the step determined by equation (5.10) in paragraph 5.4 is modified by the capillary effects related to the change of the position of the interface in the direction transverse to the displacement. Instead of x let us use n, i.e. the coordinate normal to the interface. In this new system of coordinates equation (5.10) becomes

$$\left(-\eta v - \frac{\gamma}{R}\right)\frac{d\xi}{dn} = \gamma\frac{d^2\xi}{dn^2} - \frac{dV_0}{d\xi} + \frac{\Delta\mu}{v_s} \tag{5.1}$$

where R is the radius of local curvature of the interface. This equation has the same solution as equation (5.10) in paragraph 5.4 since $-\eta v_M = -\eta v - \gamma/R$. The velocity of propagation of the curved step is written as, using equation (5.12) in paragraph 5.4,

$$v = v_M - \frac{\gamma}{\eta R} = v_M\left(1 - \frac{R_c}{R}\right). \tag{5.2}$$

A reaction-diffusion front will bend if the medium in which it propagates is inhomogeneous or if instabilities are present. The velocity of propagation is changed since instead of diffusing along one direction the reactant diffuses in all of space, shifting the equilibrium between reaction and diffusion. Let us first consider the simplest case of a single reactive species u, as in paragraph 5.5, which satisfies the following equation

$$\frac{\partial u}{\partial t} = D\Delta u + F(u). \tag{5.3}$$

Let us first consider a spherical front spreading from the centre. When the front is well-developed the D/V-thick region where the concentration changes gets far smaller than its distance from the centre. We can therefore assign a constant radius of curvature R approximately equal to the radius of the line $u = 1/2$ in this region. Thus in the diffusion layer the Laplacian can be written in spherical coordinates for a purely radial field u as : $\Delta = (\partial^2/\partial r^2)/r^2 + \kappa\ \partial/\partial r$, where $\kappa = 2/R$ is the curvature of the front. In these new coordinates the concentration in the front moving at the velocity V satisfies the following equation

$$(V + D\kappa)\frac{du}{dr} = D\frac{d^2u}{dr^2} + F(u) \tag{5.4}$$

with $u = 0$ and $u = 1$ for $r = +\infty$ and $r = 0$ as the boundary conditions. We find a problem similar to the one in paragraph 5.5, that is, if V_0 is one of the possible velocities of propagation of the plane wave, $V = V_0 - D\kappa$.

Let us consider a spherical flame already well-defined so that we can consider the curvature of the temperature and concentration lines as uniform. We obtain

$$\begin{aligned}(V + D_T\kappa)\frac{dT}{dr} &= D_T\frac{d^2T}{dr^2} + \frac{Q}{c_p\tau}(Y)^n\exp\left(-\frac{E}{RT}\right)\\ (V + D_M\kappa)\frac{dY}{dr} &= D_M\frac{d^2Y}{dr^2} - \frac{1}{\tau}(Y)^n\exp\left(-\frac{E}{RT}\right).\end{aligned} \tag{5.5}$$

Outside the reaction region the reaction term is negligible. If we multiply the second of equations (5.5) by Q/c_p, add it to the first one and integrate the sum between r_f the position of the reaction region and $r = r_\infty$, we get

$$(V + D_T\kappa)(T_0 - T_f) + \frac{Q}{c_p}(V + D_M\kappa)Y_0 = 0. \tag{5.6}$$

The advective terms are actually negligible in the reaction region. The sum of the two equations (5.5) and its integration over this region lead to a zero net flux $D_T dT/dr + D_M(Q/c_p)dY/dr$ at the end of the reaction zone. According to the assumption the front is weakly curved compared to the reaction region and once the Lewis number $Le = D_T/D_M$ is close to one,

$$T_f \approx T_0 + \frac{Q}{c_p} Y_0 (1 + (1 - Le)\kappa d). \tag{5.7}$$

Thus as in the case of a crystal the temperature of the reaction region varies proportionally to the curvature of the front but for a completely different reason. When the front wrinkles the difference of the diffusion coefficients leads to uncompensated transverse heat and reactive fluxes. This appears as an increase or a decrease at the convex points depending whether the Lewis number is smaller or larger than one. The analysis of the reaction zone is similar to the one in paragraph 5.5 and leads to heat being given out

$$\lambda \frac{dT}{dr} = \left(\frac{2\Gamma(n+1)\lambda\rho Q Y_0^n (T_b - T_0)}{Le \tau \beta^{n+1}} \right)^{1/2} \exp\left(-\frac{E}{2RT_f} \right). \tag{5.8}$$

The propagation velocity is obtained by calculating the outgoing heat flux by the first of equations (5.5) in which the reaction term has been neglected and by identifying it to equation (5.8). We obtain by replacing T_f by its expression given by equation (5.7) and by linearising the exponential since the radius of curvature is large compared to the thickness of diffusion,

$$V = V_0 \left(1 - \left(1 + \frac{\beta}{2}(Le - 1) \right) d\kappa \right) \tag{5.9}$$

where β is the reduced activation energy and $d = D_T / V_0$ the thickness of the flame. The velocity can therefore increase or decrease with the radius of curvature depending on whether the Lewis number is smaller or larger than one.

When the gas in which the flame propagates is set in motion the generated flow is not uniform in general. The resulting curvature of the flow lines affects the velocity of propagation (Karlovitz *et al.* 1953, Klimov 1963). For simplicity, in the moving frame where the flame is at rest, let us consider the point of stagnation (figure 5.14)

$$u = V - \frac{x}{\tau_0}$$
$$v = \frac{y}{\tau_0}. \tag{5.10}$$

The flame is at $x = 0$ and $x^* = V\tau_0$ is the point where the longitudinal velocity cancels.

The steady-state temperature and concentration fields satisfy the following equations

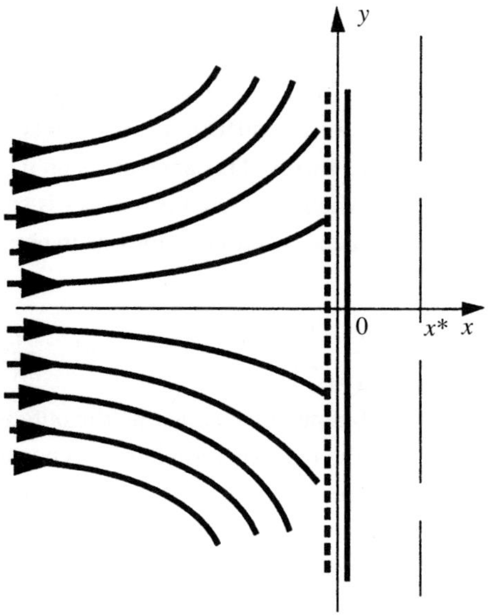

Figure 5.14 Flame in a stagnant flow.

$$\left(V-\frac{x}{\tau_0}\right)\frac{dT}{dx} = D_T\frac{d^2T}{dx^2}+\frac{Q}{c_p\tau}(Y)^n\exp\left(-\frac{E}{RT}\right)$$
$$\left(V-\frac{x}{\tau_0}\right)\frac{dY}{dx} = D_M\frac{d^2Y}{dx^2}-\frac{1}{\tau}(Y)^n\exp\left(-\frac{E}{RT}\right). \tag{5.11}$$

In the diffusion region the reaction term is negligible and we obtain

$$\frac{T-T_0}{T_f-T_0} = \frac{1-\mathrm{erf}\left(\dfrac{V\tau_0-x}{\sqrt{2D_T\tau_0}}\right)}{1-\mathrm{erf}\left(V\sqrt{\dfrac{\tau_0}{2D_T}}\right)}$$
$$Y = Y_0\left(1-\frac{1-\mathrm{erf}\left(\dfrac{V\tau_0-x}{\sqrt{2D_T\tau_0}}\right)}{1-\mathrm{erf}\left(V\sqrt{\dfrac{\tau_0}{2D_T}}\right)}\right). \tag{5.12}$$

The analysis of the reaction zone is similar to the preceding case. The net flux $D_T dT/dx + D_M(Q/c_p)dY/dx$ is thus zero at the end of the reaction region and the heat flux given out is given by equation (5.8). The application of these relations determines the velocity of propagation V as a function of the imposed velocity gradient τ_0

$$\frac{V}{V_0} = \sqrt{\pi\alpha}(1-\mathrm{erf}\sqrt{\alpha})\exp\left(\alpha + (1-Le)\beta\left(\frac{1}{2}+\alpha-\frac{\sqrt{\alpha}\exp-\alpha}{\sqrt{\pi}(1-\mathrm{erf}\sqrt{\alpha})}\right)\right) \quad (5.13)$$

where

$$\mathrm{erf}\sqrt{\alpha} = \frac{2}{\sqrt{\pi}}\int_0^{\sqrt{\alpha}} \exp(-u^2)du \quad (5.14)$$

and $\alpha = \tau_0 V_0^2/2D_T$ is the parameter characterising the intensity of the gradient. As above, the Lewis number is assumed to be close to one. When $Le = 1$, V monotonically decreases with the intensity of the gradient of velocity τ_0 until it cancels at the value of the parameter which satisfies $\sqrt{\alpha} = 1$. In fact the heat released is transversely evacuated by the convective flux instead of heating the fresh gases. The flame temperature is therefore decreased as well as the velocity of propagation. As shown in figure 5.15 the differential diffusion caused by the change in the Lewis number decreases or amplifies the effect. When $Le < 1$ the velocity increases then decreases with the gradient. When $Le > 1$ the decrease is so quick that it is no longer possible to obtain solutions with a finite velocity of propagation, related to an extinction of the flame. For a gradient of weak intensity the velocity of the flame is obtained by expanding equation (5.13) at high values of α, that is

$$V = V_0\left(1-\left(1+\frac{\beta}{2}(Le-1)\right)\frac{V_0^2}{\tau_0 D_T}\right). \quad (5.15)$$

As we can notice in equations (5.9) and (5.15) the curvature of the front has a quantitative role similar to that of a flow velocity gradient. This similar behaviour can be described in a broader framework (Clavin and Joulin 1983).

5.7 ELECTRO-DIFFUSION

In an electrolyte we must take into account the electric force acting on the ions in addition to the diffusion. The electric force $\vec{f} = -ze\vec{\nabla}\Phi$ sets the ion into motion at the velocity $\vec{v} = \mu\vec{f}$, where z is the charge of the ion, μ its mobility and Φ the electric potential. It gives rise to a net current

$$\vec{j}_s = -\rho D\vec{\nabla}c - \mu\rho zec\vec{\nabla}\Phi. \quad (5.1)$$

At thermal equilibrium this current must cancel so that

$$D\log c + \mu ze\Phi = \mathrm{cst}. \quad (5.2)$$

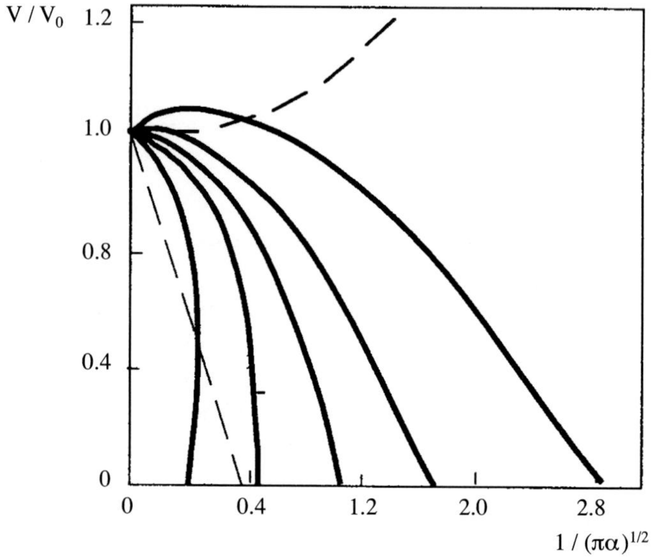

Figure 5.15 Velocity of propagation as a function of the intensity of the gradient for different values of $\beta(Le-1)$ (Zeldovich *et al.* 1985).

If we compare this expression to the expression showing that the chemical potential of the ion is constant

$$kT\text{Log}c + ze\Phi = \text{cst} \tag{5.3}$$

we deduce the Einstein relation between the coefficient of diffusion and the mobility, $D/\mu = kT$. The current of the ion can be written as

$$\vec{j}_s = -\rho D\left(\vec{\nabla}c + \frac{ze}{kT}c\vec{\nabla}\Phi\right). \tag{5.4}$$

Using the conservation equation in paragraph 1.4,

$$\frac{\partial(\rho c)}{\partial t} + \vec{\nabla}\cdot(\rho c\vec{w} + \vec{j}_s) = 0 \tag{5.5}$$

we deduce, without flow and assuming the product ρD to be constant,

$$\frac{\partial c}{\partial t} = D\vec{\nabla}\cdot\left(\vec{\nabla}c + \frac{ze}{kT}c\vec{\nabla}\Phi\right). \tag{5.6}$$

If the electrolyte contains n ions of charge number z_i and of coefficient of diffusion D_i each concentration c_i satisfies equation (5.6). Moreover the electric potential satisfies the Poisson equation,

$$\Delta\Phi = -\frac{\rho_c}{\varepsilon\varepsilon_0} = -\frac{e\sum_i z_i c_i}{\varepsilon\varepsilon_0} \tag{5.7}$$

where ρ_c is the charge density.

For simplicity, we will consider only one electrolyte consisting of two ions of opposite charges z_+ and z_-. Let us assume that the equilibrium is slightly perturbed so that the concentrations of both ions still remain close to their equilibrium values. We deduce therefore from equations (5.6) written for each ion,

$$\begin{aligned}\frac{\partial(z_+c_+ + z_-c_-)}{\partial t} &= \Delta(D_+z_+c_+ + D_-z_-c_-)\\ &\quad - \frac{e^2(D_+z_+^2c_{0+} + D_+z_-^2c_{0-})}{kT\varepsilon\varepsilon_0}(z_+c_+ + z_-c_-).\end{aligned} \tag{5.8}$$

When the two coefficients of diffusion D_+ and D_- are equal, the ratio of the two terms of equation (5.8) can be written as the square of the ratio of the two lengths $(d/L)^2$, L being the lengthscale over which the concentration fields change, and

$$d = \left(\frac{kT\varepsilon\varepsilon_0}{e^2(z_+^2c_{0+} + z_-^2c_{0-})}\right)^{1/2} \tag{5.9}$$

the Debye length. In general for usual ionic concentrations of a few mM $d << L$ so that the first term of the right-hand side of equation (5.8), of order D/L^2, is far smaller than the second one, of order D/d^2. It results that quickly, over the time scale d^2/D, $z_+c_{0+} + z_-c_{0-} \approx 0$, and the medium is almost electrically neutral. We can thus rewrite equation (5.8) in the form

$$\begin{aligned}\frac{\partial(z_+c_+ + z_-c_-)}{\partial t} &\approx D_+z_+\Delta c_+ + D_-z_-\Delta c_-\\ &\quad + \frac{e}{kT}(D_+z_+^2c_{0+} + D_-z_-^2c_{0-})\Delta\Phi\end{aligned} \tag{5.10}$$

and deduce the equation for the electric potential

$$\frac{e}{kT}\Delta\Phi = \frac{z_+(D_- - D_+)}{(D_+z_+^2c_{0+} + D_-z_-^2c_{0-})}\Delta c_+. \tag{5.11}$$

In particular when the coefficients of diffusion are equal, $\Delta\Phi = 0$, which we could have hastily and wrongly deduced from the Poisson equation (5.7) in which we would have applied the condition of exact electroneutrality. By using equations (5.6) and (5.11) we deduce that

$$\frac{\partial c_+}{\partial t} = \frac{D_-D_+(z_+^2c_{0+} + z_-^2c_{0-})}{(D_+z_+^2c_{0+} + D_-z_-^2c_{0-})}\Delta c_+ \tag{5.12}$$

showing that the cation and the anion diffuse with an effective coefficient of diffusion. In particular, when the charge numbers of the cation and the anion are equal, the coefficient of effective diffusion is $D_{\text{eff}} = 2D_-D_+/(D_- + D_+)$, which varies from the common coefficient of diffusion when these coefficients are equal, to the slowest coefficient of diffusion when they are not of the same order of magnitude. This diffusion process, known as ambipolar, comes from the fact that when the coefficients of diffusion of the two ions are different the difference of concentrations resulting from their movement creates an electric field which tends to slow down the fastest ion.

The boundary conditions for the potential and for one of the concentrations, c_+ for instance (since the other one is deduced from the first one by the electroneutrality condition), are obtained by a combination of these two relations

$$J_+ = -\rho D_\pm\left(\frac{\partial c_\pm}{\partial n} + \frac{ze}{kT}c_{0\pm}\frac{\partial \Phi}{\partial n}\right) \tag{5.13}$$

for the cation and the anion, obtained by projecting the current of equation (5.4) onto the normal to the considered interface. This combination uses the electroneutrality of the solution, that is

$$\begin{aligned} C_0\frac{e}{kT}\frac{\partial \Phi}{\partial n} &= -\left(z_+\frac{J_+}{\rho D_+} + z_-\frac{J_-}{\rho D_-}\right) \\ C_0\frac{\partial c_+}{\partial n} &= -z_-^2 c_{0-}\frac{J_+}{\rho D_+} + z_+z_-c_{0+}\frac{J_-}{\rho D_-} \end{aligned} \tag{5.14}$$

where $C_0 = z_+^2c_{0+} + z_-^2c_{0-}$. The place where this condition is met is therefore beyond the Debye layer in which the electroneutrality is not met. Since the thickness of this layer is far smaller than the size of the electrolytic aggregates we can assume that this place is the interface itself. The set of equations ends because the currents $J_\pm$ are themselves functions of the potential and of the concentrations by a Butler–Volmer-type law considered in paragraph 3.5 and paragraph 4.5.

These equations are valid when the concentrations are almost constant within the solution, which is the case at low difference of the applied potential. At higher fields we must take into account the changes of concentration in the electric contribution of the current. As shown by Chazalviel (1990) the thickness of the layer adjacent to the interface in which the condition of electroneutrality is not satisfied can become far larger than the Debye length. Nevertheless this layer remains small compared to the dimensions of the electrolytic aggregate microstructures.

5.8 SPACE CHARGED INTERFACES

Again, to keep the formulation simple we will just consider two ions, a cation of charge number z_+ and an anion of charge number z_-. The electrolyte is locally neutral at equilibrium and $z_+c_{0+} + z_-c_{0-} = 0$.

Let us first consider a planar interface charged with a surface density σ at the potential Φ_i. At equilibrium the ionic currents are zero so that in equation (5.4),

$$c_+ = c_{0+}\exp-\frac{z_+e\Phi}{kT} \; ; \; c_- = c_{0-}\exp\frac{-z_-e\Phi}{kT} . \tag{5.1}$$

The electric potential therefore satisfies the following equation

$$\varepsilon\varepsilon_0\frac{d^2\Phi}{dx^2} = -ez_+c_{0+}\left(\exp\left(-\frac{z_+e\Phi}{kT}\right) - \exp\left(-\frac{z_-e\Phi}{kT}\right)\right) \tag{5.2}$$

which we can integrate twice, after multiplying by $d\Phi/dx$,

$$\frac{\varepsilon}{2}\frac{\varepsilon_0}{kT}\left(\frac{d\Phi}{dx}\right)^2 = ez_+c_{0+}\left(\exp\left(-\frac{z_+e\Phi}{kT}\right) - \exp\left(-\frac{z_-e\Phi}{kT}\right) - 2\right). \tag{5.3}$$

In the particular case of a 1 : 1 electrolyte, $z_+ = 1$, $z_- = -1$, $c_{0+} = c_{0-} = c_0$ and

$$\frac{d\Phi}{dx} = \left(\frac{8kTc_0}{\varepsilon\varepsilon_0}\right)^{1/2}\sinh\left(\frac{e\Phi}{2kT}\right) \tag{5.4}$$

the solution of which is given by Gouy and Chapman

$$\Phi = \frac{2kT}{e}\mathrm{Log}\left(\frac{1+\gamma\exp\left(-\frac{x}{d}\right)}{1-\gamma\exp\left(-\frac{x}{d}\right)}\right) \tag{5.5}$$

where $d = (kT\varepsilon\varepsilon_0/2e^2c_0)^{1/2}$ is the Debye length, $\gamma = \tanh(e\Phi_i/4kT)$ and Φ_i the interface potential. The surface density of charge is deduced from equation (5.11):

$$\sigma = \varepsilon\varepsilon_0\frac{d\Phi}{dx}(0) = (8kT\varepsilon\varepsilon_0c_0)^{1/2}\sinh\left(\frac{e\Phi_i}{2kT}\right). \tag{5.6}$$

At low values of the interface potential equation (5.12) is written more simply with the Debye–Hückel approximation

$$\Phi = \Phi_i\exp\left(-\frac{x}{d}\right) \tag{5.7}$$

and the surface density of charge becomes $\sigma = \varepsilon\varepsilon_0\Phi_i/d$.

Let us now consider the case where the cation only deposits on the interface $(J_+ \neq 0, J_- = 0)$, the interface being at zero surface charge density $(\sigma = 0)$. If we assume that the gradients are small we can use the currents given by equation (5.13) in paragraph 5.7. Thus from $dJ_+/dx = dJ\_/dx = 0$ we deduce,

$$C_0 \frac{e}{kT} \frac{d^2\Phi}{dx^2} = -\frac{d^2}{dx^2}(z_+c_+ + z\_c\_) \tag{5.8}$$

so that using the Poisson equation,

$$\frac{d^2}{dx^2}(z_+c_+ + z\_c\_) = -\frac{1}{d^2}(z_+c_+ + z\_c\_). \tag{5.9}$$

Thus, the space charge generated by the difference in currents of the two ions at the interface decreases again over the Debye length. This decrease justifies the condition of electroneutrality used for the boundary conditions (5.14) in paragraph 5.7. Outside the Debye layer the equations (5.11) and (5.12) in paragraph 5.7 can be applied and therefore in the quasi-stationary conditions usually met, $\Delta c_+ = \Delta\Phi = 0$, $c\_$ being determined by the condition of electroneutrality. If μ_+ is the chemical potential of the cation $\Delta\mu_+ = 0$, with $\vec{v}\cdot\vec{n} = KJ_+ = -\alpha\vec{\nabla}\mu_+$ and $\mu_+ = \text{cst}$ at the interface (in fact, just at the end of the Debye layer), if we assume that the kinetics of deposition are quick enough. In that case the Butler–Volmer deposition kinetics do not explicitly intervene and the motion of the interface follows the dynamics similar to the dynamics of a rough liquid–solid interface already studied in paragraph 5.3, but with negligible capillary effects.

Let us consider a charged layer with a charge density $\rho(z)$, for instance the Debye layer mentioned above, in the proximity of a planar interface (figure 5.16). When we apply a tangential electric field $\vec{E}$, an electro-osmotic phenomenon occurs, that is the motion of the charges induces a tangential flow, the magnitude of which will be determined later (see for example Levich 1962).

When the electric field is uniform the electric potential Φ satisfies

$$\frac{d^2\Phi}{dz^2} = -\frac{\rho}{\varepsilon\varepsilon_0}. \tag{5.10}$$

The tangential electric field applies a force per unit volume $f = \rho E$, so that the fluid is set in motion with a mostly tangential velocity $v(z)$ and which satisfies

$$\eta \frac{d^2v}{dz^2} = -f \tag{5.11}$$

according to paragraph 5.1. If we replace f by its expression and if we use the Poisson equation we deduce

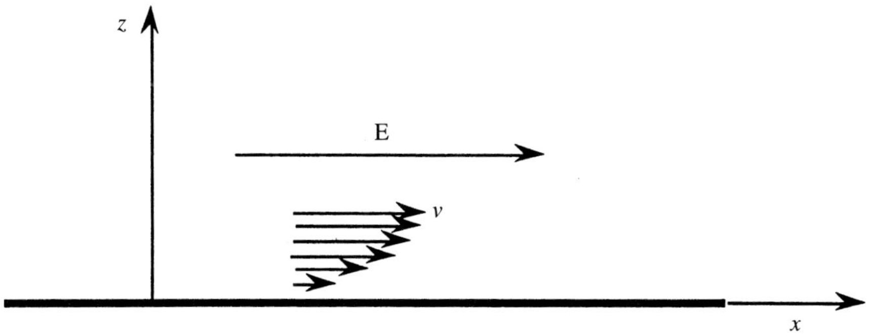

Figure 5.16 Flow induced by the action of an electric field on a space charge.

$$\eta \frac{d^2 v}{dz^2} = -\varepsilon\varepsilon_0 \frac{d^2\Phi}{dz^2} E \tag{5.12}$$

so that

$$v(z) = \frac{\varepsilon\varepsilon_0 E}{\eta}\Phi(z) + az + b\,. \tag{5.13}$$

Far away from the wall the velocity must remain finite so that $a = 0$. The velocity cancels at the wall and $b = -\varepsilon\varepsilon_0 E/\eta\ \Phi(0)$. We therefore deduce the electro-osmotic velocity far from the wall:

$$v = \frac{\varepsilon\varepsilon_0 E}{\eta}\zeta \tag{5.14}$$

where the ζ potential is written as $\zeta = \Phi(\infty) - \Phi(0)$. If the charge distribution is due to a surface density σ, $\zeta = d\sigma/\varepsilon\varepsilon_0$, and $v = d\sigma E/\eta$.

BIBLIOGRAPHY

[1] D.G. Aronson, H.F. Weinberger, 'Multidimensional Nonlinear Diffusion Arising in Population Genetics', *Advances in Mathematics*, **30**, p. 33, 1978.

[2] J.N. Chazalviel, 'Electrochemical Aspects of the Generation of Ramified Metallic Electrodeposits', *Phys. Rev.*, **A 42**, p. 7355, 1990.

[3] P. Clavin, G. Joulin, 'Premixed Flames in Large Scale and High Intensity Turbulent Flows', *J. Phys. Lett.*, **44**, p. 1, 1983.

[4] D. Dowson, C.M. Taylor, 'Cavitation in Bearings', *Ann. Rev. Fluid. Mech*, **11**, p. 35, 1979.

[5] P.G. De Gennes, 'Wetting: Statics and Dynamics', *Rev. Mod. Phys.*, **57**, p. 827, 1985.

[6] R. Hoffman, 'A study of the advancing interface. I. Interface shape in liquid–gas systems'. *J. Colloid Interface Sci.*, **50**, p. 228, 1975.

[7] J.F. Joanny, P.G. De Gennes, 'Structure statique des films de mouillage et des lignes de contact', *C.R. Acad. Sci.*, **299 II**, p. 279, 1984.

[8] G. Joulin, A. Linan, G.S.S. Ludford, N. Peters, C. Schmidt-Laine, *Siam. J. Appl. Math.*, **45**, p. 420, 1985.

[9] B. Karlovitz, D.W. Denniston Jr., D.H. Knapschaefer, F.E. Wells, 'Studies on Turbulent Flames', 4th Symp. Combust., Baltimore, p. 613, 1953.

[10] A.M. Klimov, 'Laminarnoe plamya v turbulentnom potoke', (Laminar Flames in Turbulent Flows), *Prikl. Mekh. i Teckh. Fiz*, **3**, p. 49, 1963.

[11] A.N. Kolmogorov, I.G. Petrovskii, N.S. Piskunov, 'Issledovanie uravneniya difuzii, soedinnennoi s vozrastaniem kolichestva, i ego primenie k odnoi biologicheskoi probleme', (A Study of the Diffusion Equation Associated with the Growth in the Amount of a Substance and its Application to a Biological Problem), *Byul. Mos. Gos. Un-ta*, Vol.1, **Ser. A**, Issue B, 1937.

[12] L.D. Landau, V.G. Levich, 'Dragging a Fluid by a Moving Plate', Acta Physicochimica URSS, **17**, p. 42, 1942.

[13] J.S. Langer, 'Instabilities in crystal growth', *Rev. Mod. Phys.*, **52**, p. 1, 1980.

[14] M. Lelah, A. Marmur, 'Spreading kinetics of drops on glass', *J. Colloid Interface Sci.*, **82**, p. 518, 1981.

[15] V.G. Levich, *Physicochemical Hydrodynamics*, Prentice Hall, 1962.

[16] I.M. Lifshitz, V.V. Slyozov, 'The kinetics of precipitation from supersaturated solid solutions', *J. Phys. Chem. Solids*, **19**, p. 35, 1961.

[17] V.A. Mikhel'son, 'On the Normal Ignition Rate of Fulminating Gas Mixtures', Ph. D. Thesis, University of Moscow, 1889.

[18] Ph. Nozières, 'Shape and Growth of Crystals', in *Solids far from equilibrium*, edited by C. Godrèche, C.U.P., 1992.

[19] C.W. Park, G.M. Homsy, 'Two-Phase Displacement in Hele–Shaw Cells: theory', *J. Fluid. Mech.*, **139**, p. 291, 1984.

[20] J. Quinard, thèse d'État. Limites de stabilité et structures cellulaires dans les flammes de prémélange. Étude expérimentale, Université de Provence, 1984.

[21] O. Reynolds, 'On the Theory of Lubrication and its Application to Mr. Beauchamp Tower's Experiments, Including an Experimental Determination of the Viscosity of Olive Oil', 1986.

[22] A. Sommerfeld, *Mechanics of Deformable Bodies*, Academic Press, New York, 1950.

[23] P. Tabeling, A. Libchaber, 'Film Draining and the Saffman–Taylor Problem', *Phys. Rev.*, **A 33**, p. 794, 1986.

[24] L.H. Tanner, 'The Spreading of Silicone Oil Drops on Horizontal Surfaces', *J. Phys. D: Appl. Phys.*, **12**, p. 1473, 1979.

[25] M. Van Dyke, *Perturbation Methods in Fluid Mechanics*, the Parabolic Press, Standford, 1975.

[26] Ya.B. Zeldovich, D.A. Frank-Kamenetskii, *A Theory of Thermal Flame Propagation, selected works*, Vol.1, p. 262, Princeton University Press, 1938.

[27] Ya.B. Zeldovich, 'Tsepnye reaktsii v goryachikh plamenakh-priblizhennaya teoriya skorosti plameni', (Chain Reactions in Hot Flames – an Approximate Theory of Flame Velocities), *Kinetica i Kataliz*, **11**, p. 305, 1961.

[28] Ya.B. Zeldovich, G.I. Barenblatt, V.B. Librovich, G.M. Makhviladze, The Mathematical Theory of Combustion and Explosion, Consultants Bureau, 1985.

CHAPTER 6

First instabilities

6.1 INSTABILITY OF VISCOUS FINGERING

First let us consider a linear geometry to which a uniform pressure gradient is applied along the Hele–Shaw cell. The interface is therefore flat and moving with a constant velocity V, where the displacing and displaced viscous fluids of viscosity η_2 and η_1, respectively, flow at the same speed. The stationary pressure field of the viscous fluids is determined from equations (5.3) – (5.6) in paragraph 5.2, that is

$$\bar{p}_{1,2} = p_0 - \frac{12\eta_{1,2}}{b^2}Vx \tag{6.1}$$

where $x = \xi(y) = 0$ is the position of the flat interface. The stability of the uniform displacement is examined in the following way: we superimpose a small perturbation on the stationary solution $(\vec{w}', p', \xi')$ which satisfies the equations of motion as well as their boundary conditions. The equations for $\vec{w}'$, p' and ξ' are deduced by separating the stationary and perturbed quantities in the equations of motion. Keeping only the linear terms in $\vec{w}'$, p' and ξ', we get

$$\vec{w}'_{1,2} = -\frac{b^2}{12\eta_{1,2}}\vec{\nabla}p'_{1,2} \tag{6.2}$$

and

$$\vec{\nabla}\cdot\vec{w}'_{1,2} = 0 \tag{6.3}$$

with the boundary conditions at $x = 0$

$$u'_1 = u'_2 = \frac{\partial^2\xi'}{\partial t} \tag{6.4}$$

$$p'_1 + \xi'\frac{d\bar{p}_1}{dx} - p'_2 - \xi'\frac{d\bar{p}_2}{dx} = \gamma\frac{\partial^2\xi'}{\partial x^2} \tag{6.5}$$

where u is the longitudinal component of the velocity field, and where indices 1 and 2 refer to the displaced and the displacing fluids, respectively. As in many problems of instability (Chandrasekhar 1961), since the coefficients in equations (6.2) – (6.5) are time-independent we can look for solutions proportional to the factor $\exp(\sigma t)$, where σ is the growth rate. If there is such a solution with $Re\,\sigma > 0$, its amplitude increases with time and the shape of the interface deviates from being flat. The flat interface is therefore unstable and the obtained solution only realistically describes the interface dynamics at sufficiently short times.

Using equations (6.2) and (6.3) we deduce $\Delta p'_{1,2} = 0$ so that the perturbation of the pressure field is of the form

$$\begin{aligned} p'_1 &= A\exp(\sigma t)\exp(-|k|x + iky) \\ p'_2 &= B\exp(\sigma t)\exp(|k|x + iky). \end{aligned} \tag{6.6}$$

The perturbation of the shape of the interface is written as $\xi' = C\exp(\sigma t)\exp(iky)$. The constants A, B and C are determined by equations (6.4) and (6.5) using equation (6.2) to determine the perturbed velocity field, that is

$$\sigma C = \frac{b^2}{12\eta_1}|k|A = -\frac{b^2}{12\eta_2}|k|B \tag{6.7a}$$

$$A - B = C\left(-\gamma k^2 + \frac{12}{b^2}(\eta_1 - \eta_2)V\right). \tag{6.7b}$$

These relations are compatible with non-zero solutions if the following so-called dispersion relation between the growth rate σ and the wavenumber k is satisfied (figure 6.1)

$$\sigma = \frac{(\eta_1 - \eta_2)}{(\eta_1 + \eta_2)}V|k| - \gamma\frac{b^2}{12(\eta_1 + \eta_2)}|k|k^2. \tag{6.8}$$

Thus if the displacing fluid is less viscous than the displaced fluid, wavelengths larger than $\lambda_m = 2\pi b(\gamma/(\eta_1 - \eta_2)V)^{1/2}$ are unstable and the most unstable wavelength, meaning with the fastest growth rate, is $\lambda_c = \sqrt{3}\lambda_m$. Let us assume that the viscosity of the displacing fluid is negligible ($\eta_2 = 0$). Then $\lambda_m = 2\pi b/(Ca)^{1/2}$, and so since the capillary number $Ca = \eta V/\gamma$ is small λ_m is much larger than b. It is an *a posteriori* justification of the derivation in paragraph 5.2. Moreover, if we compare it to appendix 3 in paragraph 5.2, the order of magnitude of the film and the capillary effects on the change in pressure is $(\gamma/b)Ca^{-1/3}(\eta/\gamma)\sigma\xi'$ due to the former and $\gamma\xi'/\lambda^2$ due to the latter. If we take the wavelength λ_c as the reference and the growth rate associated with it as $\sigma_c \approx V/\lambda_c$, we deduce that the capillary effects dominate the film effects proportionally to $Ca^{-1/6}$. This justifies only taking capillary effects into account in relation (6.8).

As this relation indicates the destabilising effect is the displacement of the fluid. If the interface wrinkles the pressure gradient is larger in front of the parts of the fluid further ahead. Indeed, since the interface is at an almost constant pressure, the jump of the pressure between the interface and the upstream extremity of the cell takes place over a shorter distance at the tip than at the other points. Thus, since the

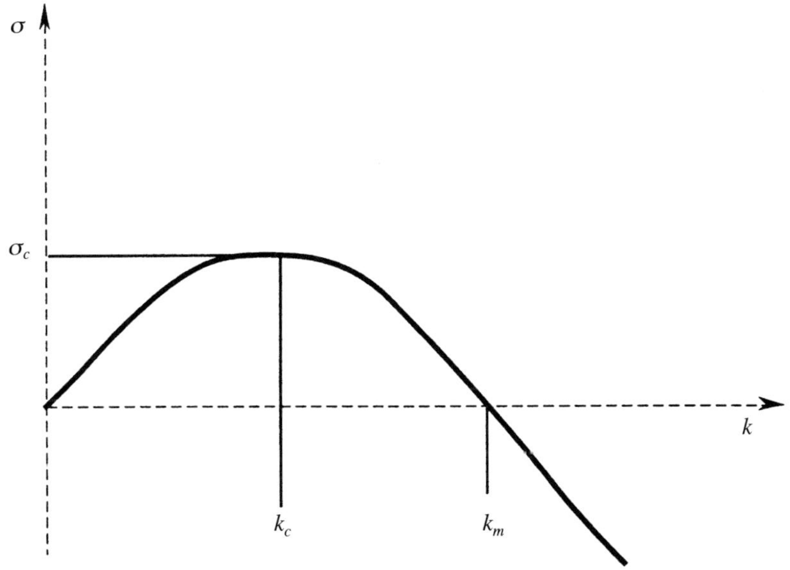

Figure 6.1 Dispersion relation $\sigma(k)$.

interface velocity is proportional to the pressure gradient at a given point, a tip tends to move faster than the parts located behind. The surface tension has the opposite effect because it tends to minimise the curvature of the interface. The existence of a wavenumber band of instability implies that the planar interface is unstable as soon as it begins to move. At short times the most unstable wavelength appears first. Then there is a process of coalescence of the tips, finally leaving only a displacing fluid finger which moves across the displaced fluid at constant velocity.

If the cell is vertical and the fluids have different densities the gravity field enters into the interface dynamics. We must simply replace the pressure p by an effective pressure $p + \rho g z$. The gravity effect affects the left hand side of the boundary conditions (6.5) where we must add the term $g(\rho_1 - \rho_2)\xi'$. The modified dispersion relation is therefore deduced as

$$\sigma = \left(\frac{(\eta_1 - \eta_2)}{(\eta_1 + \eta_2)}V + \frac{b^2}{12}\frac{(\rho_1 - \rho_2)}{(\eta_1 + \eta_2)}g\right)|k| - \gamma\,\frac{b^2}{12(\eta_1 + \eta_2)}|k|k^2\,. \tag{6.9}$$

Thus when the more viscous fluid is evacuated towards the top the interface can stay planar even if the viscosity of the displacing fluid is lower, as long as its volumetric mass is larger. The condition of instability for a planar interface is therefore simply written as: $(\eta_1 - \eta_2)V + (b^2/12)(\rho_1 - \rho_2)g > 0$. If this relation had first been determined by Hill (1952) during his work on sugar refining, the full dispersion relation (6.9) was established for the first time by Chuoke, Van Meurs and Van der Poel (1958) and by Saffman and Taylor (1958). As indicated by relation (6.9) displacement and the gravity field have equivalent effects since they are linked by the relation $\eta V = (b^2/12)\rho g$.

6.2 CIRCULAR GEOMETRY CASE

Now let us consider a radial geometry with fluid injection at the centre of the cell (Paterson 1981). Let Qb be the volume of fluid injected per unit time, assumed constant. Since the interface is circular with radius R, the growth law $R(t)$ is simply deduced from the constant flow rate $R(t)$: $R(t) = (Q(t-t_0)/\pi)^{1/2}$. From equations (5.3) and (5.4) in paragraph 5.2, the pressure field satisfies the Laplace equation $\Delta p = 0$, that is, if we only consider the radial solution

$$\frac{1}{r}\frac{\partial}{\partial r}\left(r\frac{\partial}{\partial r}\bar{p}\right) = 0. \tag{6.1}$$

Applying the boundary conditions (5.5) in paragraph 5.2 we get

$$\bar{p}_{1,2}(r) = -\frac{Q}{2\pi}\frac{12\eta_{1,2}}{b^2}\left(\mathrm{Log}\left(\frac{r}{R}\right) + \frac{\eta_1}{\eta_{1,2}}\right). \tag{6.2}$$

where media 1 and 2 are the exterior and the interior media, respectively. The radial velocity field is simply deduced from equation (5.5) in paragraph 5.2:

$$\bar{u}_{1,2} = \frac{Q}{2\pi r}. \tag{6.3}$$

The study of the stability of radial motion is carried out as follows: we describe the perturbation of the shape of the interface in the form $r' = f(t)\exp(in\theta)$ where f is much smaller than R. The perturbation of the pressure field p' satisfies, according to equations (5.3) and (5.4) in paragraph 5.2, the Laplace equation $\Delta p' = 0$. Looking for a solution in the form $p' = g(t)h(r)\exp(in\theta)$, h satisfies the equation

$$\frac{1}{r}\frac{\partial}{\partial r}\left(r\frac{\partial}{\partial r}\right) - \frac{n^2}{r^2}h = 0 \tag{6.4}$$

the solutions of which are of the form $h(r) = ar^n + br^{-n}$. Keeping only the solutions in the two media which do not diverge we obtain the perturbation of the pressure field in the form

$$\begin{aligned} p'_1 &= g_1(t)r^{-n}\exp(in\theta) \\ p'_2 &= g_2(t)r^n\exp(in\theta). \end{aligned} \tag{6.5}$$

The boundary conditions (5.5) and (5.6) in paragraph 5.2 are written as

$$u'_1 + r'\frac{\partial\bar{u}_1}{\partial r} = u'_2 + r'\frac{\partial\bar{u}_2}{\partial r} = \frac{\partial r'}{\partial t} \tag{6.6}$$

and

$$p'_1 + r'\frac{d\bar{p}_1}{dr} - p'_2 - r'\frac{d\bar{p}_2}{dr} = \frac{\gamma}{R^2}\left(r' + \frac{\partial^2 r'}{\partial\theta^2}\right) \tag{6.7}$$

where we use an expression for the curvature in polar coordinates

$$\frac{1}{R_I} = \frac{r^2 - r\frac{\partial^2 r}{\partial\theta^2} + 2\left(\frac{\partial r}{\partial\theta}\right)^2}{\left(r^2 + \frac{\partial r}{\partial\theta}^2\right)^{3/2}} \approx \frac{1}{R} - \frac{r' + \frac{\partial^2 r'}{\partial\theta^2}}{R^2}. \tag{6.8}$$

The relation (6.6) allows the expression of g_1 and g_2 as a function of f:

$$\begin{aligned} g_1 &= \frac{1}{n}\,\frac{12\eta_1 R^n}{b^2}\left(\frac{Q}{2\pi R}f + R\frac{df}{dt}\right) \\ g_2 &= -\frac{1}{n}\,\frac{12\eta_2}{b^2 R^n}\left(\frac{Q}{2\pi R}f + R\frac{df}{dt}\right) \end{aligned} \tag{6.9}$$

and from equation (6.7) we deduce the equation of evolution of f:

$$\frac{1}{f}\frac{df}{dt} = \frac{Qn}{2\pi R^2}\,\frac{\eta_1 - \eta_2}{\eta_2 + \eta_1} - \frac{Q}{2\pi R^2} - \frac{\gamma n(n^2 - 1)}{R^3}\,\frac{b^2}{12(\eta_2 + \eta_1)}. \tag{6.10}$$

When the azimuthal number n is large, the perturbed shape changes notably over a distance much smaller than R, and we can consider it as unperturbed and locally planar, moving at a velocity $V = Q/2\pi R$. The local wavenumber of the perturbation is therefore simply $k = n/R$. Neglecting the changes in the local wavenumber due to the increase in radius R, we find again $f \approx \exp(\sigma t)$ where σ is determined by the planar dispersion relation (6.8) in paragraph 6.1. Let us consider the limiting case $\eta_1 = \eta, \eta_2 = 0$ for simplification. Then,

$$\frac{1}{f}\frac{df}{dt} = (n-1)\left(\frac{Q}{2\pi R^2} - \frac{\gamma n(n+1)}{R^3}\,\frac{b^2}{12\eta}\right). \tag{6.11}$$

Each $n > 2$ corresponds to a critical radius R_n for which the perturbation is stationary:

$$R_n = \frac{2\pi}{Q}\,\frac{b^2}{12\eta}\gamma n(n+1). \tag{6.12}$$

The smallest radius R_2 is that from which the circular air bubble starts to be destabilised. If the equation (6.11) considered at constant R leads to an exponential growth of perturbations, this growth is slowed down by taking into account the increase in radius with time. Indeed, assuming that the achieved radius is large enough such that the effects of surface tension are negligible, equation (6.11) leads to

$$\frac{1}{f}\frac{df}{dt} = (n-1)\frac{1}{2t} \tag{6.13}$$

and therefore a power law growth of the perturbations, $f \approx t^{(n-1)/2}$. This is due to the fact that the wavelength of the perturbations increases with radius. As the growth rate is proportional to the wavenumber, the latter decreases as the radius increases.

6.3 INFLUENCE OF A THICKNESS GRADIENT

Let us take the example of the rotating axle in a lubricated cavity from appendix 1 in paragraph 5.2. When cavitation takes place there are gas cavities formed in the fluid which are stabilised at well-defined angles. More precisely, as we see in figure 6.2, a meniscus pointing in the opposite direction to the flow and letting only the fluid close to the rotating axle pass is stabilised in the space confined by the two pieces moving relative to each other.

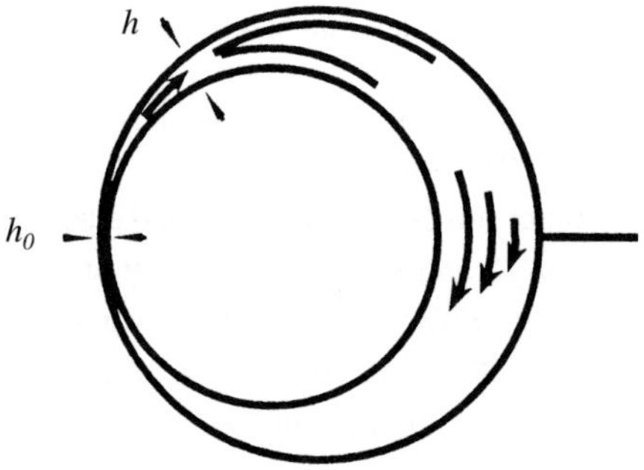

Figure 6.2 Meniscus stabilised between two rotating cylinders.

To determine the possible meniscus positions we calculate the mass of the flowing fluid per unit time and per unit length of the axle instead of the thickness h_0, where the longitudinal pressure gradient is zero. Therefore from equation (5.9) in paragraph 5.2, where the velocity profile varies linearly with the height (Pearson 1960)

$$\rho \int_0^h u(z)dz = \rho\Omega R\,\frac{h_0}{2}. \tag{6.1}$$

By conservation of mass it must be equal to the mass left behind by the film along the axle. Also, using the film thickness determined in appendix 3 in paragraph 5.2 for low capillary numbers we get

$$\rho\Omega R\frac{h_0}{2} = \pi C_0 \Omega R\,\frac{h}{2}\left(\frac{\eta\Omega R}{\gamma}\right)^{2/3}. \tag{6.2}$$

We thus deduce the thickness at which the meniscus is stabilised

$$h = \frac{1}{C_0} h_0 \left(\frac{\eta \Omega R}{\gamma} \right)^{-2/3} . \tag{6.3}$$

As h cannot exceed $2d$, the maximum value being obtained when the axle is completely decentred, the meniscus can only be stabilised if the rotation velocity is slow enough. Indeed if the capillary number of the flow is assumed small, h is much larger than h_0 from equation (6.3). The ratio h/h_0 can therefore be small only if h_0 itself is small, which is the case at low rotation velocities from the results of appendix 1 in paragraph 5.2. There are two possible positions for the meniscus at fixed h: one in which the thickness increases in the direction of the flow and the other in which it decreases. Only the position in which the thickness increases is stable. Let us assume that the meniscus is slightly going up the flow. The thickness of the space filled by the fluid diminishes, as does the film. Since a smaller quantity of fluid escapes through the film the meniscus must go down and return to its initial position. Analogous reasoning shows that the other position is unstable.

Let us examine the stability of the meniscus with respect to the perturbations in the direction parallel to the axle (Pearson 1960, Hakim *et al.*1990). When the line joining the top of the menisci wrinkles, the upstream flow is perturbed. The equation which satisfies the perturbation of the pressure field $p' = \exp(\sigma t) a(\theta) \exp(ikx)$, where x is the coordinate parallel to the axis of the axle, is obtained by linearisation of the Reynolds equation (5.10) in paragraph 5.2 around the stationary solution from appendix 1 in the same paragraph:

$$\frac{1}{R^2} \frac{d^2 a}{d\theta^2} + \frac{3}{h} \frac{1}{R} \frac{dh}{d\theta} \frac{da}{d\theta} - h^2 a = 0 . \tag{6.4}$$

Applying the Laplace law to the meniscus tip we must take into account the variation in the meniscus curvature in the transverse and parallel directions to the axle, when the meniscus tip moves by $\xi' = R\theta' = b \exp(\sigma t) \exp(ikx)$, that is

$$a + b \frac{1}{R} \frac{d\bar{p}}{d\theta} = b \left(\frac{2\gamma}{h^2} \frac{1}{R} \frac{dh}{d\theta} + \gamma k^2 \right) . \tag{6.5}$$

The linearisation of the conservation of mass of fluid is written as

$$\frac{\partial \xi'}{\partial t} h \left(1 - \frac{1}{2} Ca^{2/3} \right) = hu' + \frac{1}{R} \frac{d}{d\theta} \left(\frac{\Omega R h_0}{2} - C_0 \frac{\Omega R h}{2} Ca^{2/3} \right) b . \tag{6.6}$$

From equations (6.5) and (6.6), we deduce the growth rate of the instability

$$\sigma = \frac{1}{1 - \frac{1}{2} Ca^{2/3}} \left(-\frac{\Omega R h_0}{2h^2} \frac{1}{R} \frac{dh}{d\theta} + \frac{h^2}{12\eta} \frac{1}{Ra} \frac{da}{d\theta} \left(\frac{1}{R} \frac{d\bar{p}}{d\theta} - \frac{2\gamma}{h^2} \frac{1}{R} \frac{dh}{d\theta} - \gamma k^2 \right) \right) . \tag{6.7}$$

Let us assume that the wrinkling wavelength $1/k$ is much smaller than the distance over which the thickness varies considerably $hR/(dh/d\theta)$, so $a \approx \exp|k|R\theta$ and $(1/Ra)(da/d\theta) = |k|$. If we neglect the effects of the film thickness gradient we come back to the rate instability of uniform displacement (6.8) in paragraph 6.1. From equation (6.7) the thickness gradient has two effects which depend on its sign (divergent or convergent flow). In the case of divergent flow these effects are stabilising. The first (the first term of the right-hand side of equation (6.7)) is due to the reduction of mass of the fluid which escapes along the axle when the meniscus advances towards the more confined region. The second (the third term of the right-hand side of the same equation) corresponds to the increase in transverse capillary effects in the same situation. These effects have the opposite role in the case of convergent flow. Thus, as we have mentioned above, the meniscus places itself into an allowed position for which the flow is divergent. Above a critical velocity the instability of uniform displacement dominates the stabilising effects of the gradient and the shape of the interface is formed with a wavelength close to the first unstable wavelength from the dispersion relation (6.7).

A precise experimental study of this instability has been performed by the growth instabilities group at the École Normale Supérieure in Paris (Hakim *et al.* 1990). A rotating cylinder is submerged into a bath and drags the viscous fluid. At the cylinder base a meniscus analogous to the one mentioned above is formed in the divergent part of the space where the fluid flows. Above a critical velocity the tip of the meniscus is formed in a direction parallel to the rotation axis of the cylinder, all the more in accordance with the previously developed theory since the fluid thickness is small. The transverse structuring of that film can cause some damage, for example when a printing roller rotating too fast deposits the ink onto a surface.

6.4 FINGERING OF SURFACTANT DROPS

The spreading of a surfactant drop has particular characteristics. As shown in figure 6.3, under certain conditions, rather than spreading uniformly, the drop explodes into thin fingers which spread much faster than its main core (Lelah and Marmur 1981). The analysis of this spreading phenomenon becomes difficult because of the complexity of the system. Nevertheless, the appearance of similar phenomena in other experimental conditions enables access to the nature of the phenomenon; for example the progression by the Marangoni effect of a liquid film deposited onto a substrate undergoing a temperature gradient (Brzoska *et al.*1992), or the rise of a liquid film along a wall in the same conditions (Cazabat *et al.* 1990).

When the drop is filled with surfactant, a concentration gradient is generated between the droplet core and the upstream part of the substrate where the surfactant

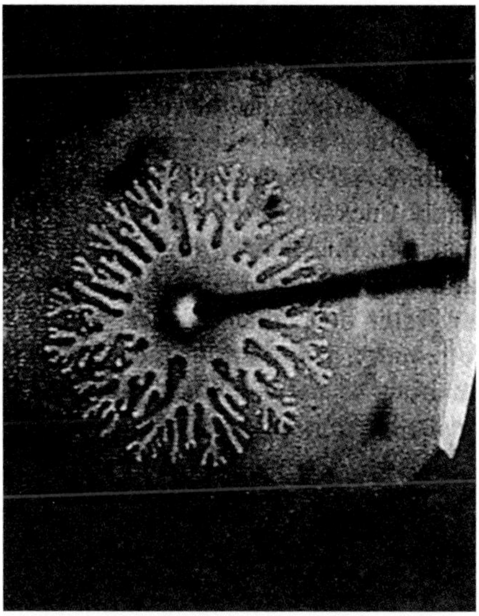

Figure 6.3 Spreading of a charged surfactant drop (Troian *et al.* 1989).

is absent. The fluid is set into motion simply by the tangential gradient of the surface tension. A tangential force acting on a unit surface of fluid

$$\vec{f}_t = \frac{\partial \gamma}{\partial c_s} \vec{\nabla}_t c_s \tag{6.1}$$

results, which modifies the movement of the underlying viscous fluid. A displacement of the contact line follows, much faster than by the effects of pressure generated by the curvature of the interfaces considered in appendix 4 of paragraph 5.2. More precisely, considering such a film of constant thickness h_0, in which there is a uniform gradient of surfactant concentration $\Gamma = dc/dx$, therefore $\eta \partial u/\partial z = (d\gamma/dc)\Gamma$ at the interface $(z = h_0)$. If we assume that $d\gamma/dc$ is constant, the film is at a uniform pressure and the velocity profile is $u(z) = (1/\eta)(d\gamma/dc)\Gamma z$. In particular, the fluid velocity on the surface is uniform and equal to $u_0 = (1/\eta)(d\gamma/dc)\Gamma h_0$. The film dynamics is limited to the evolution of the surface properties, which are the two-dimensional velocity field of components

$$\begin{aligned} u_s &= \frac{1}{\eta} \frac{d\gamma}{dc} \frac{dc}{dx} h_0 \\ v_s &= \frac{1}{\eta} \frac{d\gamma}{dc} \frac{dc}{dy} h_0 \end{aligned} \tag{6.2}$$

and the surfactant concentration field which satisfies

$$\frac{\partial c}{\partial t} + u_s \frac{\partial c}{\partial x} + v_s \frac{\partial c}{\partial y} = 0 \tag{6.3}$$

because the surface diffusion can be neglected.

The analysis of the stability of a film flowing with uniform velocity is carried out as follows: from the conservation of mass of the surface fluid

$$\frac{\partial u_s}{\partial x} + \frac{\partial v_s}{\partial y} = 0 \tag{6.4}$$

and from equation (6.1), we deduce $\Delta c = 0$. The perturbation from the stationary concentration field is therefore of the form $c' = A \exp \sigma t \exp \pm kx \exp iky$ where k is the wavenumber of the perturbation in the transverse flow direction. From equation (6.2) we deduce $\sigma = - \pm u_0 k$. Thus, the perturbations which decrease in amplitude in the flow direction, increase with the growth rate $\sigma = u_0 k$, and destabilise the film with respect to the transverse modulations (Troian *et al.* 1990).

The film is confined by boundaries upstream and downstream, where capillary effects cannot be neglected. Downstream, the contact angle imposed by the substrate forces the free surface to curve, thus generating a hillock on the longitudinal scale $l_0 \approx h_0 / Ca^{1/3}$ where $Ca = \eta u_0 / \gamma$. Upstream, the film is connected to the droplet body on the same scale. The existence of that boundary is essential in that the perturbation does not diverge upstream of the film. For example, when the concentration inside the droplet body exceeds the CMC defined in paragraph 2.1, the droplet contains an almost immobile reservoir of surfactant with a boundary in which the concentration remains constant. It is possible that the existence of this boundary is related to the observation of Lelah and Marmur (1981) that the fingering takes place only if the droplet concentration exceeds the CMC.

Experimental studies indicate that the most unstable wavelength λ_c is proportional to l_0, i.e. the smallest scale at which the capillary effects associated with the interface curvature definitely stabilise the fingers, since it is the case for the spreading of a pure liquid. The growth rate of the perturbation is then $\sigma = u_0 / l_0$, in good agreement with the experimental results. In particular Brzoska *et al.* (1992) have shown that the product λ_c / σ^2 was independent of h_0, a result which can be deduced from the expressions of σ, l_0 and u_0.

6.5 THE MULLINS–SEKERKA INSTABILITY

Let us consider the growth of a spherical germ starting from a pure liquid phase at temperature T_∞. Let us assume that the supercooling of the liquid $\Delta = c_p(T_0 - T_\infty)/Q$ is small and therefore, according to paragraph 5.4, the temperature field $u = c_p(T - T_\infty)/Q$ is quasi-stationary:

$$u = \left(\Delta - \frac{2d_{0T}}{R}\right)\frac{R}{r}, \quad r > R$$
$$u = \Delta - \frac{2d_{0T}}{R}, \qquad r < R \tag{6.1}$$

and the growth rate of the germ is written as

$$V = \frac{dR}{dt} = \frac{D_T}{R}\left(\Delta - \frac{2d_{0T}}{R}\right). \tag{6.2}$$

The analysis of the linear stability of the shape of spherical growth is carried out in an analogous way to that in paragraph 6.2 for a circular bubble. We write the perturbation of the interface shape in the form $r' = CY_{j,m}(\theta, \phi)\exp(\sigma t)$ where C is the deformation amplitude and $Y_{j,m}$ the spherical harmonic of order (j, m), the eigenfunction of the angular part L of the Laplacian

$$L(Y_{j,m}) = \left(\frac{1}{\sin\theta}\,\frac{\partial}{\partial\theta}\left(\sin\theta\frac{\partial}{\partial\theta}\right) + \frac{1}{\sin^2\theta}\,\frac{\partial^2}{\partial\phi^2}\right)Y_{j,m} = -j(j+1)Y_{j,m}\,. \tag{6.3}$$

The perturbation of the temperature field which we write as $u' = h_j(r)Y_{j,m}(\theta, \phi)\exp(\sigma t)$ satisfies the Laplace equation,

$$\frac{1}{r^2}\,\frac{\partial}{\partial r}\left(r^2\,\frac{\partial}{\partial r}\,h_j\right) - \frac{j(j+1)}{r^2}\,h_j = 0\,. \tag{6.4}$$

Thus, $h_j = A/r^{j+1}$ in the liquid, that is for $r > R$ and $h_j = Br^j$ in the solid $(r < R)$. The boundary conditions (2.4) from paragraph 2.4 and (5.8) in paragraph 5.3 linearised around the stationary solution (1) are written as

$$u'_l + r'\,\frac{\partial u_l}{\partial r} = u'_s = \frac{d_{0T}}{R^2}(2r' + L(r')) \tag{6.5}$$

and

$$\frac{dr'}{dt} = D_T\left(\frac{\partial u'_s}{\partial r} - \left(\frac{\partial u'_l}{\partial r} + r'\frac{\partial^2 u_l}{\partial r}\right)\right) \tag{6.6}$$

where L is again the angular part of the Laplacian and where we have used the expression for curvature in spherical coordinates

$$\frac{1}{R_1} + \frac{1}{R_2} \approx \frac{2}{R} - \frac{2r'}{R^2} - \frac{1}{R^2}\mathrm{L}(r'). \tag{6.7}$$

After having introduced the coefficients A, B, C in equations (6.5) and (6.6), we determine the dispersion relation

$$\sigma = \frac{(j-1)V}{R}\left(1 - (2j+1)\frac{Dd_{0T}(j+2)}{VR^2}\right). \tag{6.8}$$

Taking the limit of large radii and large j, we come back to the deformation modes of a planar interface with the wavenumber $k = j/R$. From equation (6.8) we thus obtain

$$\sigma = Vk\left(1 - 2\frac{Dd_{0\mathrm{T}}}{V}k^2\right). \tag{6.9}$$

We therefore find a dispersion relation entirely analogous to equation (6.8) in paragraph 6.1 determined by Saffman and Taylor. This is a consequence of the remarks in paragraph 5.3 about the analogies between the growth models, since we have assumed that the temperature field is almost stationary and we have neglected the effects of crystalline anisotropy. The interpretation of the instability is therefore also analogous. Firstly, let us assume that the interface is isothermic. When the interface wrinkles, it pushes back the isotherms and the temperature gradient increases in the proximity of the tips (Langer 1980) (figure 6.4). A higher solidification velocity results at the tips due to the proportionality between the normal temperature gradient and the growth rate. The capillary effects also decrease the temperature gradient by decreasing the temperature in the proximity of the tips, and therefore the solidification velocity, which tends to stabilise the original spherical shape. Thus if its radius is large enough the spherical shape becomes unstable, and this instability is responsible for many microstructures in metallurgy (Mullins and Sekerka 1964).

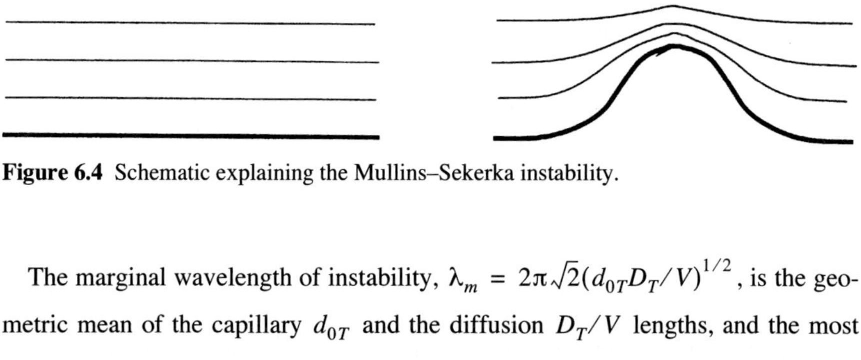

Figure 6.4 Schematic explaining the Mullins–Sekerka instability.

The marginal wavelength of instability, $\lambda_m = 2\pi\sqrt{2}(d_{0T}D_T/V)^{1/2}$, is the geometric mean of the capillary d_{0T} and the diffusion D_T/V lengths, and the most unstable wavelength is $\lambda_c = \sqrt{3}\lambda_m$. When the germ radius surpasses the critical nucleation radius $R_c = 2d_{0T}/\Delta$, its growth velocity V becomes very small and, from equation (6.8), all the arbitrary modes of j are stable. For larger radii, $V \approx D_T\Delta/R$ and from that same equation we can define for each j a radius R_j for which the growth rate cancels out:

$$R_j \approx R_c = \frac{(2j+1)(j+2)}{2}. \tag{6.10}$$

Thus, as soon as the germ reaches radius $R_2 \approx 10R_c$, it becomes unstable by the mode $j = 2$. The mode generally observed in experiments is mostly that which respects the crystalline symmetry (Oswald 1988). The principal reason for this is that, assuming the anisotropy factors of the material under consideration are weak, their equilibrium shape is a slight deformation away from the spherical shape of the mode respecting the crystal symmetry. This perturbation appears as an initial condition for the linear problem set here, which is of a much superior amplitude to the first unstable mode $j = 2$. Thus, even if its growth rate is slower this perturbation dominates the growth form at short times.

As we have mentioned in paragraph 5.8, in a simplified model the growth of an electrode surface follows the dynamics perfectly analogously to that of a rough liquid–solid interface. The instability described above must therefore be applied, as the comparisons with experiments are more difficult due to the larger number of control parameters to be considered.

6.6 TEMPERATURE GRADIENT EFFECTS

Let us take the configuration of the directional growth again, shown in paragraph 5.4. By rejecting the solute in the liquid phase, the solid phase generates a concentration profile similar to the temperature field which determines the growth in a pure liquid. The temperature gradient has a completely different role here since it has a tendency to stabilise the interface, as we will encounter later on. Let us introduce the dimensionless concentration $u = K(c - c_\infty)/c_\infty(1-K)$. The stationary concentration profile resulting from the growth of the planar interface, determined in paragraph 5.4, is then written as

$$
\begin{aligned}
u &= \exp\left(-z\frac{V}{D_c}\right), \quad z > 0 \\
u &= 0, \qquad\qquad\quad z < 0
\end{aligned}
\tag{6.1}
$$

In the moving frame where the planar interface is at rest, the perturbation of the concentration field u' in the liquid satisfies the equation

$$
\frac{\partial u'}{\partial t} - V\frac{\partial u'}{\partial z} = D_c \Delta u' \tag{6.2}
$$

with the boundary conditions (5.9) in paragraph 5.3 and (2.11) in paragraph 2.4 at the interface, linearised around the plane front

$$
u'(1-K)V + \frac{d\xi}{dt} = -D_c\left(\frac{du'}{dz} + \xi\frac{d^2u}{dz^2}\right) \tag{6.3}
$$

and

$$u' + \xi \frac{du}{dz} = -\frac{\xi}{l_T} + d_{0c}\frac{d^2\xi'}{dx^2}. \qquad (6.4)$$

We define two lengths here, the thermal length constructed with the temperature gradient G, $l_T = |m|c_\infty(1-k)/GK$ and the solute capillary length $d_{0c} = \gamma T_0 K/Qc_\infty(1-K)$. We can search for a solution of equation (6.2) in the form

$$u' = A\exp(\sigma t)\exp(-qz + ikx) \qquad (6.5)$$

where q is the root with a positive real part of the second order equation $q^2 - (1/l)q - (k^2 + \sigma/D) = 0$, and $l = D_c/V$ is the diffusion length. From boundary conditions (6.3) and (6.4) and the stationary solution (6.1), we then deduce the dispersion relation

$$1 - \frac{1}{l_T} - d_{0c}lk^2 - \frac{K + \sigma\tau}{ql + K - 1} = 0 \qquad (6.6)$$

where $\tau = D_c/V^2$ is the diffusion time. To simplify, let us consider a quasi-stationary regime for which the evolution time of the interface is slow compared to the diffusion time. We can thus neglect σ compared to $D_c k^2$ and obtain from equation (6.6)

$$\sigma\tau = \left(1 - \frac{1}{l_T} - d_{0c}lk^2\right)(ql + K - 1) - K \qquad (6.7)$$

where $ql = 1 + (1 + 4k^2l^2)^{1/2}/2$. We show this relation for different drawing velocities in figure 6.5. When V is smaller than a critical value V_c, $\sigma\tau$ is negative for all wavenumbers and the interface is planar and stable. For $V = V_c$, a critical non-zero wavenumber k_c becomes neutral, meaning that its amplitude remains stationary. For higher drawing velocities a band of wavenumbers is unstable and the interface becomes cellular.

The conditions for the instability threshold are therefore $\sigma(k_c) = 0$, $d\sigma/dk(k_c) = 0$. The critical velocity and the threshold wavenumber have a simple expression at large wavenumbers:

$$V_c = \frac{D_c GK}{|m|c_\infty(1-K)}$$
$$k_c = \left(\frac{K}{2d_{0c}l_c^2}\right)^{1/3}. \qquad (6.8)$$

These expressions are valid if $kl_c >> 1$, that is, if $V_c << D_c/d_{0c}$ in equation (6.8).

The interpretation of the instability is clearer starting from equation (6.7) considered at a large wavenumber. Indeed, we then have

$$\sigma = Vk\left(1 - \frac{1}{l_T} - d_{0c}lk^2\right) \qquad (6.9)$$

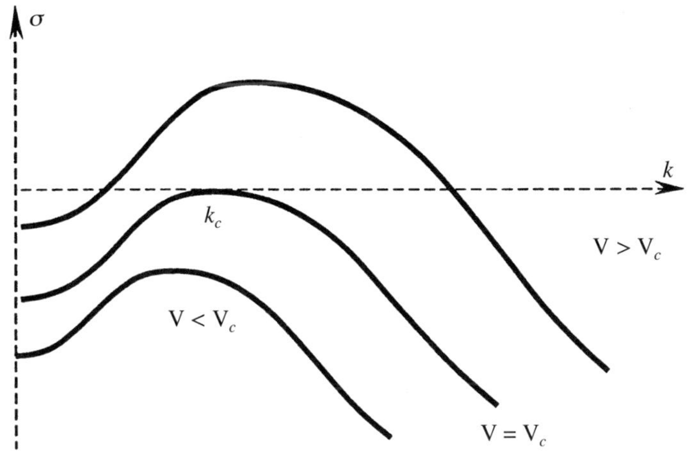

Figure 6.5 Dispersion relation for different values of growth rates.

which is a completely analogous expression to equation (6.9) in paragraph 6.5 apart from two differences. The first one, of little importance, concerns the factor 2 of the capillary term. It is due to the temperature diffusion and the absence of the concentration diffusion in the solid phase. The second one, essential in this problem, is the stabilising effect of the temperature gradient by the term $-1/l_T$. Indeed, in an analogous way to the capillary effects, the temperature gradient effect is to reduce the interfacial concentration of the tip and to therefore reduce the concentration gradient driving the instability. As we can see from equation (6.8), the threshold velocity is achieved when the thermal length becomes equal to the diffusion length, $l_c = l_T$. This relation also coincides with the criterion of 'constitutional supercooling' in the appendix of paragraph 5.4 which serves as an explanation for the generation of crystalline microstructures before the one described in this chapter (Rutter and Chalmers 1953, Tiller *et al.* 1953).

The experimental study of this instability was performed on different systems: Eshelman and Trivedi (1987) with a SCN-acetone; De Cheveigné and coworkers (1986) with CBr_4-Br_2; Oswald and coworkers (1987) in liquid crystals. According to the systems, the instability arises either in a continuous fashion (supercritical transition) or in a discontinuous fashion (subcritical transition), in which case the formed cells rapidly attain a significant amplitude. The theoretical analysis of the transition between these two behaviours was achieved by Wollkind and Segel (1970). For all these systems the agreement between theory and experiments is as satisfactory for the limits of instability as it is for the transition between the subcritical and the supercritical behaviour.

6.7 THE DARRIEUS–LANDAU INSTABILITY

Let us first suppose an infinitely thin flame which therefore propagates at a normal velocity V_0 determined by equation (5.13) in paragraph 5.5 with respect to the fresh gas. Because of the increase in temperature caused by the gas combustion, the burnt gases have a lower density ρ_b than the fresh gases ρ_0. Their ratio is the inverse of the temperatures and since the gases are flowing at a low velocity compared to the speed of sound, they can be considered at constant pressure. Thus, when the flame wrinkles, transverse flows are generated causing an intrinsic instability of the planar propagation. Satisfying the perturbed flow, on both sides of the surface of discontinuity, the Euler equations are

$$\begin{aligned}\frac{\partial u'}{\partial t} + u\frac{\partial u'}{\partial z} &= -\frac{1}{\rho}\frac{\partial p'}{\partial z}\\ \frac{\partial v'}{\partial t} + u\frac{\partial v'}{\partial z} &= -\frac{1}{\rho}\frac{\partial p'}{\partial x}\end{aligned} \tag{6.1}$$

where $u = u_1 = V_0$, $\rho = \rho_1$ for the fresh gas and $u = u_2 = \rho_1 V_2/\rho_2$, $\rho = \rho_2$ for the burnt gas. The continuity equation

$$\frac{\partial u'}{\partial z} + \frac{\partial v'}{\partial x} = 0 \tag{6.2}$$

thus allows us to deduce from equations (6.1)

$$\frac{\partial^2 p'}{\partial z^2} + \frac{\partial^2 p'}{\partial x^2} = 0\,. \tag{6.3}$$

Searching for perturbations proportional to the factor $\exp(\sigma t + ikx)$ we get

$$\begin{aligned}u'_1 &= A\exp(kz + \sigma t + ikx)\\ v'_1 &= iA\exp(kz + \sigma t + ikx)\\ p'_1 &= -A\rho_1\left(\frac{\sigma}{k} + u_1\right)\exp(kz + \sigma t + ikx)\end{aligned} \tag{6.4}$$

and

$$\begin{aligned}u'_2 &= B\exp(-kz + \sigma t + ikx) + C\exp\left(-\frac{\sigma}{u_2}z + \sigma t + ikx\right)\\ v'_2 &= -iB\exp(-kz + \sigma t + ikx) - \frac{i\sigma}{ku_2}C\exp\left(-\frac{\sigma}{u_2}z + \sigma t + ikx\right)\\ p'_2 &= -B\rho_2\left(-\frac{\sigma}{k} + u_2\right)\exp(-kz + \sigma t + ikx)\end{aligned} \tag{6.5}$$

where the wavenumber k is assumed positive. The term proportional to the constant C is a solution without the second term of equations (6.1), only present in medium 2 as we are limited to the positive real part of the growth rate σ in the search for a possible instability. At the surface of discontinuity, the conservation of mass and of momentum lead to the relations

$$\rho_1\left(u'_1 - \frac{\partial\xi}{\partial t}\right) = \rho_2\left(u'_2 - \frac{\partial\xi}{\partial t}\right)$$

$$p'_1 + 2\rho_1 u_1\left(u'_1 - \frac{\partial\xi}{\partial t}\right) = p'_2 + 2\rho_2 u_2\left(u'_2 - \frac{\partial\xi}{\partial t}\right) \tag{6.6}$$

$$\mathrm{v}'_1 + u_1\frac{\partial\xi}{\partial x} = \mathrm{v}'_2 + u_2\frac{\partial\xi}{\partial x}.$$

Furthermore, since the relative velocity of the flame is constant with respect to the fresh gas,

$$u'_1 - \frac{\partial\xi}{\partial t} = 0. \tag{6.7}$$

Introducing the expressions for the velocity field and the pressure determined by equations (6.4) and (6.5) at the boundary conditions (6.6) and (6.7), and setting $\xi = D\exp(\sigma t + ikx)$, we get a linear system for the four coefficients, *A*, *B*, *C* and *D*. The non-zero solution of this system cannot be found unless the dispersion relation

$$\sigma^2(2-\gamma) + 2\sigma V_0 k - \frac{\gamma}{1-\gamma}V_0^2 k^2 = 0 \tag{6.8}$$

is satisfied, where $\gamma = (\rho_1 - \rho_2)/\rho_1$ is the gas expansion factor. As $\gamma < 1$, this relation always has a positive square root:

$$\sigma = V_0\,\frac{-1 + \sqrt{\dfrac{1+\gamma-\gamma^2}{1-\gamma}}}{2(2-\gamma)}\,k. \tag{6.9}$$

Thus the planar flame is unstable, giving rise to the formation of rounded cells towards the fresh gases (figure 6.6) (Darrieus 1938, Landau 1944).

However, when propagating towards the bottom with a sufficiently low velocity the flame is stabilised by the effects of the flame thickness and gravity, as shown in figure 5.11 in paragraph 5.5. To take into account these effects we must essentially modify the condition which gives the pressure jump across the flame (Markstein 1951, 1964):

$$p'_1 + 2\rho u_1\left(u'_1 - \frac{\partial\xi}{\partial t}\right) - \rho_1 g\xi = p'_2 + 2\rho_2 u_2\left(u'_2 - \frac{\partial\xi}{\partial t}\right) - \rho_2 g\xi. \tag{6.10}$$

Also, from paragraph 5.6, the relative velocity of the flame with respect to the fresh gas is

$$u'_1 - \frac{\partial\xi}{\partial t} = Ld\left(\frac{\partial u'_1}{\partial z} - V_0\frac{\partial^2\xi}{\partial x^2}\right) \tag{6.11}$$

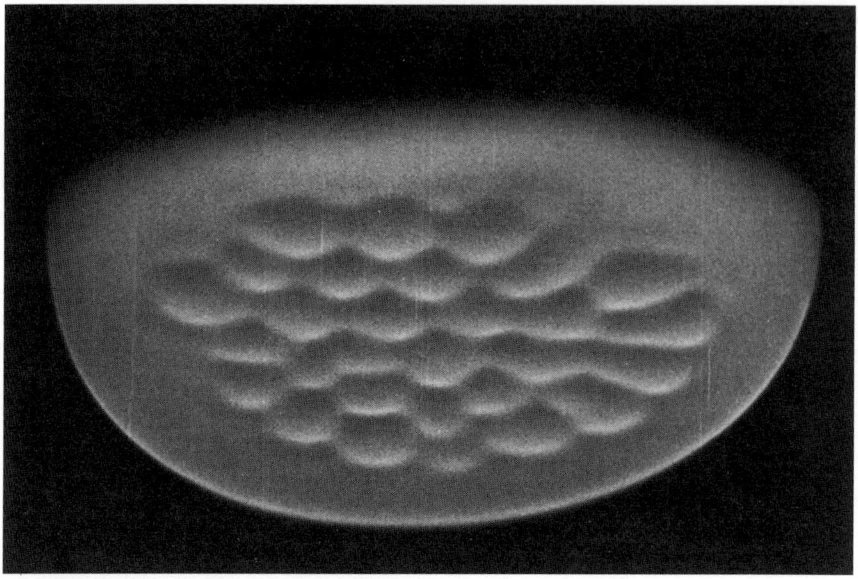

Figure 6.6 Cellular flame (Quinard 1984).

where $L = 1 + \beta(Le - 1)/2$ is the Markstein number, and $d = D_T/V_0$ the thickness of the flame. The new dispersion relation obtained is still of the form $A\sigma^2 + B\sigma + C = 0$, but with the coefficients

$$A = 2 - \gamma + \gamma\, Ldk$$
$$B = 2V_0\left(k + \frac{1}{1-\gamma}Ldk^2\right) \tag{6.12}$$
$$C = \gamma gk - \frac{\gamma}{1-\gamma}V_0^2k^2 + \frac{\gamma}{1-\gamma}\left(1 + \frac{(2+\gamma)}{\gamma}L\right)V_0^2dk^3.$$

In these expressions we have assumed a constant density throughout the thickness of the flame. Let us note the existence of a marginal wavenumber k_c, satisfying at the same time $\sigma = 0$ and $d\sigma/dk = 0$. It thus satisfies $C = 0$ and $dC/dk = 0$. Let $k_c = 2(1-\gamma)g/V_0^2$, obtained for the critical Markstein length $L_c = (\gamma/(2+\gamma))(2/k_cd - 1)$. For smaller Markstein lengths the planar flame is unstable since the solution of the dispersion relation gives a band of wavenumbers with positive growth rates. In the diagram $(\beta(Le-1), V_0)$, we thus obtain a neutral stability curve which separates the stable from the unstable region. Such a curve is shown in figure 6.7 where, contrary to the previously shown calculation, the effects of gas expansion in the thickness of the flame are taken into account (Pelcé and Clavin 1982, Clavin 1985).

In the experiments of Quinard, Searby and Boyer (1983), the flame is kept immobile in the reference frame of the laboratory in a vertical tube where the velocity of the flowing gas inside the tube is equal to the propagation velocity. A laser tomography device is used to stabilise the flame (Boyer 1981). A flattened laser beam crosses the flame. The fresh gases are sowed by fine droplets of oil which burn as they cross the front. Thus, they scatter the laser light before the front, illuminating a region bordering with the flame front itself. A regulation of the flow of fresh gas can therefore be installed in response to the laser strip surface determined by the position of the flame.

The flame velocity V_0 is modulated by dilution δ of the reaction mixture with an inert gas, nitrogen, which has an effect on the temperature of the reaction zone. The Lewis number varies according to the relative composition of the two reactants. The one considered in the calculation is that of the reactant which limits the reaction. When the mixture is rich in hydrocarbons the rate limiting reactant is oxygen, and the Lewis number is that of oxygen. In the opposite case, the mixture is said to be poor, and the Lewis number is that of the hydrocarbon, generally larger than one. For the propane–air mixture considered here, we show the obtained stability curve in figure 6.8, in the (Φ, V_0) plot. As predicted by the theory, the planar flame is unstable at high velocities. The threshold velocity is the highest at low concentration, by the fact that the Lewis number of propane is larger than one, the stabilising

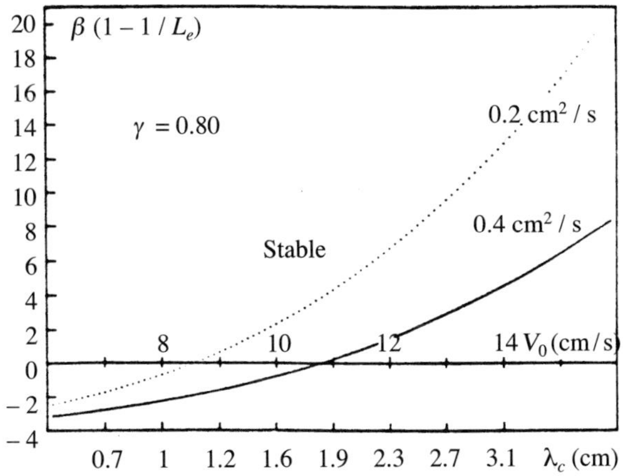

Figure 6.7 Neutral instability curve for the Darrieus–Landau instability. The solid line ($D_T = 0.4\ cm^2/s$, the thermal diffusion coefficient of the burnt gases) or the dashed line ($D_T = 0.2\ cm^2/s$) separate the stable region (above) from the unstable region; $\gamma = 0.8$ (Pelcé and Clavin 1982).

effect of diffusion is stronger and prohibits the structuring of the flame. At a higher concentration, the Lewis number of oxygen is smaller than one and the planar flame can no longer be observed since the threshold velocity is lower than the extinguishing velocity. At the stability threshold the observed wavelength is in good agreement with the prediction $k_c = 2(1-\gamma)g/V_0^2$.

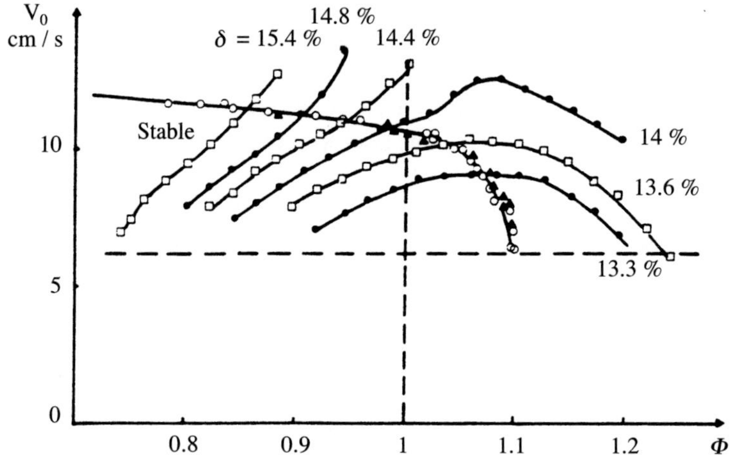

Figure 6.8 Experimental neutral stability curves for the concentration–velocity variables of propagation. For a given dilution, the change in Φ makes the neutral stability curve in the stable region (underneath) cross over to the unstable domain.For a slower propagation velocity than 6 cm/s, the flame goes out (dashed line) (Quinard *et al.* 1983).

6.8 INSTABILITIES WITH TWO DIFFUSION COEFFICIENTS

The difference in the diffusion coefficients of the reactants at the origin of the stationary reactive zones such as a flame or one developing in a strip of gel for which the kinetics is described in paragraph 4.4, leads to conceptually important instabilities of these structures. Let us consider first the case of a flame for which, from paragraph 5.5, the stationary temperature profile and the concentration of reactants is written as

$$
\begin{aligned}
T &= T_0 + (T_b - T_0)\exp\left(z\frac{V}{D_T}\right), && z<0 \\
T &= T_b, && z>0 \\
Y &= Y_0\left(1-\exp\left(z\frac{V}{D_M}\right)\right), && z<0 \\
Y &= 0, && z>0
\end{aligned}
\qquad (6.1)
$$

where the reaction zone is assumed infinitely thin. The stability analysis is carried out in an analogous way to that in paragraph 6.6, meaning that the perturbations of the temperature and concentration fields satisfy equations

$$\frac{\partial T'}{\partial t} + V\frac{\partial T'}{\partial z} = D_T \Delta T' \tag{6.2a}$$

$$\frac{\partial Y'}{\partial t} + V\frac{\partial Y'}{\partial z} = D_M \Delta Y' \tag{6.2b}$$

with the boundary conditions on the level of the reaction zone $z = \xi(x, t)$,

$$T_b'(0) = T_f'(0) + \xi\frac{d\overline{T}_f}{dz} \;;\; Y_f'(0) + \xi\frac{d\overline{Y}_f}{dz} = 0 \tag{6.3a}$$

$$\lambda\left(\frac{\partial T_b'}{\partial z} - \frac{\partial T_f'}{\partial z} - \xi\frac{d^2\overline{T}_f}{dz^2}\right) = \rho D Q\left(\frac{\partial Y_f'}{\partial z} + \xi\frac{d^2\overline{Y}_f}{dz^2}\right) . \tag{6.3b}$$

They respectively correspond to the equality of the fresh and burnt gas temperatures at the reaction zone, to the absence of the reactants in that same region and to the energy conservation obtained by integrating the appropriate linear combination of equations (5.5) and (5.6) in paragraph 5.5 across that same region, since the advective terms are negligible there. An additional relation, obtained from equation (5.12) in paragraph 5.5 determines the change in heat flux coming out of the reaction zone as a function of temperature:

$$\lambda\left(\frac{dT_f'}{dr} + \xi\frac{d^2\overline{T}_f}{dr^2} - \frac{dT_b'}{dr}\right) = \lambda\frac{d\overline{T}_f}{dr}\frac{E}{2RT_b^2}T_b' . \tag{6.4}$$

Searching for the solutions of equations (6.2) in the form

$$T_f' = A\exp(\sigma t)\exp(q_f z + ikx) \tag{6.5a}$$

$$T_b' = B\exp(\sigma t)\exp(-q_b z + ikx) \tag{6.5b}$$

$$Y_f' = C\exp(\sigma t)\exp(rz + ikx) \tag{6.5c}$$

we determine

$$q_f = \frac{V}{2D_T} + \sqrt{\frac{V^2}{4D_T^2} + k^2 + \frac{\sigma}{D_T}}$$

$$q_b = -\frac{V}{2D_T} + \sqrt{\frac{V^2}{4D_T^2} + k^2 + \frac{\sigma}{D_T}} \tag{6.6}$$

$$r = \frac{V}{2D} + \sqrt{\frac{V^2}{4D^2} + k^2 + \frac{\sigma}{D}}.$$

We thus obtain, replacing expressions (6.5) from the boundary conditions (6.3) and (6.4), a linear system of four equations for the coefficients A, B, D, E where E is the amplitude of the interface modulation defined by $\xi(x, t) = E\exp(\sigma t)\exp(ikx)$. This system has a non-zero solution if its determinant is zero, giving the dispersion relation (Barenblatt *et al.* 1962)

$$\sqrt{1+\Omega+K^2}\left(1-\frac{1}{Le}\,\frac{\beta}{2}-\sqrt{1+\frac{1}{Le}\Omega+\frac{1}{Le^2}K^2}\right)$$
$$+\frac{\beta}{2}\left(\frac{1}{Le}-1+\sqrt{1+\frac{1}{Le}\Omega+\frac{1}{Le^2}K^2}\right) = 0 \tag{6.7}$$

where $\Omega = 4D_T\sigma/V^2$ and $K = 2k\, D_T/V$ are the dimensionless growth rate and the dimensionless wavenumber. Let us first assume that the two diffusion coefficients are equal, that is, $Le = 1$ so that $\Omega = -K^2$, and the interface is stable. Indeed, in this case the reaction zone remains isothermic when it wrinkles and relaxes, hence, simply by diffusion. When the wavenumber K is small compared to 1, or when the wavelength of the wrinkling is large compared to the diffusion thickness, we deduce from equation (6.7)

$$\Omega = \left(-\frac{1}{Le}+\frac{\beta}{2}\left(\frac{1}{Le}-1\right)\right)K^2. \tag{6.8}$$

Thus, the interface can become unstable when $Le > 1$, when the reactant diffuses faster than the heat. Indeed in that case, as we have mentioned in paragraph 5.6, the temperature of the reaction zone is higher at the convex than the concave points, which increases the difference in their respective velocities (Zeldovich 1944).

Turing (1952) came up with the hypothesis that the morphogenesis of a biological tissue starts with the appearance of a stationary spatial prestructure of the concentration field: an activator of concentration X and an inhibitor of concentration Y. These variations in concentration in turn lead to the variations in the textures of biological tissues which are formed. The first experimental realisation of such a structure starting from chemical reactants in solution was achieved by Castet and coworkers (1990). The reactive medium is a strip of gel fueled with malonic acid on one side of its length and chlorite ions ClO_2^- and iodide I^- on the other. Instead of observing a uniform concentration profile in the direction of the length, as we may have expected from the homogeneity of the concentrations in that direction on both sides of the strip, we observe a well-defined periodic modulation of the wavelength (figure 6.9).

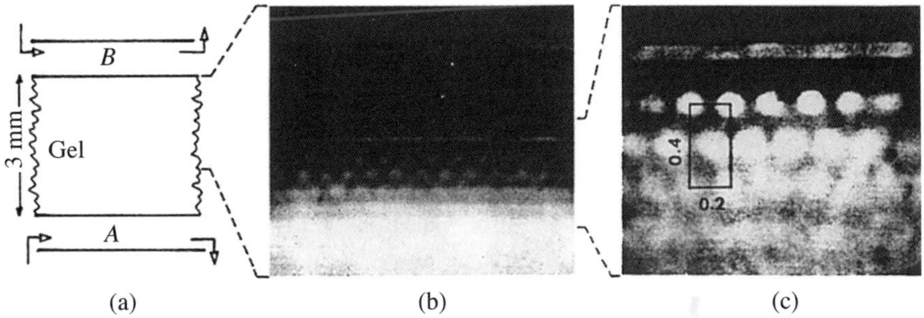

Figure 6.9 Stationary reactive structure.
(a) Schematic of the chemical reactor. A strip of gel of thickness 1 mm, fed by the reactants in reservoirs A and B, is where the reaction takes place. (b) Dark areas: reduced species; clear areas: oxidised species. The periodic modulation of the oxidation state is in the perpendicular direction to the stationary gradients of the different chemical species. (c) Enlargement of a region of (b) indicating the modulation wavelength ≈ 0.2 mm (Castet *et al.* 1990).

Using the kinetic mechanism in paragraph 4.4 (Lengyel and Epstein 1991), the concentrations X of the I^- ion and Y of the ClO_2^- ion satisfy the diffusion-reaction equations

$$\begin{aligned} \frac{\partial X}{\partial t} &= k_1' - k_2'X - 4k_3'\frac{XY}{(k_{3c}+X^2)} + D_X\Delta X \\ \frac{\partial Y}{\partial t} &= k_2'X - k_3'\frac{XY}{(k_{3c}+X^2)} + D_Y\Delta Y \end{aligned} \qquad (6.9)$$

where D_X and D_Y are the coefficients of diffusion of X and Y. In aqueous solution these diffusion coefficients are of the same order of magnitude, 10^{-5} cm^2/s. The polyacrylamide gel under consideration is not chemically neutral in relation to the species derived from the I^- ion. Thus I_2 and I_3^- can attach themselves to $-CO\,NH_2$ groups of the organic polymers making up the gel and are the temporary traps for the I^- ions. These ions can be trapped by the gel for a certain time and then released, leading to an effective diffusion of these ions which is 10 times slower than in a liquid solution. We can therefore achieve a ratio of 10 between the diffusion coefficients of the two species. The system of equations (6.9) shows a uniform stationary state of concentrations X_0 and Y_0, solutions of the system

$$
\begin{aligned}
k_1' - k_2' X_0 - 4k_3' \frac{X_0 Y_0}{(k_{3c} + X_0^2)} &= 0 \\
k_2' X_0 - k_3' \frac{X_0 Y_0}{(k_{3c} + X_0^2)} &= 0
\end{aligned}
\tag{6.10}
$$

that is, $X_0 = (k_{3c})^{1/2}\alpha/5$, $Y_0 = ((k_{3c})^{1/2}/\beta)(1 + (\alpha/5)^2)$, where α and β are the constants $\alpha = k_1'/5(k_{3c})^{1/2}k_2'$, $\beta = k_3'/((k_{3c})^{1/2}k_2')$. In the experiment the concentrations of the stationary state are not uniform throughout the thickness of the strip. Considering only the concentration variations in the plane of the strip, in the following we will assume uniform concentrations in its thickness, with stationary values equal to those determined by the system (6.10). We rewrite the system (6.9) in the form

$$
\begin{aligned}
\frac{\partial X}{\partial t} &= f(X, Y) + D_X \Delta X \\
\frac{\partial Y}{\partial t} &= g(X, Y) + D_Y \Delta Y.
\end{aligned}
\tag{6.11}
$$

Looking for perturbations of the stationary state in the form $X = X_0 + A\exp(\sigma t + i\vec{k}\cdot\vec{r})$, $Y = Y_0 + B\exp(\sigma t + i\vec{k}\cdot\vec{r})$, where $\vec{k}$ is contained within the plane of the strip, we obtain a linear system

$$
\begin{aligned}
\sigma A &= a_{11}A + a_{12}B - D_X k^2 A \\
\sigma B &= a_{21}A + a_{22}B - D_Y k^2 B
\end{aligned}
\tag{6.12}
$$

where the coefficients a_{ij} are the derivatives of the functions f and g with respect to the concentrations. A non-zero solution can be found if the dispersion relation

$$
\begin{aligned}
&\sigma^2 - \sigma(a_{11} + a_{22} - (D_X + D_Y)k^2) + \\
&a_{11}a_{22} - a_{21}a_{12} - k^2(D_X a_{22} + D_Y a_{11}) + D_X D_Y k^4 = 0
\end{aligned}
\tag{6.13}
$$

is satisfied. Since the coefficients a_{ij} generally have arbitrary signs, let us first study the necessary conditions for which the Turing structures appear. Since these structures are stationary and the wavelength is well-defined, they must correspond to marginal perturbations at $\sigma = 0$ and to a finite wavenumber k_m. In particular the homogeneous state must be stable, corresponding to growth rates with negative real parts, which imposes

$$
a_{11} + a_{22} < 0 \; ; \; a_{11}a_{22} - a_{21}a_{12} > 0 \,. \tag{6.14}
$$

Moreover, the dispersion relation $\sigma(k)$ must become positive for a single finite value k_c of the wavenumber, which means that the corresponding curve must be a tangent to the real axis at that point. This condition determines the critical wavenumber

$$k_c = \frac{1}{\sqrt{2}}\left(\frac{a_{11}}{D_X} + \frac{a_{22}}{D_Y}\right)^{1/2} \tag{6.15}$$

as soon as the condition

$$a_{11}D_Y + a_{22}D_X > 2(D_Y D_X(a_{11}a_{22} - a_{12}a_{21}))^{1/2} > 0 \tag{6.16}$$

is satisfied.

From system (4.4) in paragraph 4.4, the coefficient a_{22} is negative for all parameter values. Comparing equations (6.12) and (6.14), the instability can therefore only be produced if a_{11} is positive, which is possible if $\alpha > 5(5/3)^{1/2}$. From equation (6.14), it will be all the more likely if the coefficient D_X is small compared to D_Y. When a_{11} is positive an increase in X raises its production rate even more, meaning that I^- behaves as the reaction initiator whereas Y, ClO_2^- here, behaves as an inhibitor. The instability can only be produced if the inhibitor diffuses significantly faster than the initiator. In aqueous solution the inorganic species have approximately the same diffusion coefficient. The originality of the experiments of Castet and coworkers (1990) is to artificially reduce the diffusion coefficient of I^- ions by the presence of amidon or gel in the solution, allowing the I^- ions to be trapped and subsequently released.

BIBLIOGRAPHY

[1] G.I. Barenblatt, Ya.B. Zeldovich, A.G. Istratov, 'Diffuzionno-teplovoi ustoichivosti laminarnogo plameni', (On the Diffusional-Thermal Stability of Laminar Flames) *Prikl. Mech. i Tekh. Fiz*, **4**, p. 21, 1962.

[2] L. Boyer, 'Laser Tomographic Method for Flame Front Movement', *Combustion and Flame*, **39**, p. 321, 1980.

[3] J.B. Brzoska, F. Brochard-Wyart, F. Rondelez, 'Exponential Growth of Fingering Instabilities of Spreading Films under Horizontal Thermal Gradients', *Europhys. Lett.*, **19** **(2)**, p. 97, 1992.

[4] V. Castet, E. Dulos, J. Boissonade, P. De Kepper, 'Experimental evidence of a sustained standing Turing-type non equilibrium chemical pattern', *Phys. Rev. Lett.*, **64**, p. 2953, 1990.

[5] A.M. Cazabat, F. Heslot, S. Troian, P. Carles, 'Fingering instability of thin spreading films driven by temperature', *Nature*, **364**, p. 824, 1990.

[6] S. Chandrasekhar, *Hydrodynamic and Hydromagnetic Stability*, Oxford University Press, published by Dover in 1981, 1961.

[7] R.L. Chuoke, P. Van Meurs, C. Van der Pol, 1The Instability of Slow immiscible Viscous Liquid–Liquid Displacements in Permeable Media', *Trans. AIME*, **216**, p. 188, 1959.

[8] P. Clavin, 'Dynamical Behaviour of Premixed Flame Fronts in Laminar and Turbulent Flows', *Prog. Energy. Comb. Sci.*, **11**, p. 1, 1985.

[9] G. Darrieus, 'Propagation d'un front de flamme. Essai de théorie des vitesses anormales de déflagration par développement spontané de la turbulence'. Non-published work presented to *La technique moderne*, 1938, and in *Congrès de Mécanique Appliquée*, 1945.

[10] S. De Cheveigné, C. Guthmann, M.M. Lebrun, 'Cellular Instabilities in Directional Solidification', *J. Physique*, **47**, p. 2095, 1986.

[11] M.A. Eshelmann, R. Trivedi, 'The Planar to Cellular Transition During the Directional Solidification of Alloys', *Acta Metall*, **35**, p. 2443, 1987.

[12] V. Hakim, M. Rabaud, H. Thome, Y. Couder, 'Directional Growth in Viscous Fingering', in *New Trends in Nonlinear Dynamics and Pattern – Forming Phenomena*, P. Coullet and P. Huerre, eds, *NATO ASI Series, Vol. 237*, Plenum Press, 1990.

[13] S. Hill, 'Channeling in Packed Columns', *Chem. Eng. Sci.*, **1**, p. 247, 1952.

[14] L.D. Landau, 1On the Theory of Slow Combustion', *Acta Physicochimica*, **19**, p. 77, 1944.

[15] J.S. Langer, 'Instabilities and Pattern Formation in Crystal Growth', *Reviews of Modern Physics*, **52**, p. 1, 1980.

[16] I. Lengyel, I. Epstein, 'Modeling of Turing Structures in the Chlorite–Iodide–Malonic Acid–Starch Reaction System', *Science*, **251**, p. 650, 1991.

[17] G.H. Markstein, 'Experimental and Theoretical Studies of Flame Front Stability', *J. Aero. Sci.*, **18**, p. 199, 1951.

[18] G.H. Markstein, *Nonsteady Flame Propagation*, Pergamon Press, 1964.

[19] A. Marmur, M.D. Lelah, 'The Spreading of Aqueous Surfactant Solutions on Glass', *Chem. Eng. Commun.*, **13**, p. 133, 1981.

[20] W.W. Mullins, R.F. Sekerka, 'Stability of a Planar Interface during Solidification of a Dilute Binary Alloy', *J. Appl. Phys.*, **3**, p. 444, 1964.

[21] P. Oswald, J. Bechhoefer, A. Libchaber, 'Instabilities of a Moving Nematic–Isotropic Interface', *Phys. Rev. Lett.*, **58**, p. 2318, 1987.

[22] P. Oswald, 'Dendritic Growth of Discotic Liquid Crystal', *J. Physique*, **49**, p. 1083, 1988.

[23] L. Paterson, 'Radial fingering in a Hele–Shaw cell', *J. Fluid. Mech.*, **113**, p. 513, 1981.

[24] J.R.A. Pearson, 'The instability of uniform viscous flow under rollers and spreaders', *J. Fluid. Mech.*, **7**, p. 481, 1960.

[25] P. Pelcé, P. Clavin, 'Influence of Hydrodynamics and Diffusion upon the Stability Limits of Laminary Premixed Flames', *J. Fluid. Mech.*, **124**, p. 219, 1982.

[26] J. Quinard, 'Limites de stabilité et structures cellulaires dans les flammes de prémélange. Étude expérimentale', thèse d'Etat, Université de Provence, 1984.

[27] J. Quinard, G. Searby, L. Boyer, 'Stability Limits and Critical Size of Structures in Premixed Flames'. 9th International Colloquium on Gasdynamics of Explosions and Reactive Systems, Poitiers, France. *Progress in Astronautics and Aeronautics*, **95**, p. 129, 1983.

[28] J.W. Rutter, B. Chalmers, 'A prismatic substructure formed during solidification of metals', *Can. J. Phys.*, **31**, p. 15, 1953.

[29] P.G. Saffman, G.I. Taylor, 'The Penetration of a Fluid into a Porous Medium or Hele–Shaw Cell Containing a More Viscous Liquid', *Proc. Roy. Soc.*, **A 245**, p. 312, 1958.

[30] W.A. Tiller, K.A. Jackson, J.W. Rutter, B. Chalmers, 'The redistribution of solute during the solidification of metals', *Acta Met.* 1, p. 428, 1953.

[31] S.M. Troian, E. Herbolzheimer, S.A. Safran, 'Model for the Fingering Instability of Spreading Surfactant Drops', *Phys. Rev. Lett.*, **65**, 3, p. 333, 1990.

[32] S.M. Troian, X.L. Wu, S.A. Safran, 'Fingering Instability in thin Wetting Films', *Phys. Rev. Lett.*, **62**, p. 1496, 1989.

[33] A.M. Turing, 'The Chemical Basis of Morphogenesis', *Phil. Trans. R. Soc.*, **B 237**, p. 37, 1952.

[34] D.J. Wollkind, L.A. Segel, 'A non linear stability analysis of the freezing of a dilute binary alloy', *Phil. Trans. Roy. Soc.*, **268**, p. 351, 1970.

[35] Ya.B. Zeldovich, 'Theory of Combustion and Detonation of Gases', 1944, selected papers, Princeton University Press, Vol. I, p. 162, 1992.

CHAPTER 7

Forms without surface tension

7.1 CONFORMAL TRANSFORMATIONS

The potential of Φ velocities, the solution of the Laplace equation in two dimensions $\Delta\Phi = 0$, is said to be a harmonic function. We can therefore introduce the conjugated function Ψ defined by the Cauchy conditions

$$\frac{\partial\Phi}{\partial x} = \frac{\partial\Psi}{\partial y};\ \frac{\partial\Phi}{\partial y} = -\frac{\partial\Psi}{\partial x} \tag{7.1}$$

The function $w = \Phi + i\Psi$ is an analytic function of the variable $z = x + iy$. The derivative $dw/dz(z)$ must therefore be well-defined. We write the difference

$$\frac{w(z+h)-w(z)}{h} \approx \frac{\frac{\partial\Phi}{\partial x}(x,y)h_1 + \frac{\partial\Phi}{\partial y}(x,y)h_2 + i\left(\frac{\partial\Psi}{\partial x}(x,y)h_1 + \frac{\partial\Psi}{\partial y}(x,y)h_2\right)}{h} \tag{7.2}$$

$$= \left(\frac{\partial\Phi}{\partial x}(x,y) + i\ \frac{\partial\Psi}{\partial x}(x,y)\right)$$

where $h = h_1 + ih_2$ is a complex number of a small modulus. Thus $w(z)$ is an analytic function of the derivative $dw/dz(z) = \partial\Phi/\partial x + i\partial\Psi/\partial x$.

Such a function is a conformal transformation in which the angle between two differentiable arcs is fixed during the transformation. Let us consider two differentiable arcs $z_1(t)$ and $z_2(t)$ that meet with an angle α at a point z_0 with $dw/dz(z_0) \neq 0$. In other words if z_1 and z_2 are two points on these curves at a distance r from z_0 we have $z_1 - z_0 = r\exp(i\theta_1)$, $z_2 - z_0 = r\exp(i\theta_2)$. Then α is just the limit of the argument of the ratio $(z_2 - z_0)/(z_1 - z_0)$ at which r tends to 0. Let us now consider the same ratio but with images. We define a new angle

$$\beta = \lim_{r=0}\arg\left(\frac{w_2 - w_0}{w_1 - w_0}\right) = \lim_{r=0}\arg\left(\frac{\frac{w(z_2)-w(z_0)}{z_2 - z_0}}{\frac{w(z_1)-w(z_0)}{z_1 - z_0}}\ \frac{z_2 - z_0}{z_1 - z_0}\right). \tag{7.3}$$

Nevertheless each of the ratios $(w(z_i) - w(z_0))/(z_i - z_0)$ where $i = 1, 2$ tends to $dw/dz(z_0)$ when r tends to 0. As $dw/dz(z_0) \neq 0$, $\beta = \alpha$.

The problem of the boundary conditions which must be satisfied by the potential is also important. For instance, in the case of a Saffman–Taylor finger moving at constant velocity with a tip featuring a high enough radius of curvature so that we can neglect the surface tension effects, $\Phi = 0$ $(\mathrm{Re}\ w = 0)$ at the interface since the pressure is constant. At the cell walls the normal velocity must cancel, which leads to $\partial\Psi/\partial y = 0$, or in the same way according to the Cauchy conditions defined by equations (1) $\Psi = \mathrm{cst}$, $(\mathrm{Im}\ w = \mathrm{cst})$. Finding a solution to the equation of Laplace with boundary conditions at the interface and at the cell walls is equivalent to finding an analytic solution in a domain featuring simple values at its border.

Appendix 1: Potential flow around a corner and electric potential generated by a charged dihedral conductor.

Let us consider a potential flow around a corner of angle α. If $\alpha < \pi$ the flow occurs within the corner (figure 7.1), otherwise the flow occurs outside this angle. A solution of the equation of Laplace for the potential of velocities is written in polar coordinates as $\Phi = Ar^n\cos(n\theta)$, so that $v_r = nAr^{n-1}\cos(n\theta)$, $v_\theta = -nAr^{n-1}\sin(n\theta)$. At the wall $(\theta = 0, \alpha)$ the normal component of the velocity v_θ must cancel so that $n = \pi/\alpha$.

The stream function is simply written as $\Psi = Ar^n\sin(n\theta)$, so that the complex potential is written as $w(z) = \mathrm{A}z^n$. When $n < 1$ (the case of a flow around a convex angle) the velocity of the flow diverges as $r^{-(1-n)}$ at the vertex of the corner. When $n \geq 1$ the velocity tends to cancel at $r = 0$.

Finding the electric potential around a charged dihedral conductor is treated in a similar way. At the wall the potential must cancel so that $\Phi = Ar^n\sin(n\theta)$ with $n = \pi/\alpha$. The electric field is therefore written as $E_r = nAr^{n-1}\sin(n\theta)$, $E_\theta = nAr^{n-1}\cos(n\theta)$. Thus when $n < 1$ (the case of a sharp dihedron) the electric field diverges as $r^{-(1-n)}$ at the tip of the dihedron. When α tends to 0 as for a thin branch $n \approx 1/2$ and $\Phi \approx 1/\sqrt{r}$ in the proximity of the tip of the dihedron, the complex potential is simply written as $w(z) = -iAz^n$.

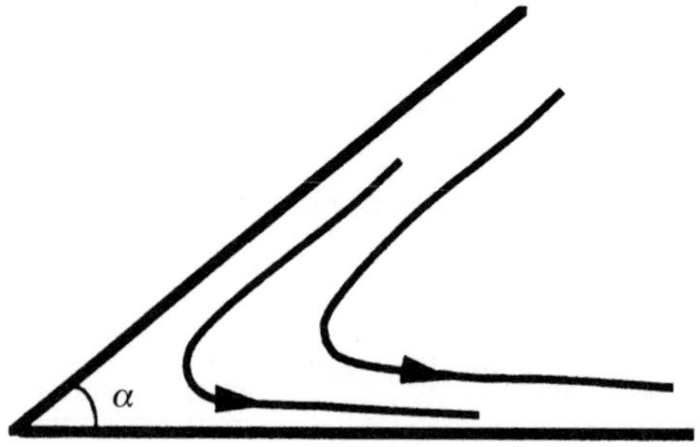

Figure 7.1 Flow in a dihedron.

Appendix 2: Electric potential generated by a charged ellipsoidal conductor.

If now $z = a\cosh w$ then $x = a\cosh\Phi\cos\Psi$ and $y = a\sinh\Phi\sin\Psi$. The equipotential lines $\Phi = \text{cst}$ are ellipses of axis lengths $a\cosh\Phi$ and $a\sinh\Phi$ (figure 7.2). The field lines $\Psi = \text{cst}$ are the hyperbolas perpendicular to the expontentials. Since the surface of a charged conductor is equipotential, the field determined above is the field generated by a charged ellipsoidal conductor.

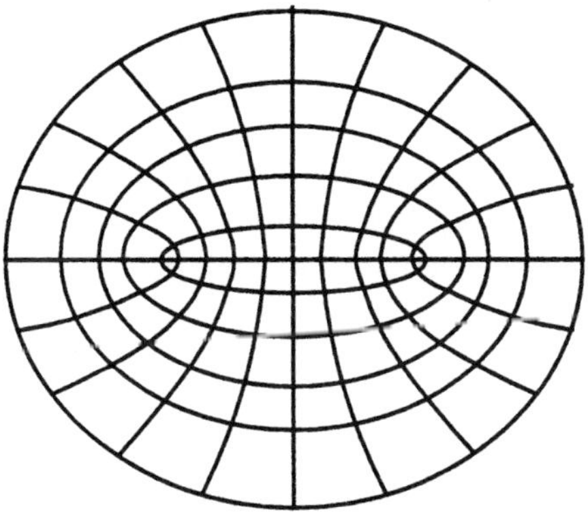

Figure 7.2 Equipotential surfaces and field lines around a charged ellipsoidal conductor.

7.2 THE SCHWARZ–CHRISTOFFEL TRANSFORMATION

In many problems of electrostatics, magnetism or hydrodynamics the domains limited by a polygonal contour are particularly important. Let us consider for instance a two-dimensional potential flow of velocity U which separates at a vertical rod of length l. The relation of conservation of mass of the fluid leads to an expression of the potential Φ as a harmonic function. Therefore, according to the above-mentioned $x = \Phi + i\Psi$, where Ψ is the stream function, it is an analytic function of the variable z defined in the flow region. At the rod the normal velocity of the fluid cancels and then $\Psi = 0$. The flow separates at the extremities of the rod, leaving behind a stagnant region where the fluid is at rest. The pressure is constant on the separation line and according to the relation of ideal fluids of Bernouilli the modulus of the velocity q is constant and equal to q_0. It is then useful to introduce the logarithm of the inverse of the complex velocity $\text{Log}\zeta = \text{Log}(q_0/q) + i\theta$, where θ is the angle between the streamline and the horizontal direction. In the plane of this variable the domain of the flow is just a rectangle with one side taken to infinity. If a connection between the variables ζ and w can be established the integration of the equation $\zeta = q_0 dz/dw$ leads to $z(w)$.

This connection can be made with an intermediary variable t for which the contour of the domain is the real axis and the domain of the flow is the upper half-plane. We just need to determine the conformal transformations of the upper half-plane inside a polygon.

Let us consider the Schwarz–Christoffel transformation defined as

$$w(z) = A\int^{z} \frac{dz}{(z-a_1)^{\alpha_1}(z-a_2)^{\alpha_2}\dots(z-a_n)^{\alpha_n}} + B \tag{7.1}$$

where $a_1, a_2 \dots$ and a_n are real numbers sorted in ascending order, $\pi\alpha_1, \pi\alpha_2 \dots$ and $\pi\alpha_n$ angles (not necessarily all positive) such that $\alpha_1 + \alpha_2 + \dots + \alpha_n = 2$, and A and B arbitrary constants. Over each real interval $[a_i, a_{i+1}]$ and for $z < a_1$ and $z > a_n$ as well the sign of $z - a_m$ does not change, so the argument dw/dz remains constant. Since this argument determines the angle of the tangent of the image of the real axis by the transformation, this image is a polygonal line. Moreover the difference in the arguments associated with two surrounding segments is

$$\arg\left(\frac{\left(\frac{dw}{dz}\right)_{z=a_{i-}}}{\left(\frac{dw}{dz}\right)_{z=a_{i+}}}\right) = \arg\left(\frac{1}{(\exp\pi i)^{\alpha_i}}\right) = -\pi\alpha_i\,. \tag{7.2}$$

The point $z = \infty$ has a finite image $w(\infty) = \mathrm{B}$ which is located on the line joining $w(a_1)$ to $w(a_n)$. Thus the image by the transformation w of the real axis which has the upper half-plane closed by infinity, is a polygon with an angle $\pi\alpha_i$ between the sides of the vertices $w(a_i)$ (figure 7.3). According to a theorem about analytic functions (Cohn 1967), $w(z)$ establishes a bi-univocal correspondence between the upper half-plane and the inside of the polygon.

This transformation can be extended to the case of polygonal lines closing at infinity. Let us assume that a_n is transformed into a point at infinity. We define the new variable ξ by $z = a_n - 1/\xi$ and

$$\begin{aligned} w(z) &= A\int_0^{\xi} \frac{1}{\left(a_n - a_1 - \frac{1}{\xi}\right)^{\alpha_1}\left(a_n - a_2 - \frac{1}{\xi}\right)^{\alpha_2}\left(-\frac{1}{\xi}\right)^{\alpha_n}\xi^2} + \mathrm{B} \\ &= A_1\int_0^{z} dz \frac{1}{(z-a_1')^{\alpha_1}(z-a_2')^{\alpha_2}(z-a_n')^{\alpha_{n-1}}} + \mathrm{B}_1 \end{aligned} \tag{7.3}$$

where the a_i' are the new constants, which is equivalent to skipping the point a_n of the Schwarz–Christoffel transformation (Nehari 1952).

As mentioned above the case of rectangular domains is more usual $(\alpha_1 = \alpha_2 = \alpha_3 = \alpha_4 = 1/2)$. When one of the sides is taken to infinity the transformation is of the form

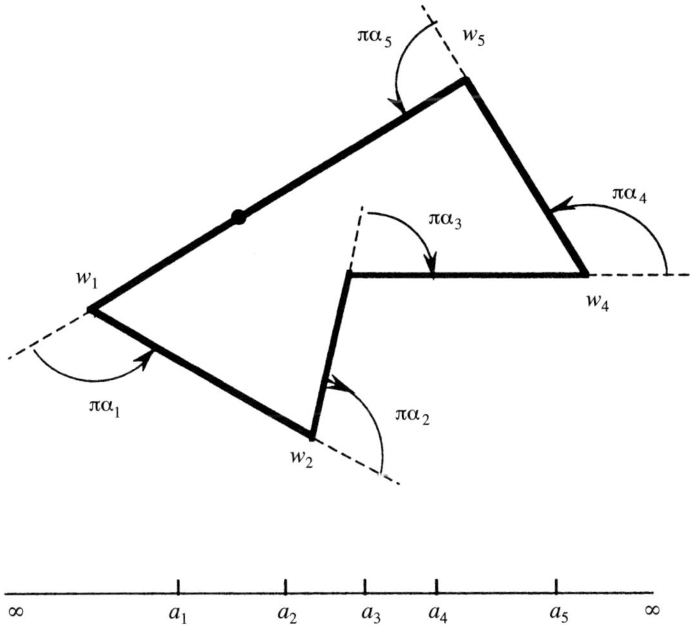

Figure 7.3 Schematic showing the Schwarz–Christoffel transformation.

$$w(z) = A\int^{z} \frac{dz}{\sqrt{(z-a)(z-b)}} \tag{7.4}$$

where a and b are fixed and the factor A sets the scale of the rectangle.

Appendix: Conformal transformation of the outside of the circle of unit radius on the outside of a rectilinear branched-star (Kurtze 1991, Peterson and Ferry 1989, Derrida and Hakim 1992).

Let us consider the n points $\exp(i\theta_i)$ of the circle of unit radius and the conformal transformation

$$w(z) = \mathrm{A}z\prod_{j=1}^{n}\left(1-\frac{\exp(i\theta_j)}{z}\right)^{\alpha_j}; \quad \sum_{j=1}^{n}\alpha j = 2$$

where A is an arbitrary constant. For a point of the unit circle $z = \exp(i\Phi)$ and

$$w(z) = -4\mathrm{A}\exp(i\Phi)\prod_{j=1}^{n}\sin\left(\frac{\theta_j-\Phi}{2}\right)^{\alpha_j}\exp i\alpha_j\left(\frac{\theta_j-\Phi}{2}\right)$$

Thus when Φ is included between two consecutive angles θ_j the sign of $\sin(\theta_j - \Phi)/2$ does not change and the argument of $w(z)$ remains constant. The image of the unit circle is therefore a collection of branches which starts from the origin. The jump in the argument $w(\exp(i\Phi))$ across the angle θ_j

$$\arg\left(\frac{w(\exp i\theta_{j+})}{w(\exp i\theta_{j-})}\right) = \arg((\exp i\pi)^{\alpha_j}) = \pi\alpha_j$$

determines the angle between two consecutive branches (figure 7.4). As $w(z)$ cancels at each angle θ_i the tips of the branches are reached at the angles Φ_i located between two consecutive angles θ. At these points $|w(\exp i\Phi)|$ is at a maximum so that the angles Φ_i are the solutions of the equation

$$\sum_{j=1}^{n} \alpha_j \cot\left(\frac{\Phi - \theta_j}{2}\right) = 0 .$$

The length of the branch l_i is determined by the modulus of $w(\exp i\Phi_i)$, that is

$$l_i = 4A\prod_{j=1}^{n}\left|\sin\left(\frac{\theta_j - \Phi_i}{2}\right)\right|^{\alpha_j} .$$

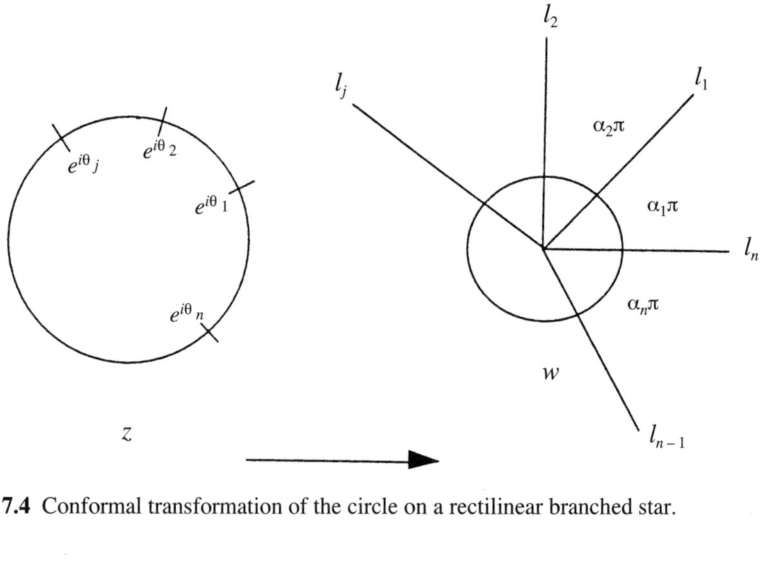

Figure 7.4 Conformal transformation of the circle on a rectilinear branched star.

7.3 POTENTIAL FLOW AROUND A VERTICAL ROD

Let us consider the example of the potential flow around the vertical rod again. The velocity field of coordinates u and v can be written in the form

$$u = \frac{\partial\Phi}{\partial x} = \frac{\partial\Psi}{\partial y}, \; v = \frac{\partial\Phi}{\partial y} = -\frac{\partial\Psi}{\partial x} . \tag{7.1}$$

According to paragraph 5.3 we can introduce the complex velocity $dw/dx = u - iv = q\exp(-i\theta)$. Let us define the variable $\mathrm{Log}\zeta = \mathrm{Log}(q_0/q) + i\theta$. As shown in figure 7.5, in the $\mathrm{Log}\zeta$ plane the rod is represented by two horizontal half-lines since $\theta = \pm\pi/2$ along the rod and $0 < q < q_0$. Along the stagnant lines of ideal fluids, according to the Bernouilli equation, the velocity modulus is constant $q = q_0$ and $-\pi/2 < \theta < \pi/2$. These lines are therefore represented in the $\mathrm{Log}\zeta$ plane by the segment $[-i\pi/2, i\pi/2]$. Thus in the $\mathrm{Log}\zeta$ variable the domain of the flow is inside a rectangle with one side taken to infinity. According to the preceding chapter this rectangle is the image of the real axis by the Schwarz–Christoffel transformation defined by equation (7.4) in paragraph 7.2 where $a = 1$, $b = -1$ by symmetry. Let t be the variable for which the domain of the flow is the upper half-plane. According to equation (7.4) in paragraph 7.2,

$$\mathrm{Log}\zeta = A\cosh^{-1} t + B. \tag{7.2}$$

As for $t = \pm 1$, $\mathrm{Log}\zeta = - + i\pi/2$, $A = 1$ and $B = -i\pi/2$.

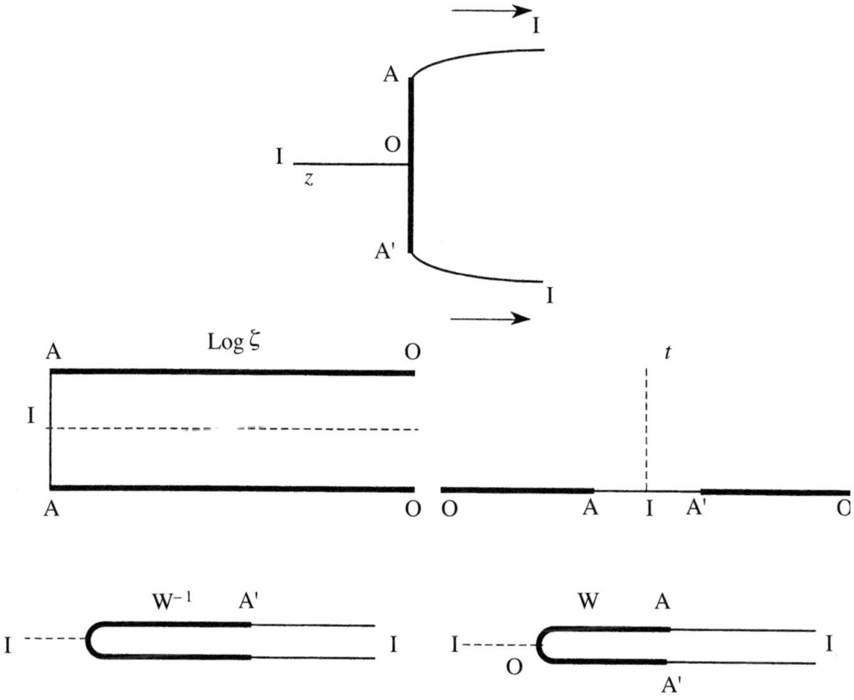

Figure 7.5 Domain of the flow for different variables.

Let us now consider the domain of w. Along each streamline which goes along the rod and which carries on by the corresponding stagnant line, $\Psi = 0$, $\Phi > 0$. Thus in the variable w the domain of the flow is outside the two half-lines folded over the positive real axis (figure 7.5). It is best to consider w^{-1} rather than w for a point at infinity to be transformed to infinity. Then we apply a Schwarz–Christoffel transformation of angle $\alpha = -\pi$ which is according to relation (7.1) in paragraph 7.2,

$$w^{-1} = \frac{1}{2}At^2 + B. \tag{7.3}$$

As for $t = 0$, $w^{-1} = 0$, $B = 0$ and $w = 2/At^2$.
If we use relations (7.1) and (7.2) and the definition of ζ we obtain

$$z = \frac{1}{q_0}\int^{w} \zeta dw = \frac{4i}{Aq_0}\int_{-\infty}^{t} dt \; \frac{(t + \sqrt{t^2 - 1})}{t^3} \tag{7.4}$$

where the origin is taken to be the middle of the rod. The upper tip of the rod corresponds to $z = il/2$ and $t = -1$. We deduce $A = 2(4 + \pi)/q_0 l$.

Along the stagnation lines $\mathrm{Log}\zeta = i\theta$ and, according to equation (7.2), $t = -\sin\theta$. Thus from equation (7.4) we deduce the equation of the upper stagnation line, whereas the other one is deduced by symmetry with the x axis,

$$\begin{aligned} x &= \frac{1}{\pi + 4}\left(\frac{1}{\sin\theta}\cot\theta - \mathrm{Log}\left(\frac{1}{\sin\theta} + \cot\theta\right)\right) \\ y &= \frac{2l}{\pi + 4}\left(\frac{1}{\sin\theta} + \frac{\pi}{4}\right) \end{aligned} \tag{7.5}$$

θ varying from $\pi/2$ to 0 (figure 7.6).

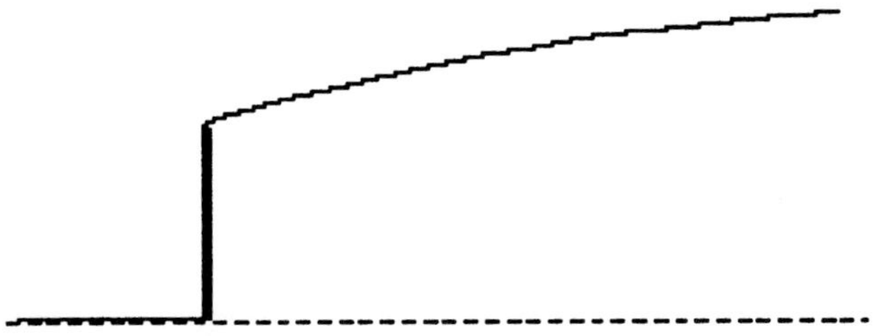

Figure 7.6 Shape of the borderline which separates the fluid at rest.

7.4 SUPERCONDUCTOR IN THE INTERMEDIATE STATE

A superconductor has the property of excluding the magnetic field as long as its intensity is lower than a critical intensity H_{cr}. When the intensity of the external field H_0 increases (by the deformation of the field lines induced by the presence of the superconductor) the critical value is not reached uniformly over the surface of the superconductor, instead an intermediate state is reached in which the regions where H exceeds H_{cr} turn into a normal state whereas the other ones remain in the superconductive state. The problem involves determining the shape of the surface of separation between the normal phase and the superconducting phase (Landau 1937).

The simplest configuration is that of a parallel-faced plate in a magnetic field H_0 perpendicular to the plate. We consider the case of a periodic planar structure where normal and superconducting layers alternate with the periodicity $\lambda = \lambda_s + \lambda_n$ – the coordinate perpendicular to this plane does not intervene in this problem (figure 7.7).

Since the curl of the magnetostatic field in the absence of current is zero this field derives from a potential Φ. Since the divergence of the field is also null Φ satisfies the equation of Laplace. The function conjugated by the relations of Laplace is in this case the vector potential $\vec{A}$ of the field. We thus determine the relations

$$H_x = -\frac{\partial \Phi}{\partial x} = \frac{\partial A}{\partial y}, \; H_y = -\frac{\partial \Phi}{\partial y} = -\frac{\partial A}{\partial x}. \tag{7.1}$$

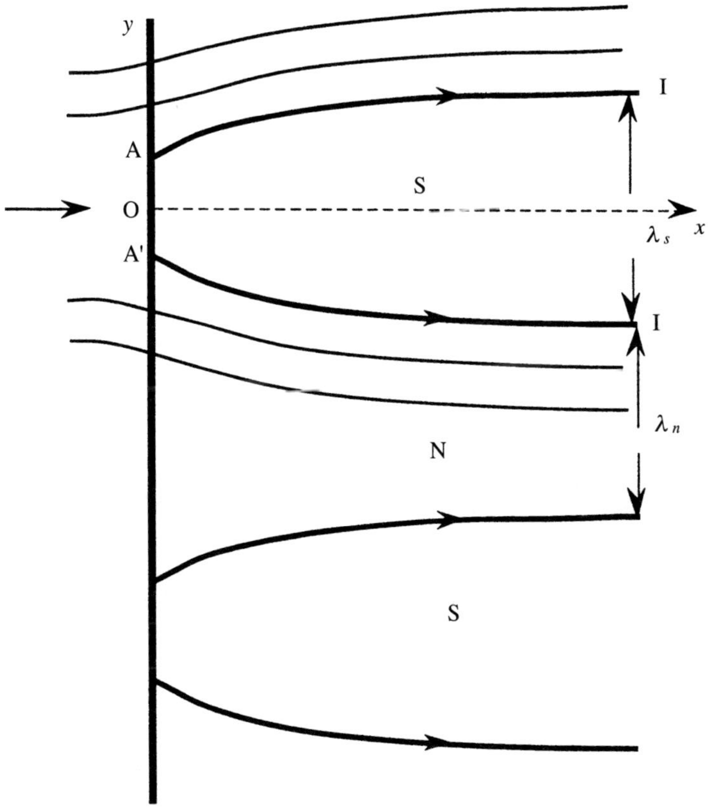

Figure 7.7 Interfaces between normal and superconductive phases in the intermediate state.

In the superconducting phase the intensity of the magnetic phase is zero, $H = 0$. Thus the field lines are excluded on the sides of the superconducting domain, a bit like the streamlines of the fluid in the preceding example move away because of the presence of an obstacle. The edge of the plate is therefore part of the form of the superconducting region. Away from the point where the field is null, along the plate, the magnitude of the field increases and reaches the critical value H_{cr}. From that point the line of separation between the normal phase and the superconducting phase gets inside the plate. Thus along the segments OA and OA', $H_x = 0$, whereas along the lines AI and A'I, $H_x^2 + H_y^2 = H_{cr}^2$. Moreover, since the structure is periodic $H_y = 0$ for $y = \pm\lambda/2$ and at infinity, upstream from the plate $H_x = H_0$. This problem is in the end similar to the problem in paragraph 7.3 if we identify the velocity field of the fluid as the magnetic field and the stream function as the vector potential, q_0 and H_{cr}. The only difference deals with the structure: here we consider a periodic succession of superconducting layers whereas in the previous case a single obstacle was taken into account.

Thus it is straightforward to introduce the analytic functions $w = \Phi - iA$, $dw/dz = -(H_x - iH_y) = -H\exp(-i\theta)$ and $\mathrm{Log}\zeta = \mathrm{Log}(H_{cr}/H) + i\theta$ where θ is the angle between the field line and the horizontal line. Let us consider the domain limited by the line $y = 0$ in the variable z, the vertical segment OA, the line of separation AI and the horizontal line $y = \lambda/2$. In the $\mathrm{Log}\zeta$ variable this domain is the rectangle determined by the two horizontal half-lines going through 0 and $\pi/2$ (figure 7.7). Rather than considering the variable w it is more convenient to consider the variable $t = \exp(-2\pi w/H_0\lambda) - 1$, since it has real values along the contour of the domain considered. On the field line $A = 0$ which, coming from infinity upstream of the plate, extends over the contour of the superconducting region, and t is a real number larger than -1. On the line IO, $A = H_0\lambda/2$ and $t < -1$ (figure 7.8). According to paragraph 7.2 we can therefore connect the variables $\mathrm{Log}\zeta$ and $u = -1/t$ by a Schwarz–Christoffel transformation. Let t_0 be the value of t corresponding to the point A where the field line $A = 0$ enters the plate. Then the point $u = 0$, corresponding to the point I, can be written according to the relation in paragraph 7.2

$$\mathrm{Log}\zeta = A\int^{u} du\frac{1}{\sqrt{u\left(u+\frac{1}{t_0}\right)}} = A\mathrm{Log}\left(\sqrt{1-\frac{t_0}{t}} - \sqrt{-\frac{t_0}{t}}\right). \tag{7.2}$$

Since at the point A, $t = t_0$ and $\mathrm{Log}\zeta = i\pi/2$, $A = -1$, we can deduce

$$\frac{dw}{dz} = -H_{cr}\left(\sqrt{1-\frac{t_0}{t}} - \sqrt{-\frac{t_0}{t}}\right). \tag{7.3}$$

Since at infinity upstream from the plate, $H_x = H_0$ and $t = -1$, $t_0 = (1/h - h)^2/4$ where $h = H_0/H_{cr}$. The ratio between the thicknesses of the normal and superconducting layers is determined by conservation of the magnetic flux. Far away upstream from the plate this flux is $H_0\lambda/2$ whereas far behind it is $H_{cr}\lambda_n/2$. We deduce that $\lambda_n/\lambda = h$ and so the intermediate state can exist only if H_0 is smaller than H_{cr}. Above this magnitude the superconducting state disappears.

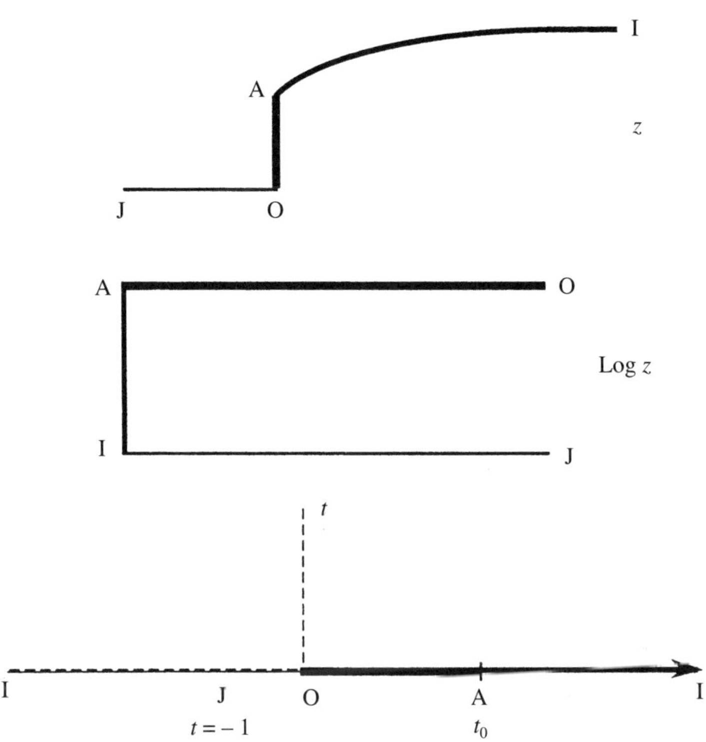

Figure 7.8 Domain of the magnetic field in different variables.

The relation $z(w)$ is obtained by integration of equation (7.3) using the relation $dt/dw = -2\pi(t+1)/H_0\lambda$:

$$z = \frac{1}{2\pi}\frac{H_0\lambda}{H_{cr}}\int^t dt \frac{1}{(t+1)\left(\sqrt{1-\frac{t_0}{t}}-\sqrt{-\frac{t_0}{t}}\right)}. \tag{7.4}$$

The shape of the separation line between the normal phase and the superconducting phase is obtained by separating the real and complex parts of relation (7.4) for $t_0 < t < +\infty$. We then obtain

$$x = \frac{h\lambda}{2\pi}\int_{t_0}^{t}\sqrt{1-\frac{t_0}{t}}\,\frac{dt}{t+1} = \frac{h\lambda}{\pi}\left(\cosh^{-1}\left(\sqrt{\frac{t}{t_0}} - \sqrt{t_0+1}\cosh^{-1}\sqrt{\frac{t(t_0+1)}{t_0(t+1)}}\right)\right)$$

$$y = \frac{\lambda(1-h)}{2} - \frac{\lambda h}{2\pi}\int_{t}^{+\infty}\sqrt{\frac{t_0}{t}}\,\frac{dt}{t+1} = \frac{\lambda}{2}((1-h) - h\sqrt{t_0}) + \frac{h\lambda}{\pi}\sqrt{t_0}\tan^{-1}\sqrt{t}. \tag{7.5}$$

In particular, the width l of the planar border of the superconductive domain is obtained by setting $t = t_0$ into the expression of y, that is $1/2 = \lambda(1-h)((1-h)/4 + (1/\pi)\tan^{-1}((1-h)/h))$. When h tends to 1 $l = \lambda(1-h)^2(4+\pi)/2\pi$. Since the planar border of a superconductive region is in this case far smaller than the wavelength of the structure, the structure of the field lines is identical to the structure of the stream lines calculated in paragraph 7.3. In particular, on the line of separation which penetrates the plate, $w = \Phi$ and $t = \exp(-\Phi(1-h)^2(4+\pi)/H_0 l) - 1 \approx -\Phi(1-h)^2(4+\pi)/H_0 l$. Besides, if we identify the imaginary parts of the two members of equation (7.3) we obtain $\sqrt{t} = \sqrt{t_0}/\sin\theta$. We therefore deduce by using equation (7.5), $y \approx (2l/(\pi+4))(\pi/4 + l/\sin\theta)$.

7.5 THE SAFFMAN–TAYLOR FINGER

In a large range of velocities the interface between two fluids moving in a Hele–Shaw cell of width $2a$ has the shape of a finger, of steady-state form with a curved tip at the scale of the width of the cell. It leaves behind a channel of moving fluid of width $2\lambda a$ where λ is by definition the relative size of the finger. It is worth noting that when the effects of surface tension are neglected exact solutions for the finger forms can be determined (Zhuralev 1956, Saffman and Taylor 1958). In this case the whole surface is free, that is, it must be determined at the same time as the potential of velocities.

As we have shown in chapter 5 the flow of a fluid in a Hele–Shaw cell is two-dimensional and derives from a potential proportional to the pressure. For simplicity, we assume that displacement is induced by air, the viscosity of which can be neglected. In that case we must solve the Laplace equation $\Delta\Phi = 0$ in the space outside the finger with boundary conditions on the interface and on the walls of the cell. The pressure and therefore the potential can be assumed to be uniform inside the finger since the viscosity of air has been neglected. The potential Φ is also constant on the interface since the effects of surface tension have also been neglected. According to relation (5.5) of paragraph 5.2,

$$\vec{\nabla}\Phi \cdot \vec{n} = \frac{\partial\Phi}{\partial x}\frac{dy}{dl} - \frac{\partial\Phi}{\partial y}\frac{dx}{dl} = U\cos\theta \tag{7.1}$$

along the interface where U is the velocity of the finger, θ the angle between the normal vector $\vec{n}$ and the direction of propagation and l the curvilinear coordinate. Using the Cauchy relation and $\cos\theta = dy/dl$ we deduce from equation (7.1) $d\Psi/dl = U\, dy/dl$, or $\Psi = Uy$, where the origin of the coordinates is fixed at the tip of the finger. At the walls of the cell, for $y = \pm 1$, the normal velocity is zero, or $\partial\Phi/\partial y = 0$. Far upstream in front on the finger the velocity of the fluid is uniform: $\Phi = Vx$.

Owing to the simple values taken by Φ and Ψ at the border of the domain it is better to determine $z = x + iy$ as a function of $w = \Phi + i\Psi$. In this variable the domain of the flow is the rectangle limited by the half-lines $\Psi = \pm V,\ \Phi > 0$ and the segment $\Phi = 0,\ -V < \Psi < V$ (figure 7.9).

$z(w)$ being analytic, $\Delta_{\Phi,\Psi} y = 0$, so that we can expand y using the eigenfunctions of the laplacian associated to the rectangle mentioned above

$$y = \frac{\Psi}{V} + \sum_{1}^{+\infty} A_n \sin\left(\frac{n\pi\Psi}{V}\right)\exp\left(-\frac{n\pi\Phi}{V}\right). \tag{7.2}$$

At the interface, $y = \Psi/U$, so that we must have

$$\frac{\Psi}{U} = \frac{\Psi}{V} + \sum_{1}^{+\infty} A_n \sin\left(\frac{n\pi\Psi}{V}\right) \tag{7.3}$$

for $-V < \Psi < V$. To determine the coefficients A_n which satisfy these conditions we multiply the two members of equation (7.3) by $\sin(m\pi\Psi/V)$ and we integrate over the interval from $-V$ to V. The term proportional to A_m is the only non-zero

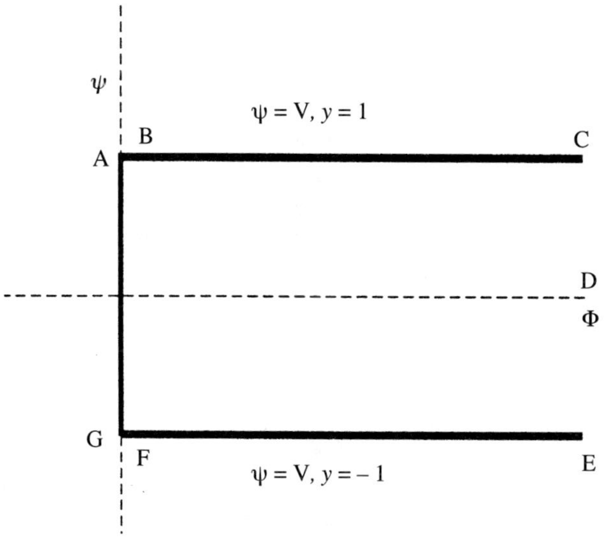

Figure 7.9 Domain of the flow in the variables Φ and Ψ.

term: $A_m = (2\cos(m\pi)/m\pi)(1 - V/U)$. We deduce from equation (7.2) and its complex conjugate

$$z = \frac{w}{V} + \frac{2}{\pi}(1 - \lambda)\mathrm{Log}\left(\frac{1 + \exp\left(-\frac{\pi w}{V}\right)}{2}\right) \tag{7.4}$$

where $\lambda = V/U$. The shape of the finger is easily obtained setting $\Phi = 0$ and $y = \Psi/U$ in the real part of equation (7.4):

$$x = \frac{(1 - \lambda)}{\pi}\mathrm{Log}\left(\frac{1 + \cos\left(\frac{\pi y}{\lambda}\right)}{2}\right). \tag{7.5}$$

When y tends to λ, x is rejected far away from the tip. Thus λ appears here as the relative width of the finger, a parameter not determined by the equations. Any relative width is therefore possible, the corresponding velocity of propagation being $U = V/\lambda$. The fingers with small λ have the shape of fine needles which propagate at high speed; the larger fingers are slower with a velocity close to the velocity of the fluid far upstream from the cell (figure 7.10).

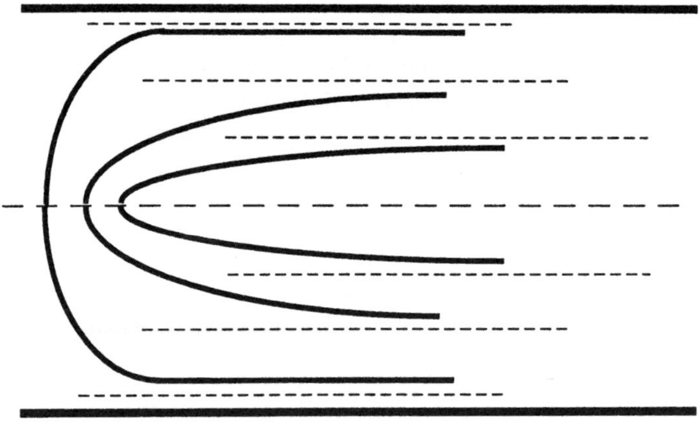

Figure 7.10 Saffman–Taylor forms.

The forms of Saffman and Taylor can also be found using a method similar to the one used in paragraph 7.2 (McLean and Saffman 1982). In the moving frame attached to the finger we introduce the dimensionless velocity field of coordinates u and v, derived from the potential $\phi = (\Phi - Ux)/(U - V)$ of the conjugated stream function $\psi = (\Psi - Uy)/(U - V)$. As in paragraph 7.2 we consider the logarithm of the inverse of the complex velocity $-\mathrm{Log}q' + i\theta'$, where $u - iv = \partial\phi/\partial x + i\partial\psi/\partial x = q'\exp(-i\theta')$. The transformation $s + it = \exp(-\pi(\phi + i\psi))$

transforms the domain of the flow of a half-cell in the upper-half plane, the borderline of the domain being the real axis. In fact on the axis of the cell upstream from the finger $-\infty < \phi < 0$, $\psi = 0$, and t is real, positive and larger than 1. Along the interface, $0 < \phi < +\infty$, $\psi = 0$ and $0 < s < 1$. On the boundary of the cell $y = 1$, $-\infty < \phi < +\infty$, $\psi = -1$ and s is real and negative (figure 7.11).

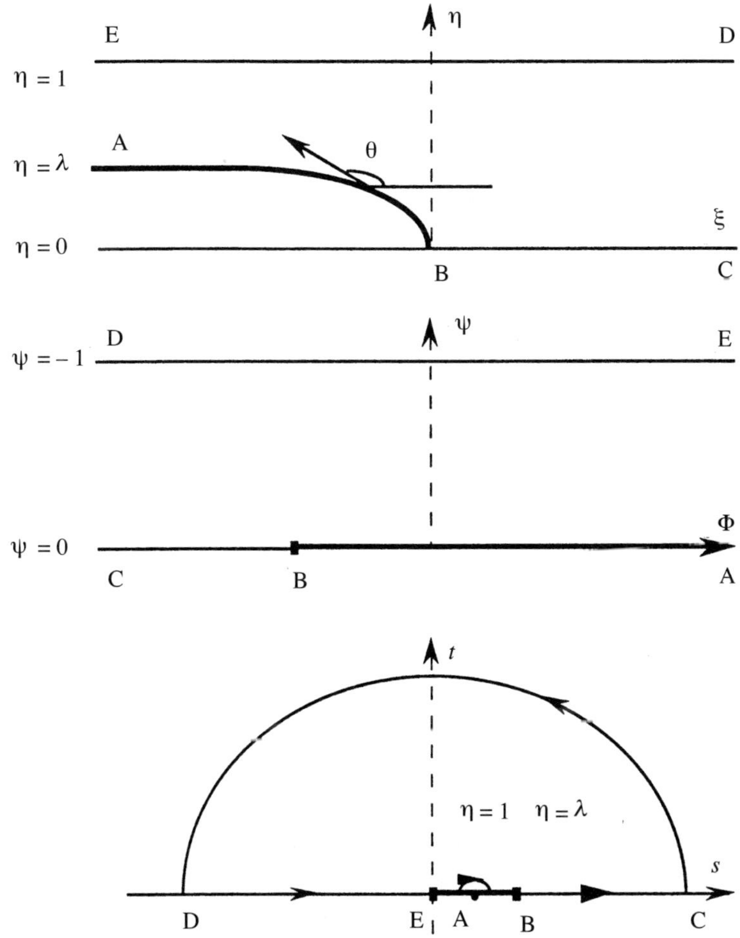

Figure 7.11 Conformal transformations of the domain of flow.

Using the residue theorem for the analytic functions over the contour shown in figure 7.11 for the function $-\mathrm{Log}\, q' + i\theta'$, at point s of segment $[0, 1]$, we obtain

$$-\mathrm{Log}(q'(s)) + i\theta'(s) = \frac{1}{\pi i}\int_C dt\, \frac{-\mathrm{Log}(q'(t)) + i\theta'(t)}{t-s} \tag{7.6}$$

or if we consider the real part only,

$$\mathrm{Log}(q'(s)) = -\frac{1}{\pi}P\int_0^1 \frac{\theta'(s')-\pi}{s'-s}ds'. \tag{7.7}$$

Another relation between q' and θ' – a local one this time – is obtained by writing $\phi = -x/(1-\lambda)$ at the interface or after derivation with respect to the curvilinear coordinate l,

$$(1-\lambda)q' = -\cos\theta' \quad 0 < s < 1 \tag{7.8}$$

where we have used $\partial\phi/\partial l = q'$. The boundary conditions to be applied to these two relations are $\theta'(0) = \pi$, $q'(0) = 1/(1-\lambda)$, $\theta'(1) = \pi/2$, $q'(1) = 0$.

To remove λ and then simplify the problem it is convenient to redefine $\theta = \theta' - \pi$ and $q = q'(1-\lambda)$; we obtain

$$\mathrm{Log}(q(s)) = -\frac{s}{\pi}P\int_0^1 ds' \frac{\theta(s')}{s'(s'-s)} \tag{7.9}$$

and

$$q = \cos\theta \tag{7.10}$$

where $\theta(0) = \pi$, $q(0) = 1$, $\theta(1) = -\pi/2$, $q(1) = 0$.

When q and θ are known we find the shape of the interface from the relation

$$z(s) = -\frac{(1-\lambda)}{\pi}\int_s^1 \frac{\exp(i\theta)}{sq}ds \tag{7.11}$$

where we have used $ds/d\phi = -\pi s$. In these variable the solution of Saffman and Taylor is written as

$$q_0 = \sqrt{\frac{1-s}{1+s\alpha}} \tag{7.12}$$

where α is the arbitrary parameter $\alpha = (2\lambda-1)/(1-\lambda)^2$. We can show that this solution satisfies equations (7.8) and (7.9).

When there is an angle between the two walls of the cell, thus forming an angular section, the shape of the interface is no longer a steady-state finger but a lobe with a self-similarly increasing radius of curvature at the tip. When the effects of surface tension are neglected a continuum of self-similar solutions can also be analytically determined (Thome *et al.* 1989, Ben Amar 1991).

7.6 IVANTSOV PARABOLOIDS

Let us introduce the parabolic coordinates ξ, η, Φ defined by

$$x = \rho\sqrt{\xi\eta}\cos\Phi, \quad y = \rho\sqrt{\xi\eta}\sin\Phi, \quad z = \frac{\rho}{2}(\eta-\xi) \tag{7.1}$$

where ρ is the radius of curvature of the paraboloid $\eta = 1$ and Φ the polar angle (figure 7.12).

Let us assume that the liquid–solid interface is actually the paraboloid $\eta = 1$, which moves at the velocity V with respect to the pure liquid. Then in the moving frame where the interface is at rest the axisymmetric temperature field satisfies the equation

$$\frac{1}{h_1} u_\xi \frac{\partial T}{\partial \xi} + \frac{1}{h_2} u_\eta \frac{\partial T}{\partial \eta} = \frac{D_T}{h_1 h_2 h_2} \left(\frac{\partial}{\partial \xi}\left(\frac{h_2 h_3}{h_1} \frac{\partial T}{\partial \xi} \right) + \frac{\partial}{\partial \eta} \left(\frac{h_1 h_3}{h_2} \frac{\partial T}{\partial \eta} \right) \right) \tag{7.2}$$

where h_1, h_2 and h_3 are the coefficient of the corresponding metrics

$$h_1 = \frac{\rho}{2}\sqrt{\frac{\eta + \xi}{\xi}}, \; h_2 = \frac{\rho}{2}\sqrt{\frac{\eta + \xi}{\eta}}, \; h_3 = \rho\sqrt{\xi\eta} \tag{7.3}$$

and $u_\xi = -(\rho/2)(V/h_1)$, $u_\eta = -(\rho/2)(V/h_2)$ the components of the uniform velocity field of intensity V in these coordinates. Since the temperature field is constant along the interface we are looking for a solution of equation (7.2) which only depends on the coordinate η. We thus have

$$-V\eta\frac{dT}{d\eta} = \frac{2D_T}{\rho} \frac{d}{d\eta} \left(\eta \frac{dT}{d\eta} \right) \tag{7.4}$$

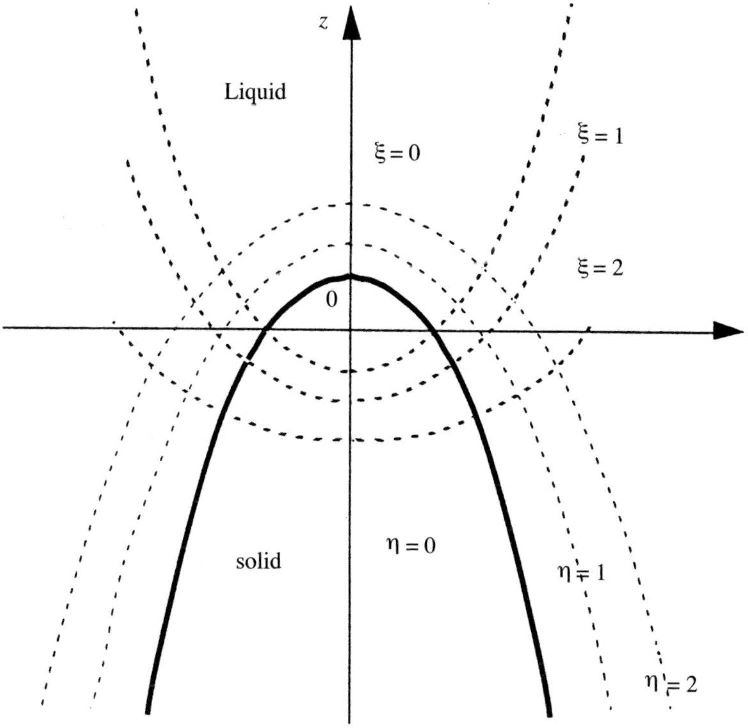

Figure 7.12 Parabolic coordinates.

with, at the interface ($\eta = 1$), the boundary conditions (5.8) of paragraph 5.3, that is

$$V\frac{Q}{c_p} = -\frac{2D_T}{\rho}\frac{dT}{d\eta}. \tag{7.5}$$

We deduce by integrating equation (7.4) once,

$$\frac{dT}{d\eta} = -P\frac{Q}{c_p}\exp P\,\frac{1}{\eta}\,\exp(-P\eta) \tag{7.6}$$

and finally by integrating between 1 and $+\infty$,

$$\Delta = P\exp P\int_1^{+\infty}\frac{1}{\eta}\exp(-P\eta)d\eta \tag{7.7}$$

where $P = \rho V/2D_T$ is the Péclet number. The paraboloids with an arbitrary radius of curvature at the tip form a set of exact solutions to the problem without surface tension (Ivantsov 1947). Only the Péclet number is determined by the supercooling of the liquid. The growth rate thus remains undetermined since the paraboloids with a large radius of curvature have a low velocity and those with a small radius of curvature a high velocity.

The equivalents of these paraboloids in two dimensions are parabolas. We introduce the two-dimensional parabolic coordinates

$$x = \rho\sqrt{\xi\eta},\ z = \frac{\rho}{2}(\eta - \xi). \tag{7.8}$$

Then in the moving frame where the parabola with a radius of curvature ρ at the tip (with equation $\eta = 1$) is at rest, the temperature field satisfies

$$\frac{1}{h_1}u_\xi\frac{\partial T}{\partial\xi} + \frac{1}{h_2}u_\eta\frac{\partial T}{\partial\eta} = \frac{D_T}{h_1 h_2}\left(\frac{\partial}{\partial\xi}\left(\frac{h_2}{h_1}\frac{\partial T}{\partial\xi}\right) + \frac{\partial}{\partial\eta}\left(\frac{h_1}{h_2}\frac{\partial T}{\partial\eta}\right)\right) \tag{7.9}$$

where

$$h_1 = \frac{\rho}{2}\sqrt{\frac{\eta+\xi}{\xi}},\ h_2 = \frac{\rho}{2}\sqrt{\frac{\eta+\xi}{\eta}} \tag{7.10}$$

are the coefficients of the corresponding metrics. We can also search for a solution of equation (7.9) depending only on the coordinate η, that is a solution of

$$-V\sqrt{\eta}\frac{dT}{d\eta} = \frac{2D_T}{\rho}\frac{d}{d\eta}\left(\sqrt{\eta}\frac{dT}{d\eta}\right) \tag{7.11}$$

with the same boundary conditions (7.5) at the interface. We deduce the relation between supercooling and the Péclet number by integrating equation (7.11) twice

$$\Delta = P\exp P\int_1^{+\infty}d\eta\ \frac{1}{\sqrt{\eta}}\exp(-P\eta). \tag{7.12}$$

By analogy to the Saffman-Taylor finger of paragraph 7.4, it is convenient for the rest of the study to find these exact solutions by the method of integral equations (Nash and Glicksman 1974, Langer 1977). Although this method may seem complicated for the present problem, it is actually necessary when the effects of surface tension are taken into account. In the moving frame where the interface is at rest the dimensionless temperature field $u = c_p(T - T_\infty)/Q$ satisfies

$$- V\frac{du}{dz} = D_T \Delta u \,. \tag{7.13}$$

Then we introduce the Green function $G(p_0|p)$ which satisfies

$$\left(- V\frac{\partial}{\partial z} + D_T\Delta\right) G(p_0|p) = - \delta(p_0 - p) \tag{7.14}$$

where $p(x, z)$ and $p_0(x_0, z_0)$ are two points in space. We write the solution of this equation simply using the Fourier transform, *i.e.*

$$G(p_0|p) = \int \frac{d^3\vec{k}}{(2\pi)^3}\hat{G}(\vec{k})\exp(i\vec{k}(\vec{r_0} - \vec{r})) \tag{7.15}$$

where
$$\hat{G}(\vec{k}) = \frac{1}{D_T k^2 - iVk_z} \tag{7.16}$$

and so after integration,

$$G(p_0|p) = \frac{1}{4\pi D_T} \frac{1}{\left|\vec{r_0} - \vec{r}\right|} \exp - \frac{V}{2D_T}(z_0 - z) \exp - \frac{V}{2D_T}\left|\vec{r_0} - \vec{r}\right| \tag{7.17}$$

where $\left|\vec{r} - \vec{r_0}\right| = ((x - x_0)^2 + (y - y_0)^2 + (z - z_0)^2)^{1/2}$ is the distance between the two points considered.

By multiplying equation (7.13) by G, equation (7.14) by u and by integrating over the volume V_T shown in figure 7.13, we obtain

$$\begin{aligned} u(p_0) &= \int_{V_T} dxdydz V\frac{\partial}{\partial z}(u(p)G(p_0|p)) \\ &+ \int_{V_T} D_T(G(p_0|p)\Delta u(p) - u(p)\Delta G(p_0|p)) \cdot \vec{n}\; dx\; dy\; dz \end{aligned} \tag{7.18}$$

where $\vec{n}$ is the vector pointing outwards from the surface S_T limiting the volume. The first integral leads to the constant $u(+\infty) = 0$ since, using equation (7.17), we deduce

$$V\int_{V_T} G(p_0|p)dxdy = \begin{cases} \exp\left(\dfrac{V}{D_T}(z - z_0)\right) & z < z_0 \\ 1 & z > z_0 \end{cases} \tag{7.19}$$

The second integral can be transformed using the Green theorem on the volume V_T, so that

$$u(p_0) = \int_{S_T} \left(D_T\left(G(p_0|p)\vec{\nabla}u(p) - u(p)\vec{\nabla}G(p_0|p)\right) \cdot \vec{n}\right)\; dx\; dy \,. \tag{7.20}$$

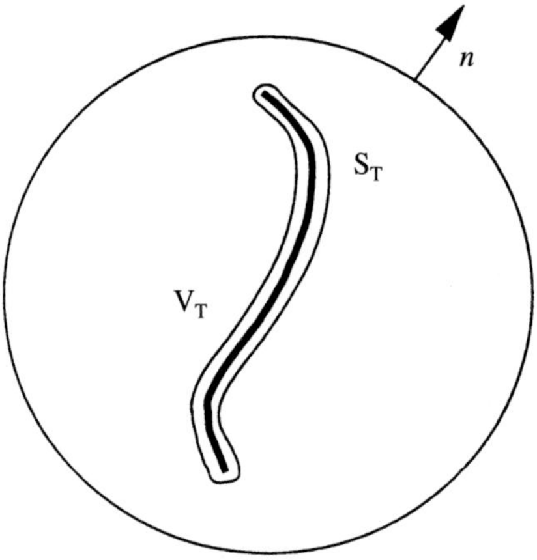

Figure 7.13 Integration contour when the diffusion coefficients of the two phases are equal.

When we deform the surface S_T along the interface the contribution of the second term of the integral is exactly zero due to the continuity of u at the interface. Using the boundary conditions (5.8) of paragraph 5.3, we deduce the integral equation

$$u(p_0) = V\int_{S_T} G(p_0|p)\,dx\,dy\ . \tag{7.21}$$

If we consider an interface contained in a plane we obtain after integrating equation (7.21) over the y coordinate

$$u(p_0) = \frac{V}{2\pi D_T}\int_{-\infty}^{+\infty} dx \exp\left(-\frac{V}{2D_T}(\xi(x)-\xi(x_0))\right) \times K_0\left(\frac{V}{2D_T}\sqrt{((\xi(x)-\xi(x_0))^2(x-x_0)^2)}\right) \tag{7.22}$$

where K_0 is the zeroth-order modified Bessel function.

If we neglect the effects of surface tension, $u(p_0) = \Delta$, and the shape of the interface is one of the Ivantsov parabolas $\xi = -x^2/2\rho$, since

$$\Delta = \frac{V}{2\pi}\int_{-\infty}^{+\infty} dx \exp\left(-\frac{V}{4D_T\rho}(x^2 - x_0^2)\right) K_0\left(\frac{V}{2D_T}\sqrt{\left(\frac{x^2}{2\rho}-\frac{x_0^2}{2\rho}\right)^2 + (x-x_0)^2}\right) \tag{7.23}$$

for all the points x_0.

Appendix 1: Integral equation for a non-axisymmetric form.

For a non-axisymmetric form in cylindrical coordinates, $z = \xi(r, \Phi)$, where r is the radial coordinate and Φ the polar angle, we deduce from equations (7.17) and (7.18)

$$u(p_0) = \frac{V}{2D_T}\int_0^{2\pi}\frac{d\Phi}{2\pi}\int_0^{+\infty} rdr\exp\left(-\frac{V}{2D_T}(\xi(r,\Phi)-\xi(r_0,\Phi_0))\right)$$

$$\times\frac{\exp\left(-\frac{V}{2D_T}\sqrt{(\xi(r,\Phi)-\xi(r_0,\Phi_0))^2+r^2+r_0^2-2rr'\cos(\Phi-\Phi_0)}\right)}{\sqrt{(\xi(r,\Phi)-\xi(r_0,\Phi_0))^2+r^2+r_0^2-2rr'\cos(\Phi-\Phi_0)}}.$$

Again if we neglect the effects of surface tension, $u(p_0) = \Delta$, and the shape of the interface is one of the Ivantsov paraboloids $\xi = -r^2/2\rho$, which allows us to obtain the rather non-trivial relation

$$\Delta = \frac{V}{2D_T}\int_0^{2\pi}\frac{d\Phi}{2\pi}\int_0^{+\infty} rdr\exp\left(-\frac{V}{2D_T}\left(-\frac{r^2}{2\rho}+\frac{r_0^2}{2\rho}\right)\right)$$

$$\frac{\exp\left(-\frac{V}{2D_T}\sqrt{\left(-\frac{r^2}{2\rho}+\frac{r_0^2}{2\rho}\right)^2+r^2+r_0^2-2rr_0\cos(\Phi-\Phi_0)}\right)}{\sqrt{\left(-\frac{r^2}{2\rho}+\frac{r_0^2}{2\rho}\right)^2+r^2+r_0^2-2rr_0\cos(\Phi-\Phi_0)}}$$

for all r_0, Φ_0.

Appendix 2: Integral equation for a liquid–solid interface of a steady-state structure with a growth in a two-dimensional space limited by the diffusion of an impurity with a negligible coefficient of diffusion in the solid phase.

Equation (7.18) in paragraph 7.6 applied to the field $u = (c-c_\infty)/\Delta C$ is still valid but applied to the surface S_c shown in figure 7.14. Transforming the volume integrals into surface integrals we obtain

$$u(p_0) = \int_{\Gamma_c} D_c\left(u(p)\vec{\nabla}G(p_0|p)-G(p_0|p)\vec{\nabla}u(p))\cdot\vec{n}-Vn_zu(p)G(p_0|p)\right)dx$$

where $\vec{n}$ is the vector normal pointing inwards the contour Γ_c limiting the surface S_T. When the point p_0 tends to the interface we must use the theorem of discontinuity of the potential of the double-layer

$$\lim(p_0\to p_{0\Gamma_c})\int_{\Gamma_c} u(p)\vec{\nabla}G(p_{0\Gamma_c}|p)\cdot\vec{n}dx = \frac{1}{2}u(p)+\int_{\Gamma_c} u(p)\vec{\nabla}G(p_{0\Gamma_c}|p)\cdot\vec{n}dx\ .$$

We thus obtain

$$\frac{1}{2}u(p_0) = \int_{\Gamma_c} D_c\left(u(p)\vec{\nabla}G(p_0|p)-G(p_0|p)\vec{\nabla}u(p)\right)\cdot\vec{n}-Vn_zu(p)G(p_0|p)\Big)dx\ .$$

Calculating the contribution of the integral coming from the part of the contour around $z = +\infty$

$$\lim_{z_0 \to +\infty} \int_{-\infty}^{+\infty} dx\left(-\frac{\partial G}{\partial z}(p_0|p) + \frac{2}{l} n_z G(p_0|p)\right)$$

$$= -\lim_{z \to +\infty} \frac{1}{2\pi l} \int_{-\infty}^{+\infty} dx \exp\left(\frac{z}{l}\right)\left(K_0\left(\frac{\sqrt{x^2+z^2}}{l}\right) + K_1\left(\frac{\sqrt{x^2+z^2}}{l}\right)\right)$$

$$= -\lim_{z \to +\infty} \frac{1}{2\pi l} \int_{-\infty}^{+\infty} dx \exp\left(\frac{z - \sqrt{x^2+z^2}}{l}\right)\left(\frac{2\pi l}{z}\right)^{1/2} = -1$$

where $l = 2D_c/V$, we finally deduce the required integral equation

$$\frac{1}{2}u(p_0) - u_\infty = \int_{\Gamma_c} D_c\left(u(p)\vec{\nabla} G(p_0|p) - G(p_0|p)\vec{\nabla} u(p)\right) \cdot \vec{n} - V n_z u(p) G(p_0|p))dx \ .$$

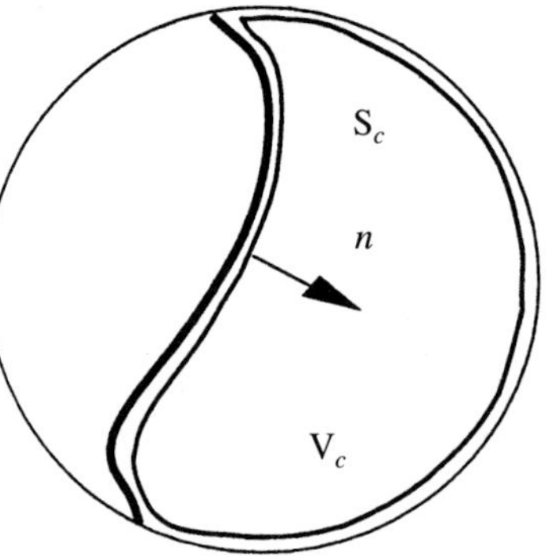

Figure 7.14 Integration contour for the asymmetric model.

Appendix 3: Relation between ρ, *V and oversaturation when impurities are present in the liquid (Karma and Langer 1984).*

When the effects of surface tension are neglected $T - T_0$ is proportional to the concentration of impurities c at the interface according to equation (2.11) in paragraph 2.4. A set of solutions for which $T = T_i$ and $c = c_i$ are constant at the interface can therefore be found. These solutions are the parabolas determined in paragraph 7.6 from which we deduce equation (7.12) for the temperature and the concentration fields, that is

$$\Delta_T = \frac{c_p(T_i - T_\infty)}{Q} = P_T \exp P_T \int_1^{+\infty} \frac{1}{\sqrt{\eta}} \exp(-P_T \eta) d\eta$$

where $P_T = \rho V/2D_T$ and $P_c = \rho V/2D_c$ are the thermal and the solute Péclet numbers. Thus for a given T_∞ and c_∞ the two preceding relations and equation (2.11) in paragraph 2.4 determine T_i, c_i and ρV. At low supercooling and oversaturation the two preceding relations become $P_T \approx \Delta_T^2/\pi$ and $P_c \approx \Delta_c^2/\pi$ so that, by the same equation (2.11),

$$P_T = (\Delta_{\text{eff}}^2/\pi)(1 + m c_\infty (1 - K)(c_p/Q)(D_T/D_c)^{1/2})^2$$

where

$$\Delta_{\text{eff}} = c_p(T_0 - mc_\infty - T_\infty)/Q$$

is the effective supercooling. We obtain a similar relation in the case with axial symmetry by using equation (7.7) of paragraph 7.6:

$$P_T = (-\Delta_{\text{eff}}\text{Log}(\Delta_{\text{eff}})/2)(1 + mc_\infty(1-K)(c_p/Q)(D_T/D_c)).$$

Appendix 4: In the case of two-dimensional growth, determination of the relation between ρ, V, and U, the velocity of a forced flow of small amplitude, in the direction opposite to the growth (Dash and Gill 1984).

We consider the configuration for which a flow resulting from an external source is forced in the axis of the crystal, far upstream from it, at the velocity U. It is worth noting that when the flow satisfies the Oseen equation

$$U\frac{\partial\vec{w}}{\partial z} = -\frac{1}{\rho}\vec{\nabla}p + \nu\Delta\vec{w}.$$

The set of Ivantsov parabolas is again the solution, still assuming that the crystal is isothermal. In fact, in the same parabolic coordinates and in the moving frame where the crystal is at rest the components of the velocity field are

$$u_\eta = -\frac{f(\eta)}{2\sqrt{\eta+\xi}}$$

$$u_\xi = \frac{\sqrt{\eta\xi}}{\sqrt{\eta+\xi}}\frac{df}{d\eta}$$

where

$$f(\eta) = 2(U+V)\sqrt{\eta} - 2Ug(\eta)$$

and

$$g(\eta) = \sqrt{\eta}\frac{\text{erfc}\left(\sqrt{\eta\frac{Re}{2}}\right)}{\text{erfc}\left(\sqrt{\frac{Re}{2}}\right)} + \sqrt{\frac{2}{\pi Re}}\frac{1}{\text{erfc}\left(\sqrt{\frac{Re}{2}}\right)}\left(\exp\left(-\frac{Re}{2}\right) - \exp\left(-\frac{Re}{2}\eta\right)\right)$$

where $Re = \rho U/\nu$ is the Reynolds number obtained from with the radius of curvature ρ of the considered parabola, and U the velocity of the flow. The temperature field still satisfies equation (7.9), that is, since it depends on the coordinate η only,

$$-\frac{f(\eta)}{2}\frac{dT}{d\eta} = \frac{2D_T}{\rho}\frac{d}{d\eta}\left(\sqrt{\eta}\frac{dT}{d\eta}\right).$$

Applying the relation giving the heat flux through the interface, we then deduce

$$\Delta = P_c\exp(P_c - P_e)\int_1^{+\infty}\frac{d\eta}{\sqrt{\eta}}\exp\left(-P_c\eta + P_e\left(2 + \int_1^{\eta}\frac{g(\eta')}{\sqrt{\eta'}}d\eta' - \eta\right)\right)$$

where $P_c = \rho V/2D_T$ is the Péclet number relative to the crystal and $P_e = \rho U/2D_T$ the Péclet number relative to the flow.

7.7 CRYSTAL IN A CAPILLARY

Let us consider the steady-state growth of a crystal from an oversaturated solution of concentration of solute c_∞ in a capillary of radius $R = \Lambda/2$. If we limit our study to the case of growth in a plane the configuration is quite similar to the configuration of a Saffman–Taylor finger moving in a Hele–Shaw cell. As will be shown here the shape of the crystal is also similar in the case where the Péclet number $P = \Lambda V/D$ is small, where V is the growth rate and D_c the diffusion coefficient of the solute (Pelcé and Pumir 1985). As in all the examples shown in this chapter we will neglect the effects of surface tension. Contrary to these examples the solution found here is not an exact solution to the posed problem of growth. This is an asymptotic solution in that its validity increases when the Péclet number decreases. We will distinguish between two scales, one being the diffusion length of thickness D_c/V, supposed far larger than the second one, of size Λ (figure 7.15).

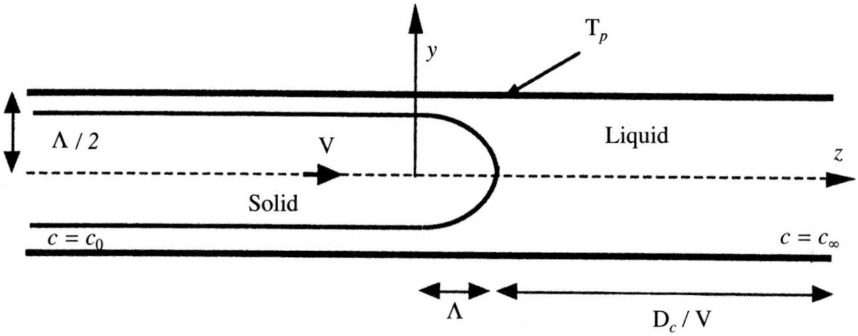

Figure 7.15 Crystal in a capillary.

The steady-state concentration field satisfies the equation

$$-V\frac{\partial c}{\partial z} = D_c \Delta c \tag{7.1}$$

with the boundary conditions

$$-D_c \vec{\nabla} c \cdot \vec{n} = c_0(1-K)V\cos\theta \tag{7.2a}$$

$$c = c_0 \tag{7.2b}$$

at the interface and $\partial c/\partial y = 0$ for $y = \pm\Lambda/2$. The wall of the capillary is at a uniform temperature T_p, so that $c_0 = (T_p - T_0)/m$.

Over the diffusion scale, upstream from the crystal, the concentration field does not change along the axis of the capillary and therefore has the same exponential profile as equation (5.8) in paragraph 5.4, that is

$$c = c_\infty + A\exp\left(-\frac{V}{D_c}z\right) \tag{7.3}$$

where A is an undetermined constant for the moment. In the region at the top, on the scale of Λ, the diffusion field is two-dimensional and changes on the order of $\delta c \approx c_0 V\Lambda/D_c$, therefore negligible compared to the diffusion region. Thus, $c \approx c_0$ in that region, $A = c_0 - c_\infty$ according to equation (7.3) and the flux given out is $-D_c dc/dz = (c_0 - c_\infty)V/D_c$. Let us introduce the effective concentration $\Phi = c - c_0$. On that scale, the advective term $-V\partial\Phi/\partial z$ of equation (7.1) is negligible so that

$$\Delta\Phi = 0 \tag{7.4}$$

with the boundary conditions at the interface

$$-D_c\vec{\nabla}\Phi \cdot \vec{n} = Vc_0(1-K)\cos\theta \tag{7.5a}$$

$$\Phi = 0. \tag{7.5b}$$

Moreover $-D_c d\Phi/dz = (c_0 - c_\infty)V$ for $z \approx +\infty$ and $\partial\Phi/\partial y = 0$ for $y = \pm 1$, where the lengths are made dimensionless with $\Lambda/2$. We thus find the same problem of the steady-state Saffman-Taylor finger again solved in paragraph 7.5. We thus deduce from equation (7.4) in the same paragraph the relation $z(w)$ where $w = u + iv$, v being the complex conjugated function of u

$$z = \frac{w}{\lambda} + \frac{2}{\pi}(1-\lambda)\mathrm{Log}\left(\frac{1 + \exp\left(-\frac{\pi w}{\lambda}\right)}{2}\right) \tag{7.6}$$

where $\lambda = Vc_0(1-K)/(c_0 - c_\infty)V = \Delta$ is the relative length of the form of the crystal, identical to that of a Saffman–Taylor finger, from equation

$$x = \frac{(1-\lambda)}{\pi}\mathrm{Log}\left(\frac{1 + \cos\left(\frac{\pi y}{\lambda}\right)}{2}\right). \tag{7.7}$$

Contrary to the case of a Saffman-Taylor finger, λ is directly determined from the control parameter $\Delta = (c_0 - c_\infty)/c_0(1-K)$, the oversaturation of the solution. On the other hand the velocity V of the crystal remains undetermined.

Using the expression of λ, we can also express the concentration flux coming out of the Saffman–Taylor region by

$$-D_c\frac{d\Phi}{dz} = \lambda V c_0(1-K). \tag{7.8}$$

7.8 HYDRODYNAMICS OF CURVED FLAMES

When the flame bends under the influence of the Darrieus–Landau instability, it also bends the streamlines. The shape which the flame adopts is therefore the result of an interaction between the growth form and the flow which we describe below (Zeldovich *et al.* 1980) (figure 7.16).

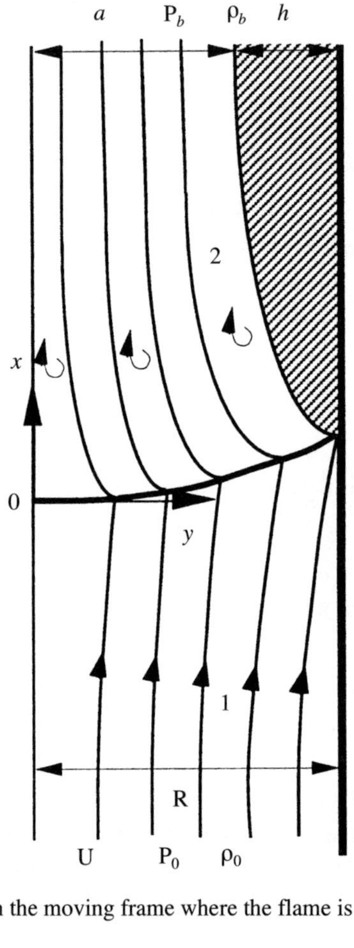

Figure 7.16 Streamlines in the moving frame where the flame is at rest.

In the moving frame where the flame is at rest the flow far in front of the flame is uniform and of propagation velocity U. Since the vorticity is conserved along the stream lines the flow is a potential flow in fresh gases. Owing to the jump in the density across the flame the flow of the burnt gases is rotational. Let us introduce the stream function Ψ which satisfies

$$\rho_b u = \frac{\partial \Psi}{\partial y}; \ \rho_b v = -\frac{\partial \Psi}{\partial x} \tag{7.1}$$

such that

$$\rho_b \Delta \Psi = -\omega(\Psi). \tag{7.2}$$

From the expression

$$\vec{w} \wedge \vec{\omega} = -\vec{\nabla}\left(\frac{p}{\rho} + \frac{w^2}{2}\right) \tag{7.3}$$

we thus deduce

$$\omega(\Psi) = -\frac{d}{d\Psi}\left(p + \rho_b \frac{w^2}{2}\right). \tag{7.4}$$

Across the flame we write the continuity conditions of the normal flux of mass, for the tangential velocity and for the normal momentum flux

$$\rho_0(\overrightarrow{w_{-f}} \cdot \vec{n}) = \rho_b(\overrightarrow{w_{+f}} \cdot \vec{n}) = \rho_0 \cdot u_L \tag{7.5a}$$

$$(\overrightarrow{w_{-f}} \cdot \vec{\tau}) = (\overrightarrow{w_{+f}} \cdot \vec{\tau}) \tag{7.5b}$$

$$p_{-f} + \rho_0(\overrightarrow{w_{-f}} \cdot \vec{n})^2 = p_{+f} + \rho_b(\overrightarrow{w_{+f}} \cdot \vec{n})^2 . \tag{7.5c}$$

Here ρ_0 and ρ_b are the densities of the fresh and burnt gases respectively and u_L is the normal velocity of flame propagation.

Since the mass flux through the flame is continuous, Ψ is continuous, such that in the flame,

$$\omega_f(\Psi) = \frac{1}{2}(\alpha - 1)\frac{d}{d\Psi}(w_{-f}^2) = \frac{1}{2}(\alpha - 1)u_L^2\frac{d}{d\Psi}(\cot^2\theta) \tag{7.6}$$

where θ is the angle between the incident streamline and the tangential direction to the front and $\alpha = \rho_b/\rho_0$. Since the flame is convex and the streamlines move away from the tip to the borders, θ decreases as we move away from the tip to the borders. The front of the curved flame thus generates the vorticity since α is smaller than 1. When the flame is planar, $\omega = 0$, since θ is constant. The presence of vorticity in the flow of the burnt gases makes the determination of the shape of the curved flame extremely difficult, and for now only the following approximation allows us to obtain the approximate shapes of the flames.

Since the upstream flow is potential, $\Delta\Phi = 0$ where $\vec{w} = \vec{\nabla}\Phi$. The flame propagates in a tube of radius R along the x direction (considered positive towards the burnt gases) and we assume the flow to be two-dimensional:

$$\Phi = Ux + \sum_{n=1}^{\infty} A_n \exp(k_n x)\cos(k_n y) \tag{7.7}$$

where $k_n = n\pi/R$, U is the required velocity of propagation sought and A_n are some coefficients to be determined. The approximation consists of keeping only the first mode k_1 and therefore determining two coefficients, U and A_I. The shape of the flame is assumed to be convex and symmetric with respect to the axis of the tube. We assume that R is far larger than the thickness of the flame so that the effects of curvature on the normal velocity of propagation can be neglected. At the tip $(x = y = 0)$, $u = u_L$ so that $A_I = -\varepsilon u_L/k_1$, where $\varepsilon = (U - u_L)/u_L$. The components of the velocity field of the fresh gases can therefore be written as

$$\begin{aligned}\frac{u}{u_L} &= 1 + \varepsilon(1 - \exp(k_1 x)\cos(k_1 y))\\ \frac{v}{u_L} &= \varepsilon\exp(k_1 x)\sin(k_1 y).\end{aligned} \tag{7.8}$$

The shape of the flame is thus deduced from the velocity field of equation (7.8) and from the condition of the normal velocity constant and equal to u_L

$$\frac{dx}{dy} = \frac{u(w^2 - u_L^2)^{1/2} - vu_L}{uu_L + v(w^2 - u_L^2)^{1/2}} \tag{7.9}$$

with $x(0) = 0$. When ε is small the flame is almost planar and from equation (7.9), we deduce

$$x = \frac{4\sqrt{\varepsilon}}{k_1}\left(1 - \frac{1}{\sqrt{2}}\,\frac{\sin(k_1 y)}{\sqrt{1 - \cos(k_1 y)}}\right) \tag{7.10}$$

whereas for ε large,

$$x = \frac{1}{k_1}\mathrm{Log}\left(\frac{k_1 y}{\sin(k_1 y)}\right). \tag{7.11}$$

As shown in figure 7.17, for intermediate values of ε, the shape is calculated by numerical integration of equation (7.9).

The problem is now to determine ε or, equivalently, to calculate the velocity of propagation U. We thus use the conservation relations for the total mass of total momentum of the fluid. If we consider the sections of the tube at $x = \pm\infty$, the conservation of mass is written as

$$\rho_b\int_0^a dy u_\infty(y) = \int_0^{\Psi(a)} d\Psi = \Psi(a) = \rho_0 UR \tag{7.12}$$

where a is the width of the section of the moving fluid far behind the flame. This width is usually smaller than the width of the tube because of the sudden decrease in density across the flame. As shown in figure 7.16 since there is a finite angle between the flame and the wall of the tube, the streamline going along the wall in

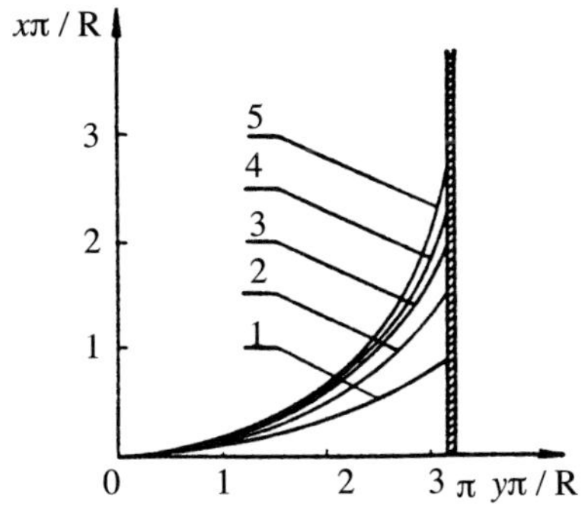

Figure 7.17 Shapes of the front for different values of ε; 1. $\varepsilon = 0.05$, 2. $\varepsilon = 0.15$, 3. $\varepsilon = 0.25$, 4. $\varepsilon = 0.35$, 5. $\varepsilon = 0.45$ (Zeldovich *et al.* 1980).

the fresh gases is deflected towards the inside when the flame passes, leaving a stagnant zone on the side. We write the conservation of momentum in the same way:

$$(p_0 - p_\infty + \rho_0 U^2)R = \int_0^{\Psi_a} d\Psi u_\infty(\Psi) \ . \tag{7.13}$$

We deduce using the Bernouilli relation along a streamline

$$u_\infty^2(\Psi) = \frac{2}{\alpha\rho_0}\left(p_{+f} + \rho_b\frac{w_{+f}^2}{2} - p_\infty\right)R \tag{7.14}$$

so that, by expressing the conditions of discontinuity (7.5) of the different quantities through the flame,

$$U^2 + u_*^2 - 2u_L^2\frac{(1-\alpha)}{\alpha} = \frac{2}{\alpha\rho_0}\int_0^{\rho_0 UR} d\Psi(u_L^2(1-\alpha^2) + u_*^2 - \alpha(1-\alpha)w_{-f}^2)^{1/2} \tag{7.15}$$

where $u_* = u(R)$. We can also calculate the width a of the section of the moving fluid using

$$a = \int_0^{\rho_0 uR} d\Psi \ \frac{1}{\rho_b u_\infty(\Psi)} = \frac{1}{\rho_0}\int_0^{\rho_0} d\Psi \ \frac{1}{(u_L^2(1-\alpha^2) + \alpha u_*^2 - \alpha(1-\alpha)w_{-f}^2)^{1/2}}. \tag{7.16}$$

As shown in figure 7.18a and figure 7.18b, equations (7.15) and (7.16) allow us to determine $\varepsilon(\alpha)$ and a.

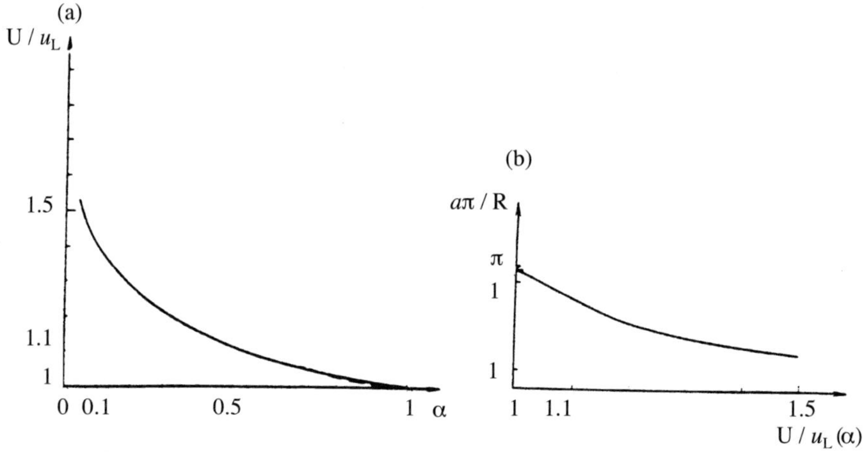

Figure 7.18 (a) U/u_L as a function of α (Zeldovich *et al.* 1980); (b) Relative thickness a as a function of U/u_L (Zeldovich *et al.* 1980).

As expected the amplitude of the flame is larger as the density gap increases, since this gap is the cause of the instability of the planar flame. Since the angle between the flame and the wall increases with the density gap the streamline going along the wall is more deflected in the centre and the radius a decreases.

Appendix: Influence of the gravity field (Pelcé 1985).

If the flame propagates in a vertical tube we must take into account the effect of the gravity field since the densities of the fresh and burnt gases are different. Let us write the gravity field in the form $\vec{g} = -g\vec{z}$, where $g > 0$ corresponds to the upwards propagation of the flame. The integral relation about the flow is written in a way similar to when gravity is absent

$$(p_0 + \rho_0 U^2)R = P_\infty R + \int_0^{\Psi_a} d\Psi u_\infty(\Psi) + (\rho_0 - \rho_b) g \int_0^R dy\xi(y)$$

by introducing the effective pressure $P = p + \rho g z$, which satisfies the condition of discontinuity

$$P_{-f} + \rho_0 u_L^2 - \rho_0 g\xi(y) = P_{+f} + \rho_b(\overrightarrow{w_{+f}} \cdot \vec{n})^2 - \rho_b g\xi(y)\ .$$

We thus deduce a relation similar to equation (7.15)

$$U^2 + u_*^2 - 2u_L^2\frac{(1-\alpha)}{\alpha} + 2(\alpha - 1)\frac{g}{R}\int_0^R dy(\xi(y) - \xi(R))$$

$$= \frac{2}{\alpha\rho_0}\int_0^{\rho_0 UR} d\Psi(u_L^2(1-\alpha^2) + u_*^2 - \alpha(1-\alpha)w_{-f}^2 - 2\alpha(1-\alpha)g(\xi(y) - \xi(R)))^{1/2}$$

and the half-width a of the region of the moving fluid as well:

$$a = \int_0^{\rho_0 UR} d\Psi \, \frac{1}{\rho_b u_\infty(\Psi)}$$

$$= \frac{1}{\rho_0}\int_0^{\rho_0 UR} d\Psi \, \frac{1}{\left(u_L^2(1-\alpha^2)+\alpha u_*^2-\alpha(1-\alpha)w_{-f}^2-2\alpha(1-\alpha)g(\xi(y)-\xi(R))\right)^{1/2}}$$

The gravity field introduces a new parameter, the Froude number $F = gR/\pi u_L^2$. Using the flow defined by equation (7.8) and the shape of the flame deduced by integration of equation (7.9), we can determine $\varepsilon(\alpha, F)$. Thus for $\alpha = 0.2$ we plot $\varepsilon(F)$ in figure 7.19. When the flame propagates downwards ε decreases as F increases until a critical value where there is no curved solution. In fact in that configuration the lighter burnt gases are above the heavy fresh gases and the gravity field tends to flatten the flame. When the flame propagates upwards the light burnt gases are below and tend to rise quicker, making the flame more convex and therefore ε larger, as a bubble of warm gas (Von Lavante and Strehlow 1983). The velocity of propagation should thus be the same as for a gas bubble, *i.e.* $\approx \sqrt{gR}$ (Davies and Taylor 1950).

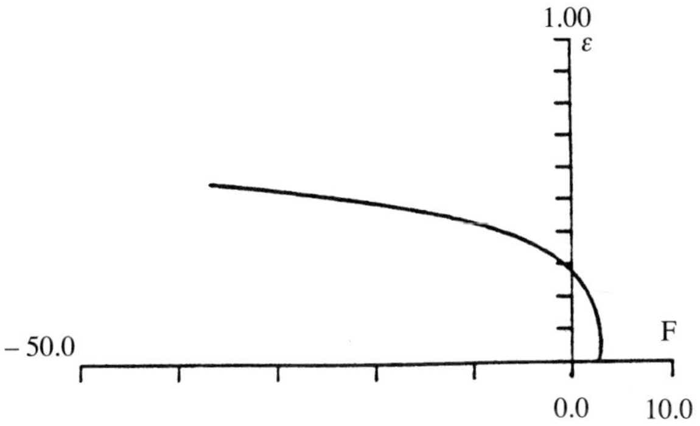

Figure 7.19 ε as a function of F for $\alpha = 0.2$ (Pelcé 1985).

BIBLIOGRAPHY

[1] M. Ben Amar, 'Exact Self-Similar Shapes in Viscous Fingering', *Phys. Rev.*, **A43**, p. 5724, 1991.

[2] S.K. Dash, W.N. Gill, 'Forced Convection Heat and Momentum transfer to dendritic structures', *Mass Heat Transfer*, **27**, p. 1345, 1984.

[3] R. M. Davies, G.I. Taylor, 'The Mechanics of Large Bubbles Rising through Extended Liquids and through Liquids in Tubes', *Proc. Roy. Soc.*, **A 200**, p. 375, 1950.

[4] B. Derrida, V. Hakim, 'Needle Models of Laplacian Growth', *Phys. Rev.*, **A45**, p. 8759, 1992.

[5] H. Cohn, *Conformal Mapping on Riemann Surfaces*, Dover, 1967.

[6] G.P. Ivantsov, 'Temperature Field around Spheroïdal, Cylindrical and Acicular Crystal Growing in a Supercooled Melt', *Translated from Doklady Akad. Nauk*, **58 (4)**, p. 567, 1947.

[7] A. Karma, J.S. Langer, 'Impurity effects in dendritic solidification', *Phys. Rev.*, **A 30**, p. 3147, 1984.

[8] D. Kurtze, 'Bifurcation in Singular Needle Growth', *Phys. Rev*, **A 43**, p. 7066, 1991.

[9] L. Landau, 'On the Theory of Superconductivity', collected papers, D. Ter. Haar Gordon and Breach ed., p. 217, 1937.

[10] J.S. Langer, 'Studies in the theory of interfacial stability. II. Moving symmetric model', *Acta Metall.*, **25**, p.1121, 1977.

[11] J.W. McLean, P.G. Saffman, 'The Effect of Surface Tension on the Shape of Fingers in a Hele–Shaw Cell', *J. Fluid. Mech.*, **102**, p. 455, 1981.

[12] G.E. Nash, M.E. Glicksman, 'Capillarity-Limited Steady-State Dendritic Growth', *Metall.*, **22**, p. 1283, 1974.

[13] Z. Nehari, *Conformal Mapping*, Dover, New York, 1952.

[14] P. Pelcé, 'Effect of Gravity on the Propagation of Flames in Tubes', *J. Physique*, **46**, p. 503, 1985.

[15] P. Pelcé, Y. Pomeau, 'Dendrites in the small undercooling limit', *Studies in applied maths*, **74**, p. 245, 1986.

[16] P. Pelcé, A. Pumir, 'Cell Shape in Directional Solidification in the Small Péclet Number Limit', *J. Crystal Growth*, **73**, p. 337, 1985.

[17] M.A. Peterson, J. Ferry, 'Spontaneous Symmetry Breaking in Needle Crystal Growth', *Phys. Rev.*, **A 39**, p. 2740, 1989.

[18] P.G. Saffman, G.I. Taylor, 'The Penetration of a Fluid into a Porous Medium or Hele–Shaw Cell Containing a More Viscous Liquid', *Proc. Roy. Soc.*, **A 245**, p. 312, 1958.

[19] H. Thome, M. Rabaud, V. Hakim, Y. Couder, 'The Saffman–Taylor Instability: from the Linear to the Circular Geometry', *Phys. Fluids*, **A 1 (2)**, p. 224, 1989.

[20] E. Von Lavante, A.R. Strehlow, 'The Mechanism of Lean Limit Flame Extinction', *Comb. and Flame*, **49**, p. 123, 1983.

[21] Ya.B. Zeldovich, A.G. Istratov, N.I. Kidin, V.B. Librovich, 'Flame Propagation in Tubes: Hydrodynamics and Stability', *Comb. Sci and Tech.*, **24**, p. 1, 1980.

[22] P.A. Zhuralev, *Zap Leningr. Com. Inst.*, **33**, p. 54, 1956.

CHAPTER 8

Pattern formation and growth rate

8.1 EXPERIMENTAL RESULTS

We will limit ourselves to the experiments determining the velocity of a Saffman–Taylor finger and a free dendrite, for which precise and reproducible results have been obtained. These experiments indicate that a curved shape, such as a Saffman-Taylor finger or a dendrite, grow at a well-defined velocity determined by the control parameters. More precisely, as long as those parameters are fixed, the final velocity of propagation of the curved shape does not depend on the initial conditions of the experiment.

The first characterisations of shapes and the determinations of the velocities of propagation of the fingers in a Hele–Shaw cell were achieved by Saffman and Taylor (1958). Using different oils as viscous fluids displaced by water or air, they observe that over a wide range of capillary number variation $Ca = \eta U/\gamma$, a single finger of stationary shape propagates in the cell at a constant velocity U. The relative width of the finger λ appears as a well-defined function of the capillary number (figure 8.1). At low velocity the finger is thick and occupies almost all the cell width, $\lambda \approx 1$. λ decreases with the velocity and unexpectedly saturates at a value close to $1/2$ at high velocities. At even higher velocities the stationary propagation is no longer stable. If the observed shapes at low velocities deviate from the forms without surface tension at the same λ calculated in paragraph 7.1, they are very similar at high velocities.

In an analogous fashion, the dendrites grow with a well-defined velocity as a function of the supercooling or the oversaturation of the liquid. Figure 8.2 shows the dimensionless velocity $\mathrm{v} = d_{0T}V/2D_T$ in logarithmic coordinates of thermal dendrites of ice and succinonitrile measured by different experimentalists as a function of the supercooling of the liquid Δ, where d_{0T} is the capillary length mentioned in paragraph 2.4, and D_T is the coefficient of thermal diffusion. It is remarkable that the different experimental data fit on the same curve over a rather large range of the supercooling and the velocity. The solid line is the theoretical curve of Langer and Müller-Krumbhaar (1978) determined from the set of Ivantsov paraboloids and the constant $\sigma = 2d_{0T}D_T/\rho^2 V \approx 0.025$, which we will discuss later on.

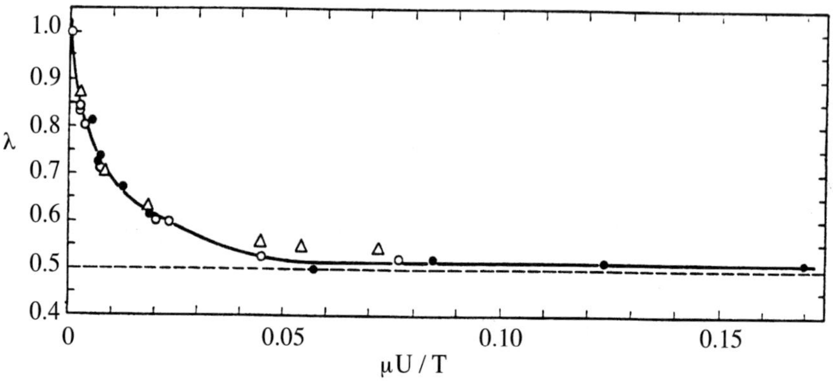

Figure 8.1 λ as a function of *Ca* when water penetrates oil, Diala (triangles and white circles) Talpa (black circles) (Saffman and Taylor 1958).

On the other hand, we note that in general the quantity $\rho^2 V$ is constant over a large range of supercooling.

These experimental results contradict those from the previous chapter, where the stationary curved structures determined in the absence of surface tension have an undetermined velocity. In fact, as we will show in this chapter, the effects of surface tension have a crucial role in determining the growth rate of these curved structures. Along a Saffman–Taylor finger for example, the change in pressure due to the flow is of the order of $\Delta p \approx \lambda a(\eta/b^2)V$, where a is the half-width of the cell and V is the velocity of the fluid far in front of the finger. The change in pressure due to the surface tension is of the order of $\Delta \mathrm{p} \approx \gamma/\lambda a$. Equating those two quantities we get $\lambda^2 V \approx (\gamma/\eta)(b/a)^2$, and therefore a decreasing relative thickness with the velocity. Even if this argument is of little value, particularly since it does not describe the saturation of the relative thickness of a Saffman–Taylor finger at high velocities, it indicates that an additional relation between the parameters can be obtained when the effects of surface tension are taken into account.

Analogously, the capillary length d_{0T} allows the construction of a dimensionless velocity $v = d_{0T}V/2D_T$ and thus its connection to supercooling Δ. Without surface tension and with the dimensions of the parameters in the problem, it is impossible to link the growth rate to the supercooling and therefore to explain the absence of the selection of velocities of the Ivantsov paraboloids. According to the Gibbs–Thomson relation, the change in temperature at the tip of a dendrite is of the order of $\delta T \approx T_0(\gamma/\rho)$, whereas that due to the release of heat flow is $\delta T \approx (Q/c_p)\rho V/D_T$. We thus deduce $\rho^2 V = \mathrm{cst}$, the additional relation observed in experiments.

The determination of that additional relation has a long history, from criteria without real theoretical justification, such as the criterion of a maximum velocity and then passing to certain principles of out-of-equilibrium systems, such as that of

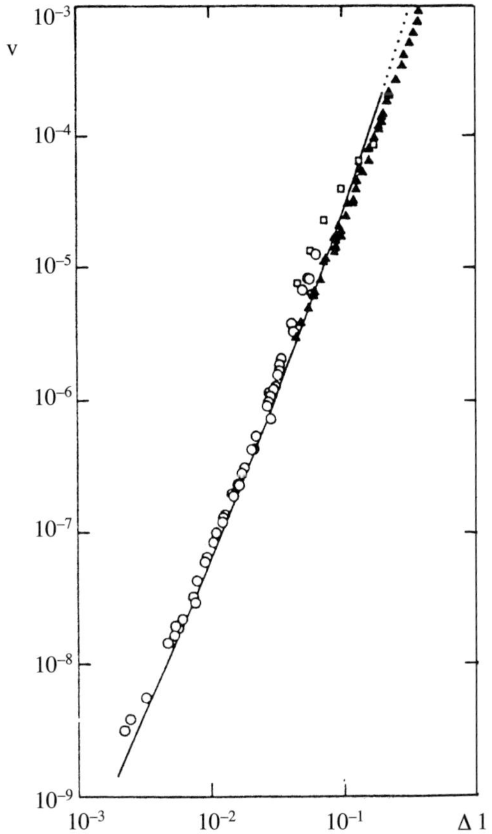

Figure 8.2 v as a sunction of Δ for different materials: ice (white circles), succinonitrile (black triangles), or other (white squares). The solid line is determined by eliminating the radius of curvature of the tip from the Ivantsov relation in paragraph 7.6 and from $\sigma = 0.025$ (Langer 1980).

marginal stability (Langer and Müller-Krumbhaar 1980). The problem is complicated by the fact that the growth form is not always steady-state, as for a dendrite, which renders the mathematical analysis very complicated. First we can limit ourselves to the steady-state structures and solve the so-called selection problem, that is to examine the way in which the capillary effects break down the continuum solutions determined in the previous chapter. This situation is different to that of a structure analogous to equilibrium. If we consider the case of a superconductor for instance, in the intermediate state discussed in paragraph 7.4, when the wavelength of the lattice is kept constant (which comes to the same as keeping the width of a Hele–Shaw cell constant), a single structure is determined. The surface tension of the interface between the normal and the superconducting phases only sets this wavelength by minimisation of the total free energy of the system (Landau 1937).

8.2 PROBLEM OF SELECTION

The geometric model of Brower and coworkers (1983) shown in paragraph 10.1 allows the understanding of the nature of the selection problem. It consists of determining the stationary structures growing at a constant velocity V in the direction Oz, which are therefore the solutions of the equation

$$V\cos\theta = \kappa + \gamma \frac{d^2\kappa}{dl^2} \tag{8.1}$$

that is

$$V\cos\theta = \frac{d\theta}{dl} + \gamma \frac{d^3\theta}{dl^3} \tag{8.2}$$

where θ is the angle between the normal to the interface and the direction of propagation. We search for the needle-like morphologies, symmetric with respect to the growth direction for which $\theta = \pm\pi/2$ far away from the tip. At the tip $(l = 0)$, $\theta = 0$ and $d^2\theta/dl^2 = 0$, since the sought solution is symmetric. The important property of the model is that it recreates in a simpler way the conditions met in the models of viscous fingering and crystal growth, meaning that in the absence of surface tension (or the parameter γ which represents it) a solution can be determined at every growth rate V. Indeed at $\gamma = 0$, we get the set of solutions

$$\theta_0 = \cos^{-1}(\mathrm{sech}(Vl)) \tag{8.3}$$

where $\mathrm{sech}\, x = 1/\cosh x$. As shown in figure 8.3a, these structures are fingers of width π/V which advance at a velocity V. We need to study how these solutions are modified when γ is non-zero. Equation (8.2) contains only one parameter; indeed, if we redefine the curvilinear abscissa by $s = Vl$ and the surface tension by the parameter $\nu = \gamma V^2$, equation (8.2) becomes

$$\cos\theta = \frac{d\theta}{ds} + \nu \frac{d^3\theta}{ds^3}. \tag{8.4}$$

In general, we cannot solve equation (8.4) with the aforementioned boundary conditions. However, when ν is small, an approximation method inspired by the method of Wenzel–Kramers–Brillouin (WKB) and applied to this non-linear equation by Kruskal and Segur (1985) shows that the tip of the finger develops a singularity (figure 8.3b). As we show in appendix 1, it occurs as an exponentially small amplitude of the curvature derivative at the tip $|d^2\theta/ds^2|(0) \approx \nu^{-5/4} \exp-\pi/2\sqrt{\nu}$, which is incompatible with the existence of symmetric solutions, since one of the conditions stated above is not satisfied. Note however that the sought condition is almost verified because the term which should be zero is in fact exponentially small. The physical relevance of this weak difference will be discussed in paragraph 8.10.

Figure 8.3 (a) The regular shape; (b) the shape with a singularity at the tip.

Here we present an approximation method, certainly less rigorous but simpler, which captures the essentials of the unexpected properties of the solutions of this equation. In addition however, as we will see later, it gives a more physical form to the problem of selection, applicable to realistic configurations found in experiments. This method applied by Langer (1986) and Shraiman (1986) to the above-mentioned problem consists of searching for a solution to the equation in the vicinity of θ_0. Introducing ξ such that $\nu\xi = d\theta/ds - \cos\theta$, we get, after linearisation, the equation with the second term

$$\nu \frac{d^2\xi}{d\theta^2} + \nu p(\theta) \frac{d\xi}{d\theta} + q(\theta)\xi = r(\theta) \tag{8.5}$$

where

$$p(\theta) = -2\tan\theta; \quad q(\theta) = \frac{1}{\cos^2\theta}$$
$$r(\theta) = -\frac{1}{\cos\theta}(1 - 2\cos^2\theta) \tag{8.6}$$

where we have kept the terms of leading order in ν. For $\theta = \pm\pi/2$, $\xi = 0$. Equation (8.5) has a solution only if the condition for solvability is satisfied. Indeed, ζ is an element of the kernel of the adjoint linear operator of the left-hand term of equation (8.5), meaning that it satisfies

$$\nu \frac{d^2\zeta}{d\theta^2} - \nu p(\theta) \frac{d\zeta}{d\theta} + q(\theta)\zeta = 0 \tag{8.7}$$

so, by integration by parts, we find that the integral

$$\int_{-\pi/2}^{-\pi/2} \zeta(\theta) r(\theta) d\theta = 0 \tag{8.8}$$

is null. At small values of ν, according to the WKB approximation, we search for solutions of equation (8.7) proportional to $\exp(S(\theta)/\sqrt{\nu})$, and we expand S in powers of $\sqrt{\nu}$. Thus, the solutions of equation (8.7) are of the form

$$\zeta(\theta) = \frac{1}{q_0^{1/4}} \exp\left(\pm \frac{i}{\sqrt{\nu}} \Psi(\theta) + \frac{1}{2} \int_0^\theta p_0(\theta') d\theta' \right) \tag{8.9}$$

where we define

$$\Psi(\theta) = \int_0^\theta \sqrt{q_0(\theta')} d\theta' \tag{8.10}$$

the symmetric combination of which leads to the condition for solvability

$$\int_{-\pi/2}^{\pi/2} d\theta \, \frac{r_0(\theta)}{q_0(\theta)^{1/4}} \cos\left(\frac{\Psi(\theta)}{\sqrt{\nu}} \right) \exp\left(\frac{1}{2} \int_0^\theta p_0(\theta') d\theta' \right) = 0 \,. \tag{8.11}$$

The calculation of the integral (8.10) achieved in appendix 2 of this paragraph shows that this condition is proportional to $\nu^{-5/4} \exp(-\pi/2\sqrt{\nu})$, of the same order of magnitude as $d^2\theta/ds^2(0)$. The condition for solvability therefore cannot be satisfied, which explains the absence of solutions to equation (8.4) for non-zero values of ν. Note also that since the value of the integral is exponentially small, the condition is almost satisfied.

Appendix 1: Kruskal and Segur's method (1985)

If the parameter ν is assumed small, we determine the solutions of equation (8.4) similar to the form (8.3). We therefore write $\theta = \theta_0 + \theta_1$, θ_1 being a small correction, and therefore a solution of

$$-\sin\theta_0 \theta_1 = \frac{d\theta_1}{ds} + \nu \frac{d^3\theta_1}{ds^3} \,.$$

This expansion is valid away from the singularities of $\cos\theta_0$, that is, away from the points $s = \pm i\pi/2$. In the WKB approximation, we look for a solution of that equation in the form $\theta_1 \approx \exp(S(s)/\sqrt{\nu})$, and expand S in powers of $\sqrt{\nu}$. Thus, up to the first order of the expansion, $\theta_1 \approx a_\pm (\exp s + \exp - s)^{1/2} \exp(\pm is/\sqrt{\nu})$ where $a_\pm$ are constants. Around the point $s = -i\pi/2$, we redefine a new variable varying on the order of unity $\zeta = (s + i\pi/2)/\sqrt{\nu}$, as well as $\theta = -\pi/2 + i\mathrm{Log}(2/\sqrt{\nu}) + i\Phi(\xi, \sqrt{\nu})$. Equation (8.4) is then written as

$$\frac{d\Phi}{d\xi} + \frac{d^3\Phi}{d\xi^3} = (\exp\Phi - (\nu/4)\exp - \Phi) \approx \exp\Phi$$

independent of all parameters. To connect the two solutions near the singularity, we rewrite the WKB solution in the new variable $\zeta = (s + i\pi/2)/\sqrt{v}$, that is, up to a constant prefactor, $\theta_1 \approx a_{\pm} v^{1/4} \exp(\pm\pi/2\sqrt{v}) \exp(\pm i\xi)$. From the preceding argument this solution must be independent of ν. Thus, $a_{\pm}$ must be, up to a factor of order unity, of the order of $v^{-1/4} \exp(-\pm\pi/2\sqrt{v})$. Close to the tip, at $s \approx 0$, the – component is exponentially large and must be cancelled out, $a_{-} = 0$. The other component is exponentially small and contributes to a derivative of the curvature of the point of the order $v^{-5/4} \exp(-\pi/2\sqrt{v})$. When the crystalline anisotropy factor is taken into account this component can be cancelled out for a discrete ensemble of values of ν, giving rise to possible solutions of needle-like crystals.

Appendix 2: Calculation of the integral

$$I(x) = \int_0^1 dt \exp{-4xt^2} \cos(5xt - xt^3)$$

at large values of x (Bender and Orszag 1978).

We can rewrite the integral in the form

$$I(x) = \frac{1}{2}\int_{-1}^{1} dt \exp(-4xt^2 + 5ixt - ixt^3)$$

$$= \frac{1}{2}\exp(-2x)\int_{-1}^{1} dt \exp(x\rho(t))$$

where $\rho(t) = -(t-i)^2 - i(t-i)^3$. On the segment $[-1, 1]$, the imaginary part of ρ is non-zero and the integral oscillates rapidly. To avoid the interferences of these oscillations we deform the contour of integration to a so-called steepest descent contour, on which the imaginary part of ρ is constant. This contour must pass through the saddle point of the complex plane $t = i$ where $\rho'(t) = 0$, where the integrand part is at a local maximum (figure 8.4) and the neighbourhood of which gives the most important part of the integral. In the vicinity of the saddle point, the steepest descent contour is the straight line $t = i + u$ and $\rho(t) \approx -u^2$. Thus,

$$I(x) \approx \frac{1}{2}\exp(-2x)\int_{-\varepsilon}^{\varepsilon} du \exp(-xu^2)$$

$$\approx \frac{1}{2}\exp(-2x)\sqrt{\frac{\pi}{x}}$$

when x is large.

Appendix 3: Calculation of integrals of the form

$$I_p(v) = \int_{-\infty}^{+\infty} d\xi (\operatorname{sech}\xi)^p \exp\frac{i\xi}{\sqrt{v}}$$

when ν is small (Langer 1986).

By changing the variable $\eta = \tan\theta$, we rewrite

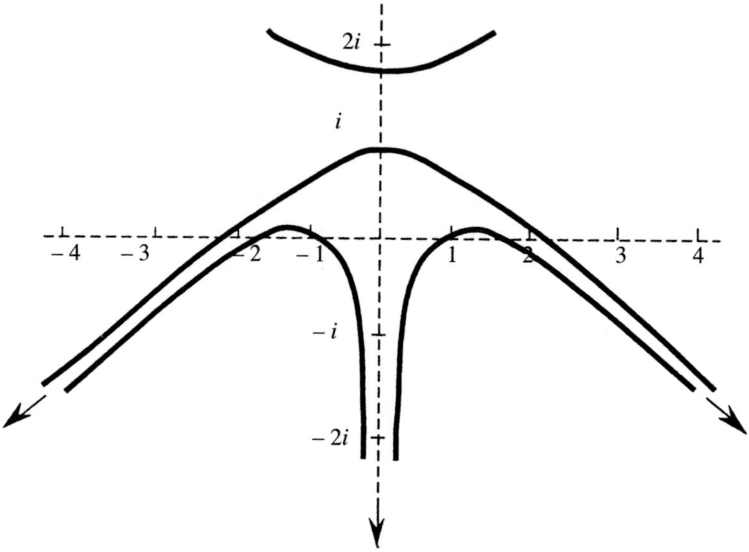

Figure 8.4 Contour of the steepest descent.

$$I_p(\nu) = \int_{-\infty}^{+\infty} d\eta \; \frac{1}{(1+\eta^2)^{(1+p)/2}} \exp\frac{i\psi(\eta)}{\sqrt{\nu}}$$

where

$$\psi(\eta) = \int_0^{\theta} d\theta' \; \frac{1}{\cos\theta'} = \int_0^{\eta} d\eta \; \frac{1}{(1+\eta^2)^{1/2}}.$$

The steepest descent contour passes through the point $\eta = i$, close to which

$$\psi(\eta) = \int_0^{i} d\eta \; \frac{1}{(1+\eta^2)^{1/2}} + \int_0^{w} dw' \; \frac{1}{(2iw')^{1/2}} = i\;\frac{\pi}{2} - i(2iw)^{1/2}$$

where $\eta \approx i + w$. Thus,

$$I_p(\nu) \approx \exp\left(-\frac{\pi}{2\sqrt{\nu}}\right) \int_{-\infty}^{+\infty} dw \; \frac{1}{(2iw)^{(1+p)/2}} \exp\left(\frac{2iw}{\nu}\right)^{1/2}$$
$$\approx \frac{2\pi}{\nu^{(p-1)/2}} \exp\left(-\frac{\pi}{2\sqrt{\nu}}\right).$$

8.3 THE EQUATIONS OF MCLEAN AND SAFFMAN

In this paragraph, let us determine the equation of the shape of a Saffman–Taylor finger when the effects of surface tension are taken into account. The velocity potential takes the value $\Phi = (\gamma/12\eta)(b/a)^2(1/R)$, where R is the radius of curvature of the interface made dimensionless with the half cell-width. At the interface we still have $\partial\Phi/\partial n = U\cos\theta$ and far in front of the finger $d\Phi/dz = V$. We thus deduce

$$\phi = -\frac{x}{1-\lambda} + \frac{\gamma}{12\eta U(1-\lambda)}\left(\frac{b}{a}\right)^2\frac{1}{R} \tag{8.1}$$

at the interface, meaning that for $\phi_0 < \phi < +\infty$, where ϕ_0 is the value of the potential at the tip of the finger, $(x = 0)$. We introduce, as in paragraph 7.5, the variable $s + it = \exp{-\pi((\phi - \phi_0) + i\psi)}$ for which the domain of the flow is the upper half-plane. Using expressions $1/R = d\theta'/dl = (\partial\theta'/\partial\phi)(d\phi/dl) = q'(\partial\theta'/\partial\phi)$ and $d\phi/ds = -1/\pi s$, we get after the derivation of equation (8.1) with respect to l,

$$\frac{\gamma\pi^2}{12\eta U}\left(\frac{b}{a}\right)^2 q's\frac{d}{ds}\left(q's\frac{d\theta'}{ds}\right) - (1-\lambda)q' = \cos\theta' \tag{8.2}$$

for $0 < s < 1$. Redefining the variables $q = q'(1-\lambda)$ and $\theta = \theta' - \pi$, we deduce the system of equations

$$\mathrm{Log}(q(s)) = -\frac{s}{\pi}P\int_0^1 \frac{\theta(s')}{s'(s'-s)}ds' \tag{8.3}$$

unchanged with respect to paragraph 7.7 and

$$\kappa qs\frac{d}{ds}\left(qs\frac{d\theta}{ds}\right) - q = -\cos\theta \tag{8.4}$$

where κ is the control parameter

$$\kappa = \frac{\gamma\pi^2}{12\eta U(1-\lambda)^2}\left(\frac{b}{a}\right)^2 \tag{8.5}$$

with the boundary conditions $\theta(0) = 0$, $q(0) = 1$, $\theta(1) = -\pi/2$, $q(1) = 0$.

Equations (8.3) and (8.4) are solved by numerical methods. In a first attempt, McLean and Saffman found only one branch of solutions. For a given κ, a solution $q(s)$, $\theta(s)$ is determined, and the corresponding relative width λ is deduced from equation (7.6) in paragraph 7.5 applied to the point $s = 0$

$$\mathrm{Log}(1-\lambda) = \frac{1}{\pi}P\int_0^1 \frac{\theta_0(s')}{s'}ds'. \tag{8.6}$$

Figure 8.5a shows the curve $\lambda(\kappa)$ obtained in this way. It solves the given problem. On the one hand, using the definition of κ, it determines the relative finger velocity as a function of capillary number. On the other hand, the obtained values of λ are in good agreement with experiment. In particular, at high velocities and so for small values of κ, λ tends to $1/2$, the minimal value obtained. At low velocities and so for large values of κ, λ tends to 1.

In fact, these equations allow for other solutions. Roméro (1982) determines a second branch of possible solutions and Vanden-Broeck (1983) an infinite , but countable number of solutions (we can number the branches of solutions). All these branches behave in the same way at small and large values of κ (figure 8.5b). The selection problem therefore arises again. We will see in the next chapter that the properties of the stability of these solutions alone allow the determination of the solution observed in experiments. Indeed, only the first branch of solutions, that determined by McLean and Saffman, is stable, and should therefore be associated with the observed stationary fingers.

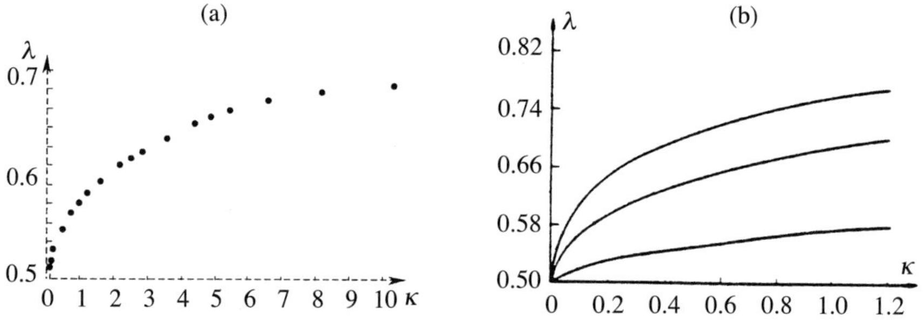

Figure 8.5 (a) McLean and Saffman branch (1982); (b) Vanden-Broeck branch of solutions (1983).

In an angular section, analogous equations to equations (8.3) and (8.4) can be determined (Ben Amar 1991). The equivalent of parameter κ is $\sigma = (b/R)^2(\gamma/12\eta V\theta_0^2)$, where R and V are the radius and the velocity of the tip of the lobe at time t, respectively, and θ_0 is the angle of the section. The numerical solution of these equations leads to a discrete number of branches $\lambda(\sigma)$. However, these branches group in pairs, connecting themselves at small values of σ. We can thus pass in a continuous fashion from the first to the second, third to the fourth, etc.

Appendix: Shape of a Saffman–Taylor finger when $\lambda \approx 1$ (Pomeau 1985, Dombre and Hakim 1987).

When $\lambda \approx 1$, the finger leaves a thin trail of liquid on the sides. This configuration is reminiscent of the lubrication flow we already met in appendices 2 and 3 in paragraph 5.2. In the region of the tip, the flow can be assumed to be of constant velocity U since the proportion of the liquid lost along the finger is negligible. Thus, $\Phi = Ux$, therefore using $\Phi = (\gamma/12\eta)(b/a)^2(1/R)$ at the interface, we deduce

$$\alpha \frac{d^2\theta}{dl^2} = -\sin\theta$$

where $\alpha = (\pi^2\gamma/12\eta U)(b/a)^2$. This equation is the same as that which describes the deviation of a pendulum with respect to the vertical, the solution of which is in the form

$$x(\theta) = -(\alpha/2\pi^2)^{1/2}\int_0^\theta d\theta' \frac{\sin\theta'}{(\cos\theta' + C)^{1/2}}$$

$$y(\theta) = -(\alpha/2\pi^2)^{1/2}\int_0^\theta d\theta' \frac{\sin\theta'}{(\cos\theta' + C)^{1/2}}$$

where we have only retained the tangential solutions at the cell walls. C is a constant of integration determined as a function of α (a parameter which is still undetermined at that stage) by the condition that the meniscus completely fills the width of the channel,

$$y(\theta = \pi/2) - y(\theta = 0) = \left(\frac{\alpha}{2\pi^2}\right)^{1/2}\int_0^{\pi/2} d\theta \frac{\cos\theta}{(\cos\theta + C)^{1/2}} = 1 .$$

Behind the tip, the shape of the finger goes along the cell walls. We can therefore evaluate the relative fluid flux in the channel

$$j(x) = \int_{y(x)}^1 dy'\left(\frac{\partial\Phi}{\partial x} - U\right) \approx Uh(x)\left(\frac{\alpha}{\pi^2}\frac{d^3h}{dx^3} - 1\right) = j(-x) .$$

We have used the fact that, the interface going along the wall, the integral is the product of the local width of the channel $h(x) \approx 1 - y(x)$ by the integral evaluated at the interface, that $dh/dx << 1$ and therefore that $\kappa \approx d^2h/dx^2$. Redefining the coordinates $x = ((1-\lambda)\alpha/\pi^2)^{1/3}X$ and $h = (1-\lambda)H$ in that region we deduce

$$\frac{d^3H}{dx^3} = \frac{H-1}{H}$$

an equation comparable to those determined in appendices 2 and 3 in paragraph 5.2. Again, the ratio between the two forms is carried out by identifying the expansion of $(1-\lambda)H(X)$ for large and positive X, and of $h(((1-\lambda)\alpha/\pi^2)^{1/3}X)$ for small $((1-\lambda)\alpha/\pi^2)^{1/3}X$. As $H(X) \approx X^3/6$ for large X, we deduce $dh/dx\ (x=0) = \cot\theta_0$, $d^2h/dx^2\ (x=0) = \kappa_0/\sin^3\theta_0$, $d^3h/dx^3\ (x=0) = 3\cos\theta_0\kappa_0^2/\sin^5\theta_0 + (\pi^2/\alpha)/\sin^3\theta_0$, where the quantities characterising the shape of the meniscus at the point of connection are indexed by 0. The application of these conditions of connection leads to

$$\cot g\theta_0 \approx -C \approx 3\left(\frac{\pi^2}{\alpha}\right)^{1/3}(1-\lambda)^{2/3}$$

and therefore

$$1-\lambda \approx (U-U_c)^{3/2}$$

where U_c is the critical velocity for which α reaches its maximum value $\alpha^* \approx 13.75$ and $C = 0$.

8.4 THE CONDITION FOR SOLVABILITY

The reduction of the continuum of solutions by the effects of surface tension can be understood for small κ in an analogous way to that in paragraph 8.2. This problem has been the focus of many mathematical studies which are too long and too specific to be presented here. Different perturbation methods are used, the most rigorous of which is the asymptotic expansion of an extension to the equations of McLean and Saffman in the complex plane (Combescot, Dombre, Hakim, Pomeau and Pumir 1987), extending the method of Kruskal and Segur (1987) mentioned in paragraph 8.2. We show here a most simple method of approximation, inspired by the work of Dombre, Hakim and Pomeau (1986), Hong and Langer (1986) and Shraiman (1986), analogous to that developed in the previous paragraph.

At high velocities the effects of surface tension are quantitatively weak and we can look for a solution of equations (8.3) and (8.4) in paragraph 8.3, close to a solution without surface tension. In order to do that we write the solution in the form $q(s) = q_0(s) + \kappa q_1(s)$, $\theta(s) = \theta_0(s) + \kappa\theta_1(s)$, where q_1 and θ_1 are small in front of q_0 and θ_0, respectively, and we linearise equations (8.3) and (8.4) around the solution q_0, θ_0. We thus obtain

$$\kappa s q_0 \frac{d}{ds}\left(s q_0 \frac{d\theta_1}{ds}\right) - \theta_1 \sin\theta_0 + q_0(s)\frac{s}{\pi}P\int_0^1 \frac{\theta_1(t)}{t(t-s)}dt = -\, s q_0 \frac{d}{ds}\left(s q_0 \frac{d\theta_0}{ds}\right) \tag{8.1}$$

which we can again write in the form

$$\nu \frac{\partial^2\theta_1}{\partial l^2} + \cos\theta\theta_1 + \sin\theta\Lambda(\theta_1) = -\frac{1}{R_0}(1) \tag{8.2}$$

where $\nu = (\gamma/12\eta U)(b/a)^2$ is assumed small, and spread the domain to the two unperturbed sides of the interface. Here, θ is the angle between the normal to the unperturbed interface and the propagation direction, l the curvilinear abscissa made dimensionless with a, which goes from $-\infty$ to 0 as s goes from 0 to 1, and $1/R_0$ the unperturbed curvature of the interface. Λ is the linear operator which, as a function of $\Phi(l)$, is given by

$$\Lambda(\Phi) = \frac{1}{\pi}P\int_{-\infty}^{+\infty} dl' \frac{\Phi(l')}{l'-l} \tag{8.3}$$

since applying the residue theorem of analytic functions gives

$$\frac{s}{\pi}P\int_{-\infty}^{+\infty} dl' \frac{\Phi(l')}{l'-l}\,\frac{l'-l}{t(t-s)}\,\frac{dt}{dl'} = \frac{1}{\pi}P\int_{-\infty}^{+\infty} dl' \frac{\Phi(l')}{l'-l}\,. \tag{8.4}$$

This trasformation is defined more simply when we apply the Fourier components $\exp(ikl)$ to it. Indeed, still by application of the residue theorem of analytic functions, $\Lambda(\exp(ikl)) = i\,\mathrm{sgn}(\mathrm{Re}k)\exp(ikl)$, since according to the sign of k, the contour of integration passes through the upper or lower half-plane.

Rather than solving equation (8.2) completely, we determine the condition for solvability which reflects the existence of a non-zero solution of the equation. For that we calculate the scalar product $\langle\Psi|L\theta_1\rangle$ in two ways, where L is the linear operator $L = \nu\partial^2/\partial l^2 + \cos\theta + \sin\theta\Lambda$, and Ψ is an element of the kernel of the adjoint operator $L^+ = \nu\partial^2/\partial l^2 + \cos\theta - \sin\theta\Lambda$. It is firstly equal to

$$I = \int_{-\infty}^{+\infty} dl' G_0(l')\Psi(l') \tag{8.5}$$

and also to $\langle\theta_1|L^+\Psi\rangle = 0$ by definition of Ψ. We therefore deduce the condition for solvability

$$\int_{-\infty}^{+\infty} dl' G_0(l')\Psi(l') = 0 \tag{8.6}$$

an additional condition which must satisfy the parameter λ. We will now show that this condition can only be verified for certain values of λ, which thus solves the problem of selection.

All the difficulty lies in the determination of the Ψ function. Due to the structure of the operator L, it is analogous to determining an element Φ in the kernel of L, that is, satisfying $L(\Phi) = 0$. Ψ is obtained from Φ by changing θ to $-\theta$. We must therefore solve the linear differential equation with variable coefficients

$$\nu\frac{\partial^2\Phi}{\partial l^2} + \cos\theta\Phi + \sin\theta\Lambda(\Phi) = 0\,. \tag{8.7}$$

To simplify, let us first suppose that $\theta = 0$. So Φ is constant, or $\Phi \approx \exp(ik_m l)$ where $k_m = \pm 1/\sqrt{\nu}$ is the marginal dimensionless wavenumber of the plane front determined in paragraph 6.1. The non-trivial solution is therefore the marginal stationary sinusoid of the plane front.

Now consider θ constant, or the case where the unperturbed interface is an inclined plane front relative to the propagation direction. So Φ is constant, or $\Phi \approx \exp(ik_m(\theta)l)$, where $k_m(\theta)$ satisfies the equation

$$\mathrm{sgn}(\mathrm{Re}k_m)(\cos\theta - \nu k_m^2) + i\sin\theta = 0\,. \tag{8.8}$$

It is once again a marginal sinusoid of the inclined plane front which is moving at a normal velocity $U\cos\theta$. In fact the relation (6.8) in paragraph 6.1 is derived for the immobile perturbations with respect to the normal direction of propagation. But, here, a marginal perturbation at constant l moves with respect to the normal direction of propagation at a tangential velocity $-\sin\theta$ (figure 8.6). A marginal solution at constant l will therefore be proportional to $\exp(-i\, k \sin\theta t)\exp(ik(l+\sin\theta t))$, that is, with an equivalent growth rate $\sigma = -i\sin\theta k$ given by the aforementioned relation (8.8). The solution of equation (8.7) is simply determined by $k_m(\theta) = \pm(1/\sqrt{\nu})\exp\pm i\theta/2$.

Now consider the case where θ, given by the stationary shape of the unperturbed interface, varies. The WKB method allows the determination of approximate solutions of equation (8.6) which vary on a much faster scale than that of a stationary shape. As in paragraph 8.2, this method leads to searching for a solution in the form $\Phi(l) = \exp(iS(l)/\sqrt{\nu})$. We therefore obtain

$$\cos\theta + i\sin\theta\,\mathrm{sgn}\left(\frac{dS}{dl}\right) + \nu\left(\frac{i}{\sqrt{\nu}}\frac{d^2S}{dl^2} - \frac{1}{\nu}\left(\frac{dS}{dl}\right)^2\right) = 0\,. \tag{8.9}$$

Expanding S in powers of $\sqrt{\nu}$, $S = S_0 + \sqrt{\nu}S_1 + \ldots$, we get, to zeroth order of the expansion,

$$\left(\cos\theta - \left(\frac{dS_0}{dl}\right)^2\right) + i\sin\theta\,\mathrm{sgn}\left(\frac{dS_0}{dl}\right) = 0 \tag{8.10}$$

that is,

$$S_0(l) = \pm\int^l k_m(\theta)dl = \pm\int^l \exp\left(\pm i\frac{\theta}{2}\right)dl\,. \tag{8.11}$$

In the next order of the expansion, S_1 satisfies

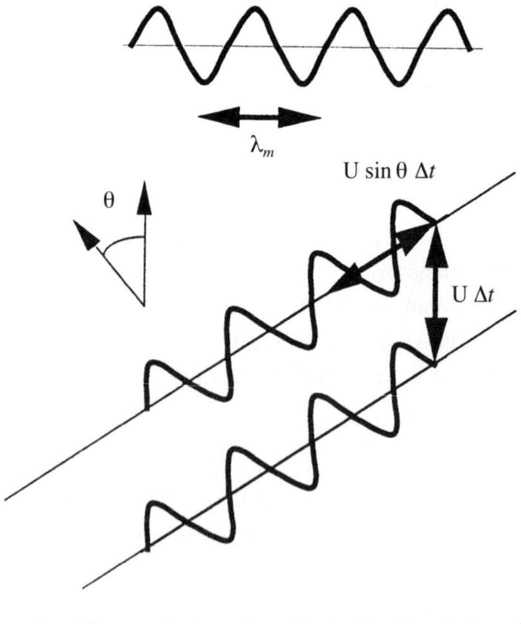

Figure 8.6 Schematic of the marginal mode on the inclined local plane front.

$$2\ \frac{dS_1}{dl}\ \frac{dS_0}{dl} = i\ \frac{d^2 S_0}{dl^2}. \tag{8.12}$$

After integration, we first deduce $S_1 \approx -(1/4)\theta(l)$, then the form of the kernel elements of the adjoint operator

$$\Phi(l) = \exp\left(-\pm\frac{i}{\sqrt{\nu}}\int_0^l \exp\left(\pm i\frac{\theta}{2}\right)dl + i\frac{\theta}{4}\right) \tag{8.13}$$

where, as mentioned above, θ used in relation (8.7) must be changed to – θ. Thus, the condition for solvability (8.5) can be written in the form

$$\int_{-\infty}^{+\infty} dl G_0(l)\exp\left(-\frac{i}{\sqrt{\nu}}\int_0^l \exp\left(i\frac{\theta}{2}\right)dl' + i\frac{\theta}{4}\right) = 0 \tag{8.14}$$

Before evaluating the integral (8.14), it is necessary to determine the relation $\theta(l)$ for the Saffman–Taylor finger. The relations (7.9) and (7.11) in paragraph 7.5, and from $dl/ds = (1-\lambda)/\pi sq$, we obtain

$$l = \frac{(1-\lambda)}{\pi}\left(\mathrm{Log}\left(\frac{1-\sin\theta}{1+\sin\theta}\right) - 2\sqrt{\alpha}\ \tan^{-1}(\sqrt{\alpha}\sin\theta)\right). \tag{8.15}$$

Using the variable $\eta = \tan\theta$ instead of l, we can rewrite the integral for solvability (8.14) in the form

$$\int_{-\infty}^{+\infty} d\eta\ F(\eta)\exp\left(\frac{\Psi(\eta)}{\sqrt{\kappa}}\right) = 0 \tag{8.16}$$

where,

$$\Psi(\eta) = 2i(1+\alpha)\int_0^\eta d\eta\ \frac{(1+i\eta)^{3/4}(1-i\eta)^{1/4}}{1+(1+\alpha)\eta^2} \tag{8.17}$$

and

$$F(\eta) = \frac{(1-\lambda)}{\pi\sqrt{1+\alpha}(1+\eta^2)}\ \frac{((1+\alpha)(1+\eta^2)-3\alpha)}{\sqrt{(1+\alpha)\eta^2+1}}\left(\frac{1+i\eta}{1-i\eta}\right)^{1/8} \tag{8.18}$$

Integral (8.16) is of the same form as the one found by Hong and Langer (1986), but differs in the expression of the prefactor $\mathrm{F}(\eta)$. The reason for this discrepancy is that the simplified presentation of the calculation of the element of the kernel of the adjoint linear operator only gives the zeroth order of the WKB expansion. The first order of the expansion modifies the curvature term with an exponential prefactor, and therefore the term given by equation (). However, the calculation of this oscillating integral essentially only takes into account the phase of the exponential given by equation (8.17), minimising the role of the prefactor when the parameter κ is small. The calculation of integral (8.16) is carried out in the same way as that of integral (8.2) in paragraph 8.2. The $d\Psi/d\eta$ derivative cancels out at the points $\eta = \pm i$. The most important contribution to integral (8.16) comes from the point

$\eta = i$. We deform the initial contour (the real axis) into a steepest descent contour which passes close to the point $\eta = i$. We must therefore distinguish between the two cases, according to whether λ is smaller or larger than $1/2$ (figure 8.7).

If λ is smaller than $1/2$, the contribution near the point $\eta = i$ simply gives a term proportional to $\exp(\Psi(i)/\sqrt{\kappa})$ where $\Psi(i)$ is the negative real number

$$\Psi(\eta) = -2(1+\alpha)\int_0^1 dx \, \frac{(1-x)^{3/4}(1+x)^{1/4}}{1-(1+\alpha)x^2}. \tag{8.19}$$

Integral (8.16) is therefore an exponentially small negative real number and cannot be cancelled. Thus, as is observed in experiments, fingers of relative width smaller than $1/2$ cannot be selected.

If λ is larger than $1/2$, the steepest descent contour must bypass the $\Psi(\eta)$ singularity at the point $\eta = i/\sqrt{(1+\alpha)}$. The part of the contour which lies along the imaginary axis introduces an oscillating factor of the form $\cos(\alpha^{3/4}/\sqrt{\kappa})$, and the integral (8.16) can be cancelled for the following discrete values of λ

$$\lambda - \frac{1}{2} \approx n^{3/4}\kappa^{2/3} \tag{8.20}$$

where n is an arbitrary integer, in good agreement with the numerical results of Vanden-Broeck (1983).

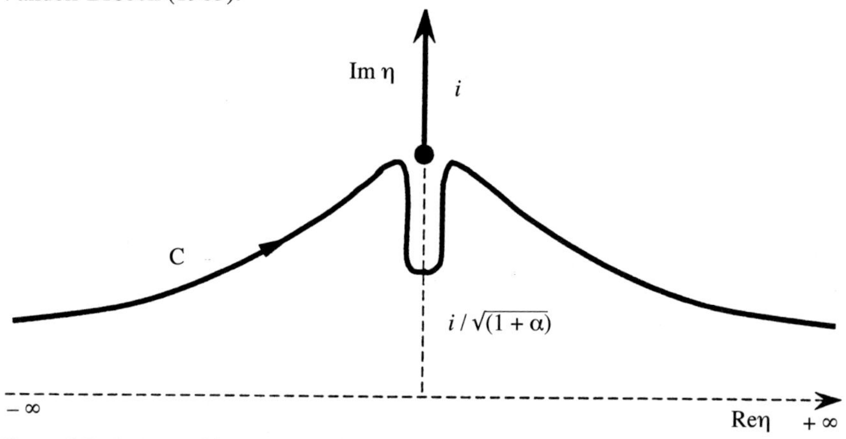

Figure 8.7 Contour of integration of equation (8.16).

We can give a relatively simple form to the condition for solvability which strays from the framework of the propagation of a Saffman–Taylor finger. Indeed, if we rewrite equation (8.6) using equation (8.11) in the form

$$\int_{-\infty}^{+\infty} dl \, \frac{1}{R_0}(l) \exp i \int_0^l k_m(l')dl' = 0 \tag{8.21}$$

the specificity of the Saffman–Taylor problem disappears, since only the exact curvature of the solution without surface tension, $1/R_0$, and the local marginal wavenumber of the dispersion relation in the plane front, $k_m(l)$, play a role. Its region of validity is at small ν, or, since $\rho = R_0(0) \approx a$ and $\lambda_m \approx 1/k_m \approx b/\sqrt{Ca}$, $\lambda_m << \rho$. Thus, relation (8.21) is applicable to other cases of curved interfaces such as crystalline dendrites for which an analogous selection problem arises. It should be noted that only the leading order in ν is retained from this relation, as the next order in ν is taken into account incorrectly. In particular, as mentioned above, the curvature term is generally not exact, since it is changed by the first order of the WKB expansion of the kernel element of the adjoint operator. However, the asymptotic calculation of this oscillating integral essentially only includes the phase of the exponential, correctly evaluated by equation (8.21).

The interest in the form (8.21) of the condition for solvability is to bring out the important physical factors which appear in this condition. The original papers of strong mathematical character, presenting these calculations for the first time, are completely unappreciated by experimentalists. In particular, it turns out that the form without surface tension as well as the stability of the planar interface play an important role in the selection of the finger and dendrite velocities. All external perturbations likely to modify them are therefore also important in this selection. This can be considered in relation to the criterion of marginal stability, initially proposed by Langer and Müller-Krumbhaar (1978). The essential difference is that in this last criterion, only the marginal wavenumber of the tip plays a role, whereas in the condition for solvability it is the integral of this wavenumber over the whole interface which is the determining factor.

The selection of the finger-widths at high velocities has a very specific character. When the condition for solvability is non-zero it is exponentially small, suggesting that the infinitesimally small perturbations of the form could lead to its accomplishment. For instance, the initial data of Saffman and Taylor (1958) were recently corrected by Tabeling, Zocchi and Libchaber (1987) who find that the relative width of a finger, instead of saturating at high velocities at $\lambda = 1/2$, passes slightly underneath that line (figure 8.8).

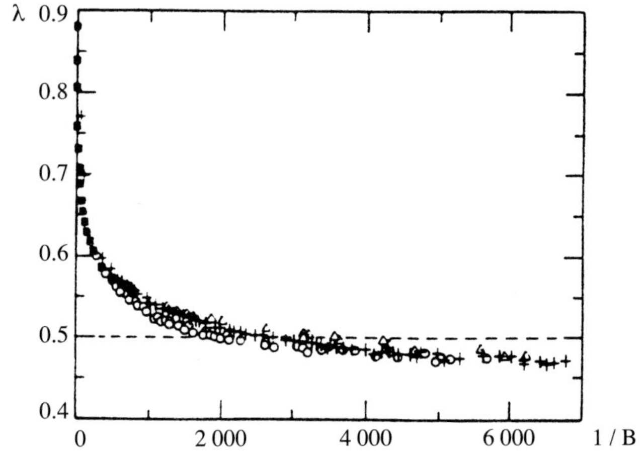

Figure 8.8 λ as a function of $1/B = 12(w/b)^2 Ca$, where w is the cell-width for three aspect ratios: $w/b = 112.5$ (triangles), $w/b = 65$ (crosses), $w/b = 34$ (circles) (Tabeling *et al.* 1987).

This difference is attributed to the effects of thickness (Reinelt and Saffman 1986) which, from appendix 3 in paragraph 5.2, modifies the jump in pressure at the interface as a function of the capillary number. These effects clearly must dominate the discrepancies of the exponentially small forms discussed above and lead to different values of selected λ.

Let us place a small air bubble at the tip of a Saffman–Taylor finger (Couder *et al.* 1986). The stationary fingers can always propagate at a constant velocity, but only with relative sizes which are smaller than $1/2$ (figure 8.9). These fingers are perfectly superimposed with the exact solutions of fingers with no surface tension, determined in paragraph 7.5.

In another experiment (Zocchi *et al.* 1987), a thread is stretched along a linear Hele–Shaw cell and a stationary Saffman-Taylor finger crossed by the thread can propagate at a constant velocity with a relative size smaller than $1/2$. These experiments illustrate the fragility of forms at high velocities, since a small-amplitude perturbation contributes to the modification of their form in a significant way. In order to explain this phenomenon more quantitatively, Hong and Langer (1987) consider that a bubble placed at the tip of a finger leads to a singularity in the given effective slope $\Delta\theta$ on the scale of the cell width. The fingers with a relative width smaller than $1/2$ have a cusp at the tip of the angle $\Delta\theta(\lambda)$. It suffices to render that angle equal to the imposed perturbation to determine the corresponding value of λ. A simplified a model allows the correct inclusion of the new dependence $\lambda(Ca)$.

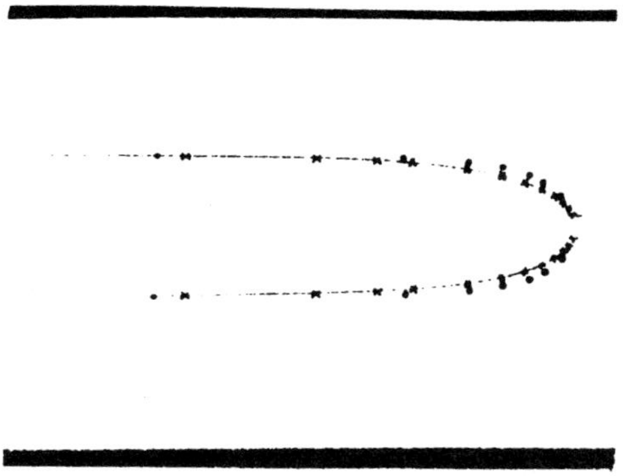

Figure 8.9 Finger with a bubble at the tip. The crosses correspond to the Saffman–Taylor solution for $\lambda = 0.32$ (Couder *et al.*, 1986).

In all cases, the agreement between theory and experiments for the forms is perfect, from the rounded forms at low velocities to the sharper forms at higher velocities. The latter are similar to those which we can determine exactly by neglecting the effects of surface tension. Thus the model for viscous fingering in a Hele–Shaw cell quantitatively describes the observed Saffman–Taylor fingers in a very precise way. For the growth forms, this model is a prototype on which we can rely in order to understand other forms generated by more complex mechanisms.

8.5 THE EQUATION OF NASH AND GLICKSMAN

As in paragraph 8.3, relative to the Saffman–Taylor finger, we can determine the integral equation for the form of a needle-like crystal. We limit ourselves to the symmetric model for which the heat diffusion coefficients are the same in the liquid and the solid phase. In fact, we can directly use the integral equation (7.21) in paragraph 7.6,

$$u(p_0) = V\int_S G(p_0|p)dxdy \tag{8.1}$$

where here the value of $u(p_0)$ depends on the curvature by the effect of surface tension. If we consider a two-dimensional crystal of squared symmetry,

$$u(p_0) = \Delta - \frac{d_{0T}}{R}(1-\beta\cos 4\theta) \tag{8.2}$$

where, from appendix Appendix: in paragraph 2.4, $\beta = 15\varepsilon/(4+3\varepsilon)$ and $d_{0T} = \gamma_0(1+(3/4)\varepsilon)c_pT_0/Q^2$, ε being the anisotropy factor.

From equation (7.17) in paragraph 7.6 and from equation (8.1), we deduce the equation of Nash and Glicksman (1974):

$$\Delta-\frac{d_{0T}}{R}(1-\beta\cos(4\theta))$$

$$=\frac{V}{2\pi D_T}\int dx\exp\left(-\frac{V}{2D_T}(\xi(x)-\xi(x_0))\right)K_0\left(\frac{V}{2D_T}\sqrt{((\xi(x)-\xi(x_0))^2+(x-x_0)^2)}\right)$$

(8.3)

where K_0 is a zeroth order modified Bessel function. It is an integral equation, containing an eigenvalue, the velocity of the crystal, to be determined as a function of supercooling Δ. When $d_{0T} = 0$, we find equation (7.22) in paragraph 7.6, the solutions of which are the Ivantsov parabolas. When d_{0T} is not 0, in general the equation must be solved numerically. The structure of equation (8.4) suggests making the lengths dimensionless by $l = 2D_T/V$. We thus obtain

$$\Delta-\frac{v}{R}(1-\beta\cos(4\theta))$$
$$=\frac{1}{\pi}\int_{-\infty}^{+\infty}dx\exp(-(\xi(x)-\xi(x_0)))K_0(\sqrt{((\xi(x)-\xi(x_0))^2+(x-x_0)^2)}\ ) \quad (8.4)$$

where $v = d_{0T}V/2D_T$, ξ and x are the dimensionless quantities here. Far from the tip, a solution of equation (8.4) must tend to an Ivantsov solution and at the tip, the slope $d\xi/dx$ must cancel out. At a given Δ, the possible parameters v must be determined, for which there are existing solutions. The first numerical solutions (Nash and Glicksman 1974) are incorrect. They confirm that the continuum solutions exist, or that at a given Δ, a solution can be found for all values of v. The numerical solutions of Kessler, Koplik and Levine (1986) and of Meiron (1986) show that, just as for a Saffman–Taylor finger (McLean and Saffman 1982), the effects of surface tension reduce the number of possible solutions, but that solutions exist only if $\beta \neq 0$. The properties of equation (8.4) are somehow analogous to those of equations (8.3) and (8.4) in paragraph 8.3. Indeed, as shown in figure 8.10, the amplitude of the singularity $\mathrm{Log}(-f(v))$ increases linearly with $1/\sqrt{v}$, at fixed $P = \rho V/2D_T$ (here $P = 0.25$), where ρ is the radius of curvature of the tip. This leads to a relation $f \approx \exp(-C(P)/\sqrt{v})$, exponentially small at small v, or at low surface tensions.

When $\beta \neq 0$, a discrete number of solutions can be found. If the problem seems to be solved from a mathematical point of view, we can ask for the physical relevance of the parameter β. It plays no role in the first instability, but it becomes essential in the determination of the crystal velocity without being able to attribute a simple physical reason to it. We will look deeper into the discussion of the role of this parameter later on, after we have studied the influence of other effects on the growth, such as impurities, flow, dimensionality of the medium and confinement.

Equation (8.4) is simplified at low Péclet number $P = \rho V/2D_T$ if we make the hypothesis that at long distances away from the tip the sought solution tends towards

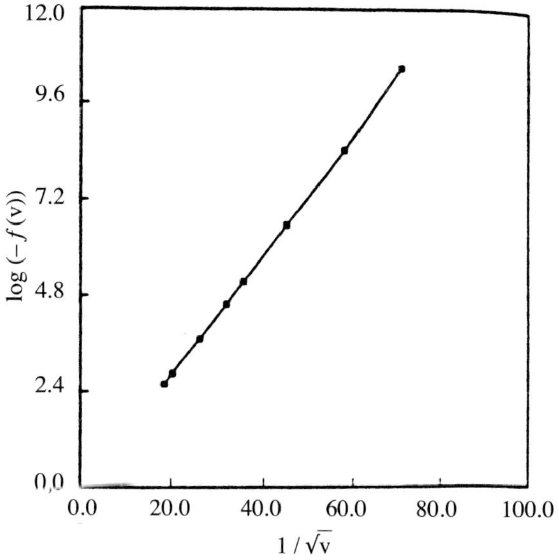

Figure 8.10 Log(−f(v)) as a function of $l/\sqrt{v}$ at $P = 0.25$ (Kessler *et al.* 1986).

an Ivantsov parabola of the radius of curvature ρ. This hypothesis is reinforced by the fact that, as the interface curvature becomes smaller and smaller as we get further away from the tip, the effects of surface tension become negligible and the interface isothermic. Deduce from equation (8.4) the same equation but at $d_{0T} = 0$, for which the solution is an Ivantsov parabola of the same radius of curvature. If we make the lengths dimensionless by ρ and we make the Péclet number tend to 0, we get the equation

$$\frac{\dfrac{\partial^2\xi}{\partial x^2}}{\left(1+\left(\dfrac{\partial\xi}{\partial x}\right)^2\right)^{3/2}}(1-\beta\cos(4\theta)) = -\frac{2C}{\pi}\int_{-\infty}^{+\infty} dx_1 \mathrm{Log}\left(\frac{(\xi(x)-\xi(x_1))^2+(x-x_1)^2}{(x^2-x_1^2)^2+(x-x_1)^2}\right). \tag{8.5}$$

where $C = \rho^2 V/2D_T d_{0T}$ (Pelcé and Pomeau 1986). The change of notation is not by chance. Let us note that C is simply $1/\sigma$, where σ is called the stability constant, a name not well adapted to the role which this parameter plays here. Thus, we observe that the form of a crystal with a given anisotropy is associated with a value of C which is independent of the supercooling. This gives an interpretation of the experimentally observed law $\rho^2 V = \mathrm{Cst}$. The numerical integration of equation (8.5) shows that it only has a solution if the surface tension is anisotropic $(\beta \neq 0)$

(Ben Amar and Moussalam 1987). The possible values of the constants form a discrete ensemble.

The kinetic effects described in paragraph 3.2 can be taken into account by adding a term proportional to the normal velocity of the interface growth to the left-hand term of equation (8.3) (Brener 1989, Brener and Mel'nikov 1991). However, in this case $\rho^2 V$ is not constant. This effect could explain the observed variations in $\rho^2 V$ with Δ in certain cases, such as for the dendrites of xenon and krypton (Livescu *et al.* 1987).

8.6 SELECTING THE CONSTANT $\rho^2 V$

Different methods of perturbation are used there as well (Caroli *et al.* 1986, Barbieri *et al.* 1987), of which the most rigorous is the asymptotic expansion of an extension to the equation of Pelcé and Pomeau (8.5) in paragraph 8.5 in the complex plane (Ben Amar et Pomeau 1986), inspired by the method of Kruskal and Segur (1987) mentioned in paragraph 8.2. The reader is referred to the reviews of Langer (1986), Brener and Mel'nikov (1991), Pomeau and Ben Amar (1992) and the book of Kassner (1996). A simple analytic determination of these values can be achieved for low anisotropy by using the condition for solvability (8.21) in paragraph 8.4 which we write again here. Indeed,

$$\int_{-\infty}^{+\infty} dl \, \frac{1}{R_0}(l) \exp i \int_0^l k_m(l') dl' = 0 \tag{8.1}$$

where $k_m(l)$ is the marginal wavenumber of the dispersion relation (6.9) in paragraph 6.5, and $1/R_0$ is the unperturbed curvature of the Ivantsov parabola. There as well, the relation is only valid if $\rho >> \lambda_m$, where ρ is the radius of curvature of the tip of the parabola, and $\lambda_m = 2\pi/(2D_T d_{0T}/V)^{1/2}$ is the marginal wavelength of the planar interface. Taking into account the effect of the angle between the normal to the interface and the propagation front in the same way as in equation (8.8) in paragraph 8.4 we deduce the equation for k_m

$$\mathrm{sgn}(\mathrm{Re} k_m)\left(\cos\theta - \frac{1}{C}(1 - \beta\cos(4\theta))k_m^2\right) + i\sin\theta = 0 \tag{8.2}$$

where the lengths are made dimensionless with the radius of curvature ρ of the parabola, and C is the constant defined above. Before evaluating the integral (8.1), it is necessary to determine the relation $\theta(l)$ for the Ivantsov parabola

$$l = -\frac{1}{2}\left(\frac{\tan\theta}{\cos\theta} + \mathrm{Log}\left(\frac{1}{\cos\theta} + \tan\theta\right)\right) \tag{8.3}$$

where l is the curvilinear abscissa, again made dimensionless with the radius of curvature of the tip of the parabola. Using the variable $\eta = \tan\theta$ instead of l, we can rewrite the integral of solvability (8.1) in the form

$$\int_{-\infty}^{+\infty} d\eta G(\eta)\exp(\sqrt{C}\Psi(\eta)) = 0 \tag{8.4}$$

where,

$$\Psi(\eta) = \frac{i}{2}\int_0^{\eta} d\eta \frac{(1+i\eta)^{7/4}(1-i\eta)^{5/4}}{\sqrt{(1+\eta^2)^2(1-\beta)+8\beta\eta^2}}. \tag{8.5}$$

The integral calculation (8.4) is carried out just like that of integral (8.16) in paragraph 8.4. The $d\Psi/d\eta$ derivative cancels out at the points $\eta = \pm i$. The most important contribution to integral (8.4) comes from the point $\eta = i$. We deform the initial contour (the real axis) into a steepest descent contour which passes close to the point $\eta = i$ and which bypasses the gap from the point $\eta = i(1-\sqrt{2\beta})$ (figure 8.11). On the part of the contour which goes along the gap Ψ is imaginary, which gives the exponentially small integral (8.4) an oscillating factor of the form $\cos(\sqrt{C}\beta^{7/4})$. This factor cancels for values of C

$$C \approx \frac{n^2}{\beta^{7/4}} \tag{8.6}$$

where n is an arbitrary integer.

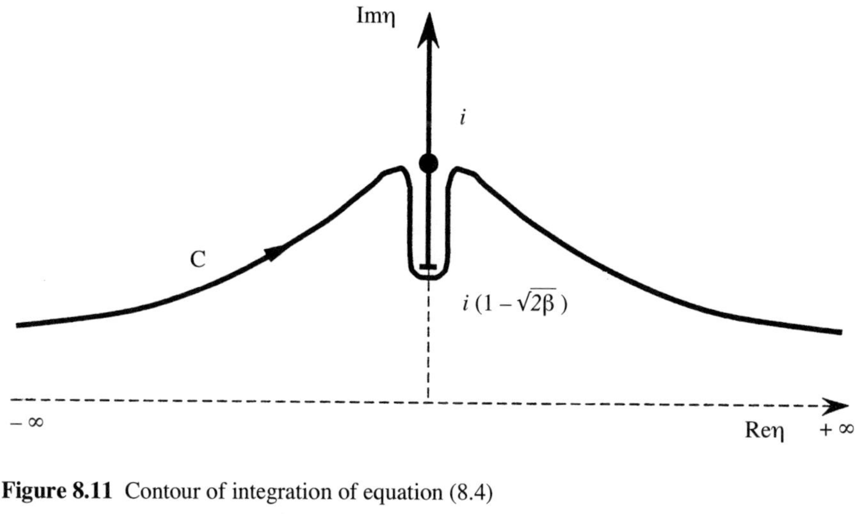

Figure 8.11 Contour of integration of equation (8.4)

Once C is determined, the velocity of the crystal is obtained by eliminating the radius of curvature ρ between the relations $C = \rho^2 V/2D_T d_{0T}$ and the Ivantsov law, $P_T = \rho V/2D_T = f(\Delta)$ determined from equation (7.7) in paragraph 7.6.

Thus, at small supercooling Δ and in two dimensions, the velocity of growth of a crystal is written as

$$V = \frac{1}{2C}\frac{D_T}{d_{0T}}\Delta^4 . \tag{8.7}$$

8.7 THREE-DIMENSIONAL DENDRITES

For simplicity, we have considered two-dimensional dendrites, which have a curved interface. Real dendrites grow in a three-dimensional medium and the liquid–solid interface is then a surface, and according to paragraph 2.4 we must use the Gibbs–Thomson relation in the form

$$T - T_0 = -\frac{T_0}{Q}\left(\frac{1}{R_1}\left(\gamma + \frac{\partial^2\gamma}{\partial\theta_1^2}\right) + \frac{1}{R_2}\left(\gamma + \frac{\partial^2\gamma}{\partial\theta_2^2}\right)\right) . \tag{8.1}$$

If we consider a cubic crystalline phase the surface tension, according to appendix 1 in paragraph 2.4, is written as

$$\gamma = \gamma_0'(1 + \varepsilon'(\sin^4\theta(1 - 2\sin^2\Phi\cos^2\Phi))) \tag{8.2}$$

where θ and Φ are the polar and azimuthal angles of the spherical coordinates respectively, and the polar axis is represented by the z axis. Taking into account the variations of the surface tension with respect to θ and Φ makes the problem far more complicated and like Barbieri and Langer (1989), we will consider an anisotropy averaged over the angle Φ, that is

$$\gamma = \overline{\gamma_0}\left(1 + \frac{\beta}{15}\cos 4\theta - \frac{\beta}{15}\sin^4\theta\right) \tag{8.3}$$

where $\beta/15 = \varepsilon/(4 + 3\varepsilon)$.

Again we make the assumption that far away from the tip the form tends to the Ivantsov paraboloid of radius of curvature ρ. This assumption is justified by the fact that, the form being axisymmetric, its curvature tends to 0 far away from the tip and the interface becomes isothermic. We generalise the condition for solvability (8.6) in paragraph 8.4 to

$$\int_{S_0} dSG_0(M)\Psi(M) = 0 \tag{8.4}$$

where G_0 is the curvature of the unperturbed paraboloid and Ψ is an element of the adjoint operator of the dynamics linearised in the proximity of the paraboloid. In the same way as in paragraph 8.4, we are searching for Ψ in the form

$$\Psi(M) \approx \exp im\Phi \exp i\int^{l} k_{//}dl . \tag{8.5}$$

A calculation of the dispersion relation similar to the one in paragraph 8.5, for the marginal wavenumber k_m with components $k_{//}$ and m/r along the polar and azimuthal directions, leads to an equation of the form

$$i\sin\theta\; k_{//} + \sqrt{k_{//}^2 + \frac{m^2}{r}}\left(\cos\theta - \frac{1}{C}((1+a(\theta))k_{//}^2) + b(\theta)k_{//}\frac{m}{r} + \left(1 + c(\theta)\frac{m^2}{r}\right)\right) = 0 \tag{8.6}$$

where a, b and c are small compared to 1. Thus, each azimuthal number m is associated with a function Ψ. The difficulty brought by the three-dimensional nature of the problem is in the multiplicity of the possible elements Ψ, each of which leads to a relation (8.4). However, there is only one parameter to be determined: the growth rate of the crystal. In the simplified case considered here $k_{//}$ depends only on θ, and the condition for solvability with $m = 0$ alone allows the determination of the constant C (after integration over Φ, the other relations are satisfied in the same way). When $m = 0$, $\theta_1 = \theta$, and the coefficient $a(\theta)$ is determined by the first term of equation (8.1)

$$1 + a(\theta) = \frac{1}{\gamma_0}\left(\gamma + \frac{\partial^2\gamma}{\partial\theta^2}\right). \tag{8.7}$$

Since the growth is assumed to take place along the $[1, 0, 0]$ axis, θ is the angle between the normal of the unperturbed Ivantsov paraboloid, and the direction of propagation. Thus, $\cos\theta = 1/(1+r^2)^{1/2}$, $\sin\theta = r/(1+r^2)^{1/2}$ and $dl/dr = (1+r^2)^{1/2}$. The condition for solvability associated to $m = 0$ is written as

$$\int_0^{2\pi} d\Phi \int_0^{+\infty} dr G(r)\exp(\sqrt{C}Z(r)) = 0 \tag{8.8}$$

where

$$Z(r, \Phi) = i\int_0^r dr\; \frac{(1-ir)^{1/4}(1+ir)^{3/4}}{\sqrt{A(r)}} \tag{8.9}$$

and

$$A(r) = 1 - \frac{4\beta}{5} + \frac{\beta}{5}\,\frac{(34r^2-1)}{(1+r^2)^2}. \tag{8.10}$$

As in the preceding cases, when C is large, integral (8.8) is singular and is dominated by the part of the contour close to the point of stationary phase $r = i$. Again the anisotropy factor introduces a branch cut on the imaginary axis from the point $r_c = i(1-(7\beta)^{1/2}/2)$ when β is small. Bypassing this cut leads to oscillatory behaviour of the form $\exp(-\beta^{7/8}\sqrt{C})\cos(\beta^{7/8}\sqrt{C})$, which leads to the relation

$$C \approx \frac{n^2}{\beta^{7/4}} \tag{8.11}$$

that is, to within numerical factors, to a law similar to the two-dimensional case.

The problem becomes more difficult if we take into account the anisotropy effects in the plane [100]. Indeed since the shape is no longer axisymmetric its curvature is no longer necessarily small far away from the tip since high curvatures can be allowed in the azimuthal direction. We cannot stick to the resolution of an equation of the same type as equation (8.5) anymore and we must go back to the initial equation

$$\Delta - \frac{d_0(\theta, \Phi)}{R} = \left(\frac{V}{2D_T}\sqrt{((\xi(x) - \xi(x_0))^2(x - x_0)^2)}\right)$$
$$\times \frac{\exp - \frac{V}{2D_T}\sqrt{(\xi(r, \Phi) - \xi(r', \Phi'))^2 + r^2 + r'^2 - 2rr'\cos(\Phi - \Phi')}}{\sqrt{(\xi(r, \Phi) - \xi(r', \Phi'))^2 + r^2 + r'^2 - 2rr'\cos(\Phi - \Phi')}}. \tag{8.12}$$

The form of the needle-like crystal is suggested by an expansion in powers of r

$$\xi(r) = -\frac{r^2}{2\rho} + \sum_m A_m r^m \cos(m\Phi) \tag{8.13}$$

of a solution of equation (8.12) linearised about an Ivantsov paraboloid. Kessler and Levine (1987) notice that, by truncating the expansion to the first non-axisymmetric term $A_4\cos(4\Phi)$, the values of A_4 and of $C = \rho^2 V/D_T d_{0T}$ can be determined, the value of C being close to the value of the axisymmetric solution. By the method of the asymptotic expansion of the extension of the integral equation into the complex plane, Ben Amar and Brener (1993) show that these two values are actually determined, the first one being a number independent of the anisotropy $A_4 \approx (1/44\rho^2)$, the second one being determined by the two-dimensional relation (8.6) in paragraph 8.6. This relation is obviously valid at distances far smaller than the radius of curvature above which the form is no longer close to the Ivantsov paraboloid. This form can be related to another form which will be discussed in paragraph 10.6.

When the constant C is determined, the growth rate of the crystal is obtained by eliminating the radius of curvature of the tip of the crystal ρ from the relation which defines the constant C using the relation of Ivantsov which is written, at small supercooling, as $\rho V/D_T = -2\Delta\log\Delta$. We thus obtain the velocity of the crystal

$$V = \frac{1}{C}\frac{D_T}{d_{0T}}\left(\frac{\Delta}{\mathrm{Log}\Delta}\right)^2. \tag{8.14}$$

8.8 EFFECTS OF IMPURITIES

Let us now consider the growth of a crystal from a binary mixture, where the concentration of the solute is c_∞. It appears that for a given oversaturation, the growth rate first increases with c_∞, reaches a maximum and then decreases at higher concentrations (Lipton *et al.* 1984).

Again, according to appendix 3 in paragraph 7.6, when we neglect the effects of surface tension, we have a set of parabolas of arbitrary radius growing at constant velocity. To select the growth rate we use the extra relation (Ben Amar and Pelcé 1989)

$$\int_{-\infty}^{+\infty} dl \; \frac{1}{R_0}(l) \exp i \int_0^l k_m(l')dl' = 0 . \tag{8.1}$$

As in paragraph 8.6, $k_m(l)$ is the marginal local wavenumber of a steady-state perturbation of the Ivantsov parabola in the moving frame where the parabola is at rest. Since the wavelength of the perturbation is assumed to be small compared to the radius of curvature of the parabola we can assume that the unperturbed interface is locally planar and that there is an angle θ between its normal and the direction of propagation. The growth rate of the steady-state perturbation in the local frame is deduced from the growth rate of the planar interface moving in the direction at a rate $\sigma(k) + iV\sin\theta k$, because of the relation between the tangential velocity $V\sin\theta$ in the local moving frame and the fluid at rest. According to the appendix mentioned above a marginal mode k_m therefore satisfies

$$\begin{aligned} &\operatorname{sgn}(\mathrm{Re}k_m)\left(V\cos\theta\left(1 + \frac{2mc_p}{Q}c_i(1-K)\frac{D_T}{D_c}\right)\right) \\ &- 2D_T d_{0T}(1 - \beta\cos(4\theta)k_m^2) - iV\sin\theta k_m\left(1 + \frac{2mc_p}{Q}c_i(1-K)\frac{D_T}{D_c}\right) = 0 \end{aligned} \tag{8.2}$$

so that
$$k_m = k_m^0\left(1 + \frac{2mc_pc_i(1-K)}{Q}\frac{D_T}{D_c}\right)^{1/2} \tag{8.3}$$

where k_m^0 is the marginal wavenumber associated with the growth in the pure liquid determined by equation (8.2) in paragraph 8.6. Thus integral (8.1) is the same as integral (8.4) in the same paragraph if we redefine the constant C by

$$C = \frac{\rho^2 V}{2D_T d_{0T}}\left(1 - \frac{2mc_\infty(1-K)c_p}{Q}\frac{D_T}{D_c}\right) . \tag{8.4}$$

Indeed, the possible values of C are determined by equation (8.6) of the same paragraph. At high concentration, by using the definition of the chemical capillary constant d_{0c} of paragraph 5.4 we have $C = \rho^2 V/D_c d_{0c}$. We thus find the definition of the pure thermal case again, to within a factor of 2, which comes from the fact that the impurities diffuse only trough the liquid (Misbah 1987). After eliminating the radius of curvature ρ from equation (8.4) using the equation in appendix 3 of paragraph 7.6 which determines the Péclet number, we obtain the velocity of the crystal as a function of the concentration of impurities

$$V = V_0 \frac{\left(1 - \frac{2mc_\infty(1-K)c_p}{Q}\frac{D_T}{D_c}\right)^2}{\left(1 - \frac{mc_\infty(1-K)c_p}{Q}\left(\frac{D_T}{D_c}\right)^{1/2}\right)^4} \qquad (8.5)$$

which is in qualitative agreement with the experimental observations.

Appendix: Growth rate of a perturbation of the planar interface when the effects of latent heat and impurities are taken into account simultaneously.

In order to determine the marginal wavenumber $k_m(l)$ we determine the dispersion relation of the stability of the planar interface where the dynamics of the temperature and concentration fields are taken into account. Let us consider a planar interface moving at a constant velocity V in the z direction. The steady-state temperature and concentration profiles are, respectively,

$$\bar{T}_l = T_\infty + \frac{Q}{c_p}\exp\left(-z\frac{V}{D_T}\right); \quad \bar{T}_s = T_\infty + \frac{Q}{c_p}$$

$$c = c_\infty + c_i(1-K)\exp\left(-z\frac{V}{D_c}\right).$$

We consider perturbations of the form $\xi(x) = A\exp(\sigma t + ikx)$ of the interface and

$$T_l' = B\exp(\sigma t + qz + ikx)$$

$$T_s' = C\exp(\sigma t + q'z + ikx)$$

$$c' = D\exp(\sigma t + q''z + ikx)$$

for the temperature and the concentration, σ and k being the growth rate and the wavenumber of the perturbation, respectively. Here q and q' are the square roots of the negative real part and the positive real part of the second degree equation $\sigma - Vq = D_T(q^2 - k^2)$, respectively. By applying the boundary conditions at the interface determined by equations (2.11) in paragraph 2.4 and (5.9) in paragraph 5.3 we obtain the relations

$$\sigma = -Vq + V(q'-q)\frac{mc_p}{Q}c_i(1-K)\frac{D_T}{D_c} - d_{0T}(1-\beta\cos(4\theta))D_T k^2(q-q')$$

$$-\frac{V^2}{D_T} + D_T\frac{mc_p}{Q}(q-q')h$$

$$\sigma = -h\left(\frac{V}{c_i} + \frac{D_c q''}{c_i(1-K)}\right) - K\frac{V^2}{D_c}$$

where h is the coupling parameter $h = D/A$. The marginal wavenumber k_m is obtained by eliminating h in the two preceding equations and by setting: $\sigma = 0$

$$\sigma = -Vq + V(q'-q)\frac{mc_p}{Q}c_i(1-K)\frac{D_T}{D_c} - d_{0T}(1-\beta\cos(4\theta))D_Tk^2(q-q')$$

$$-\frac{V^2}{D_T} + D_T\frac{mc_p}{Q}(q-q')h$$

$$\sigma = -h\left(\frac{V}{c_i} + \frac{D_cq''}{c_i(1-K)}\right) - K\frac{V^2}{D_c}.$$

Since, in the experimental conditions, $k_m >> V/D_c$, $q' \approx -q \approx -q'' \approx k\,\mathrm{sgn}\,\mathrm{Re}(k)$ and the growth rate of the perturbations is obtained by eliminating h from the two equations

$$\sigma = k\,\mathrm{sgn}(\mathrm{Re}k)\left(V\left(1 + \frac{2mc_p}{Q}c_i(1-K)\frac{D_T}{D_c}\right) - 2D_Td_{0T}(1-\beta\cos(4\theta))k^2\right).$$

8.9 SOLIDIFICATION IN A CAPILLARY

Usually crystalline microstructures, such as cells or dendrites, grow to form a lattice. The growth rate of each structure is therefore modified by the surrounding structures. As in paragraph 7.7 the simplest configuration associated with this confinement effect is a crystal growing along the axis of a capillary (Honjo and Sawada 1985). Then, the problem consists of determining the growth rate of the crystal as a function of the oversaturation of the liquid and the radius of the capillary $\Lambda/2$. Furthermore, let us assume that the velocity of the crystal is constant and equal to V and let us change Λ. At a very large radius the crystal is no longer influenced by the walls of the capillary and we must find the free growth of dendrites. At small radius the Péclet number $P = \Lambda V/D_c$ is small and we again find the regime already studied in paragraph 7.7 where the shape of the crystal must be close to a Saffman–Taylor finger. In the region of the tip let us introduce the effective concentration $\Phi = c - c_0$ again. Over that scale the advective term $-V\partial\Phi/\partial z$ from equation (7.1) in paragraph 7.7 is negligible, such that

$$\Delta\Phi = 0 \tag{8.1}$$

with the boundary conditions at the interface

$$\vec{\nabla}\Phi \cdot \vec{n} = -\frac{V}{D_c}\frac{\Lambda}{2}c_0(1-K)\cos\theta \tag{8.2a}$$

$$\Phi = (1-\beta\cos(4\theta))\frac{2\gamma T_0}{mQR\Lambda}. \tag{8.2b}$$

Moreover, $-D_c d\Phi/dz = (c_0 - c_\infty)V\Lambda/2$ at $z \approx +\infty$, and $\partial\Phi/\partial y = 0$ at $y = \pm 1$, where the lengths are made dimensionless with the half-width of the capillary. We find again the problem of the steady-state Saffman–Taylor finger solved in paragraph 8.3, with crystalline anisotropy effects in addition (Dorsey and Martin

1987). In appendix 1 we determine the corresponding condition for solvability. We thus obtain a set of curves $\lambda_n(\kappa, \beta)$, where

$$\kappa = \frac{4D_c d_{0c}}{\Lambda^2 V} \frac{\pi}{(1-\lambda)^2} \tag{8.3}$$

is analogous to the parameter defined for the Saffman–Taylor finger. When $\beta = 0$, we come back to the curves of McLean, Saffman and Vanden–Broeck, which has several consequences: the first one is that the relative width of a crystal cannot be smaller than $1/2$. However, according to paragraph 7.7, $\lambda = \Delta$. If there is no crystalline anisotropy the solutions at small Péclet number can only exist at oversaturation larger than $1/2$. The second is that λ is an increasing function of κ, therefore a decreasing function of the velocity and of the size Λ of the capillary. We obtain the unexpected result that, at a given Λ, the velocity of the crystal must decrease when the oversaturation increases and that, for a given oversaturation, the velocity decreases as Λ increases. We will see in paragraph 9.9 that this property is associated with an instability of the steady-state growth in the same conditions. In the presence of anisotropy, $\beta \neq 0$, λ still increases with κ, but tends to 0 for large κ. More precisely, when λ tends to 0,

$$\lambda^2 \approx n \frac{\kappa^{1/2}}{\beta^{7/8}}. \tag{8.4}$$

But, for a Saffman–Taylor finger at small λ, $\rho \approx a\lambda^2$, where ρ is the radius of curvature of the tip, returning to the law of dendrites $\rho^2 V/D_c d_{0c} \approx n^2/\beta^{7/4}$. Nevertheless the considered form is not a free dendrite. Indeed the relation $\rho/a \approx \lambda^2 = \Delta^2$ is different from the law of Ivantsov $\rho V/D_c \approx \Delta^2$. The branch of solutions obtained here, which is not feasible from a physical point of view since it corresponds to crystals with a decreasing velocity when the oversaturation increases, does not therefore describe the free dendrite as a limiting case. It comes from the fact that the Péclet number calculated from the capillary radius is assumed to be small here, whereas for the free dendrite it does not satisfy this assumption.

No exact solution for dendritic form without surface tension at large channel width is known as yet. We write an approximation of this form without surface tension as

$$\xi(x) = \frac{1}{s} \mathrm{Log} \cos \frac{\pi x}{\Delta} \tag{8.5}$$

which is a Saffman–Taylor-type structure when $s = \Delta$, and we calculate the corresponding condition for solvability in appendix 2 (Brener et al. 1988). We then find the free dendrite branch by making Δ small and P large, that is

$$s \approx \frac{\pi^2}{P}; \ \lambda^2 s^2 \approx n \frac{\kappa^{1/2}}{\beta^{7/8}} \tag{8.6}$$

that is
$$\frac{\rho V}{D_c} \approx \Delta^2; \; \frac{\rho^2 V}{D_c d_{0c}} \approx \frac{n^2}{\beta^{7/4}}. \tag{8.7}$$

For a given Δ the growth rate of a crystal increases with the diameter of the capillary Λ at the dendrite branch. As shown in figure 8.12, the two branches are again connected at a minimum diameter of the channel depending on the oversaturation. Besides, at a given Λ, the velocity increases with Δ at the dendrite branch and decreases with Δ at the Saffman–Taylor branch (Brener *et al.* 1988).

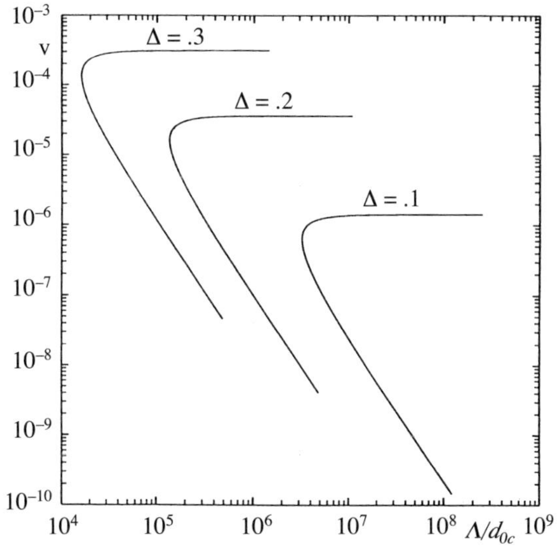

Figure 8.12 $v = Vd_{0c}/2D_c$ as a function of Λ/d_{0c} at different values of Δ (Brener *et al.* 1988).

The origin of the reconnection of the branches can be seen in the first order analysis of the Péclet number for the problem of the crystal in the capillary (Ben Amar and Brener 1995). At this order the diffusion thickness deforms the Saffman–Taylor tip, as a small gap between the walls of a Hele–Shaw cell would (i.e. an angular section with a small angle). The reconnection of the Saffman–Taylor branches and dendrite of the capillary would therefore be analogous to the reconnection of the first two branches of solutions of the angular section.

In their experiments, Becchoeffer (1988) and Molho (1990) and their coworkers use 6 cm-long capillary of rectangular section $200/20\ \mu\text{m}^2$. Two materials of different anisotropy factors are used, succinonitrile and pivalic acid. The temperature is regulated to within a few mK around the liquid–solid equilibrium temperature. After nucleation, the crystal grows very slowly at first, at a velocity of a few $\mu\text{m/s}$, then the growth accelerates up to higher velocities. During the slow phase which lasts several minutes the crystal occupies a substantial part of the capillary and, as shown in figure 8.13, its shape is very similar to a Saffman–Taylor finger.

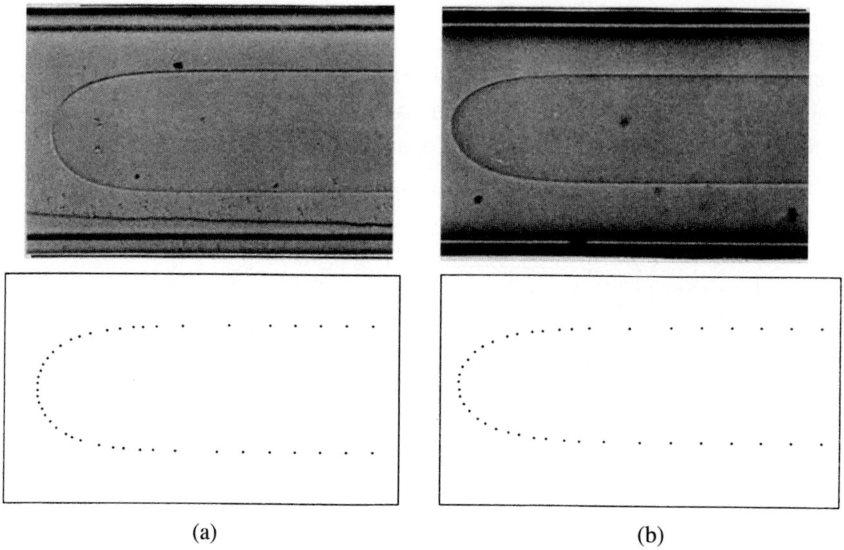

Figure 8.13 Shape of the crystal at low velocity: (a) succinonitrile, (b) pivalic acid (Molho *et al.* 1990).

As shown in figure 8.14, the relative half-width of the crystal decreases down to values lower than $1/2$ with the velocity, and the higher the crystalline anisotropy the lower the values, in qualitative agreement with equation (8.3).

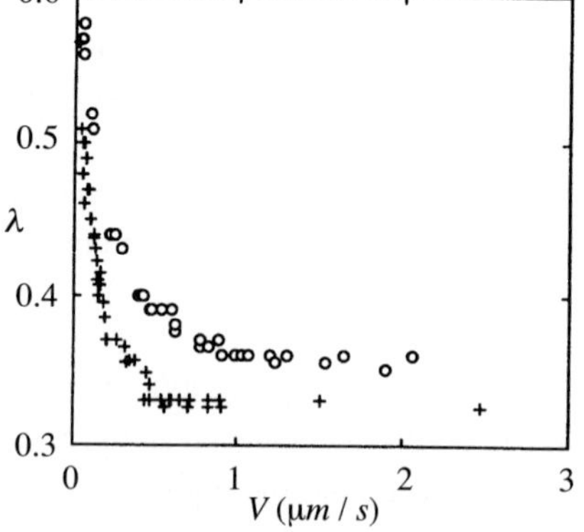

Figure 8.14 Relative width of the crystal as a function of the velocity (Molho *et al.* 1990).

At higher velocities, as shown in figure 8.15, the crystal looks really like a free dendrite. The crystal can therefore get both limiting growth forms discussed in the paragraph. Their respective stability will be discussed in the next chapter.

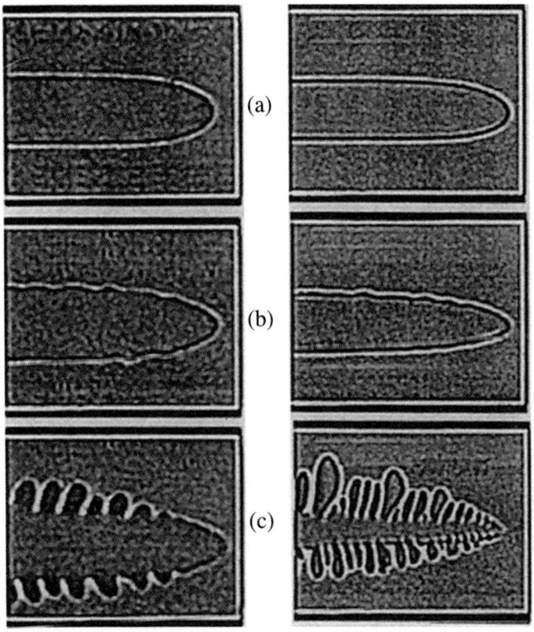

Figure 8.15 From the Saffman–Taylor form to the dendritic from. (a) $V = 14.9$ μm/s, P = 3, (b) $V = 39.4$ μm/s, $P = 7.9$, (c) $V = 72$ μm/s, $P = 14.4$ for succinonitrile, (a') $V = 1.5$ μm/s, $P = 0.3$, (b') $V = 9$ μm/s, $P = 1.8$, (c') $V = 38.8$ μm/s, $P = 7.8$ for pivalic acid (Molho *et al.* 1990).

Appendix 1: Integral for solvability at small Péclet number.

For this last case we still apply the condition for solvability (8.2) from paragraph 8.6 where the forms without surface tension are Saffman–Taylor fingers. The marginal wavenumber satisfies the equation

$$\operatorname{sgn}(\operatorname{Re}k_m)\left(\cos\theta - \frac{1}{C}(1-\beta\cos(4\theta))k_m^2\right) + i\sin\theta = 0$$

where $C = \Lambda^2 V/4D_c d_{0c}$ and where the lengths are made dimensionless with $\Lambda/2$.

The relation $\theta(l)$ is the same as for the Saffman–Taylor finger, that is

$$l = \frac{(1-\lambda)}{\pi}\left(\operatorname{Log}\left(\frac{1-\sin\theta}{1+\sin\theta}\right) - 2\sqrt{\alpha}\tan^{-1}(\sqrt{\alpha}\sin\theta)\right)$$

where $\alpha = (2\lambda - 1)/(1-\lambda)^2$. We deduce the expression for the condition for solvability

$$\int_{-\infty}^{+\infty} F(\eta)\exp\left(\frac{\Psi(\eta)}{\sqrt{\kappa}}\right)d\eta = 0$$

where,

$$\Psi(\eta) = 2i(1+\alpha)\int_0^{\eta} d\eta \frac{(1+i\eta)^{7/4}(1-i\eta)^{5/4}}{(1+(1+\alpha)\eta^2)\sqrt{(1+\eta^2)^2(1+\beta)+8\beta\eta^2}}$$

and $\kappa = (4D_c d_{0c}/\Lambda^2 V)(\pi/(1-\lambda))^2$ is the analogue of the control parameter of McLean and Saffman.

Appendix 2: Branches of solutions.

No exact solution for needle-like crystals without surface tension in a capillary for an arbitrary Péclet number is known as yet. However, even if it is not justified from a mathematical point of view, the following procedure allows us to describe qualitatively all the branches of solutions. We start from a form without surface tension

$$\xi(x) = \frac{1}{s}\text{Log}\cos\frac{\pi x}{\Delta}$$

which gives again the Saffman–Taylor form when $s = \pi/(1-\Delta)$. We determine s by identifying the forms at large distances from the tip given by equation (8.5), and the form calculated from the equations of crystallisation, by searching for the concentration field in the form $c(x, y) = c_0 - A(x)\exp(sy)$. We thus determine

$$s = \left(P^2 + \frac{\pi^2}{(1-\Delta)^2}\right)^{1/2} - P$$

where $P = \Lambda V/D$ which gives, $s = \pi/(1-\Delta)$ when P is small. To calculate the condition for solvability we need the relation $\theta(l)$ which we determine from equation (8.5), that is

$$l = \frac{1}{s}\left(\text{Log}\left(\frac{1-\sin\theta}{1+\sin\theta}\right) - 2\sqrt{\frac{\lambda^2 s^2}{\pi^2} - 1}\tan^{-1}\left(\sqrt{\frac{\lambda^2 s^2}{\pi^2} - 1}\sin\theta\right)\right).$$

We obtain a condition similar to the condition in appendix 1, except that

$$\Psi(\eta) = 2i\frac{\lambda^2 s^2}{\pi^2}\int_0^{\eta} d\eta \frac{(1+i\eta)^{7/4}(1-i\eta)^{5/4}}{\left(1+\frac{\lambda^2 s^2}{\pi^2}\eta^2\right)\sqrt{(1+\eta^2)^2(1-\beta)+8\beta\eta^2}}$$

and $\kappa = (D_c d_{0c}/\Lambda^2 V)s^2$.

8.10 EFFECT OF A TEMPERATURE GRADIENT

When a temperature gradient is applied we find the configuration of directional growth from paragraphs 5.4 and 6.6. When the growth rate exceeds a critical order of magnitude of the instability threshold of the planar interface a lattice of cells of high amplitude and of well-defined wavelength can be observed, as shown in figure 8.16 for instance (Georgelin and Pocheau 1998).

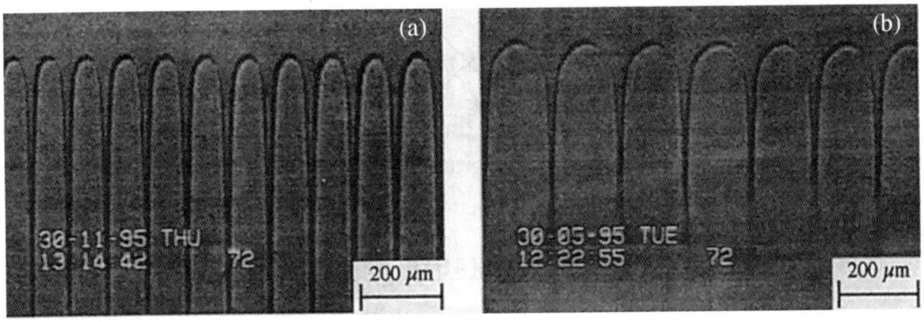

Figure 8.16 Lattice of succinonitrile cells, (a) $G = 140$ K / cm, V = 12 μm / s, $\Lambda = 100$ μm, (b) $G = 36$ K / cm, $V = 8$ μm / s, $\Lambda = 150$ μm (Georgelin and Pocheau 1998).

The position of the tip of the cells in the temperature gradient z_0 can be evaluated approximately (Burden and Hunt 1974). The concentration field around a cell can be written as the sum of a longitudinal component and a radial one: $c = c_l + c_r$. Behind the tip the cell almost goes along the direction of growth. The concentration field is therefore almost longitudinal, with the gradient $dc_l/dz \approx G/m$, since the effects of surface tension are negligible. On average, this gradient must be equal to the gradient upstream from the cell balancing advection and diffusion: $dc_l/dz \approx (c_\infty - c_l(z_0))V/D_c$. We evaluate the order of magnitude of the radial concentration by using the condition of flux given out $-D_c \partial c/\partial n = c(1-K)V$, which leads to $c_r(z_0) \approx -c_\infty(1-K)RV/D_c$, where R is the radius of curvature of the tip. We thus evaluate $c(z_0) - c_\infty \approx -GD_c/mV - c_\infty(1-K)RV/D_c$, and therefore according to the Gibbs–Thomson condition,

$$\Delta T = T_0 - T_i = \frac{GD}{V} - \frac{mVR}{D}(1-K)c_\infty + \frac{\gamma T_0}{QR} \tag{8.1}$$

or even by defining the dimensionless oversaturation of the tip as $\Delta = K(c_i - c_\infty)/(c_\infty(1-K)) = 1 - z_0/l_T$,

$$\Delta = \frac{D}{Vl_T} - \frac{K}{1-K} + \frac{KVR}{D} - \frac{d_{0c}}{R}. \tag{8.2}$$

More precisely the problem is indeed close to the case of a crystal growing in a capillary apart from a few differences. In the latter configuration the size of the cap-

8.11 DISCUSSION

In the case of dendrites the agreement between theory and experiment is not as good as in the case of the viscous fingering in a Hele–Shaw cell. More precisely, as shown in figure 8.22 with logarithmic coordinates, the orders of magnitude are in good agreement. For a large number of materials, in a wide range of supercooling and oversaturation, the product $\rho^2 V$ is constant (Glicksman *et al.* 1976, Huang and Glicksman 1981, Muschol *et al.* 1992). Curiously it is the same for the dendrites of liquid crystals (Oswald 1988), states of matter which exhibit specific features (De Gennes 1974, De Gennes and Prost 1992). On the other hand, as shown in figure 8.22, the expected dependence of the factors $\rho^2 V/2D_T d_{0T}$ or $\rho^2 V/2D_T d_{0c}$ with respect to the factor of anisotropy is not met.

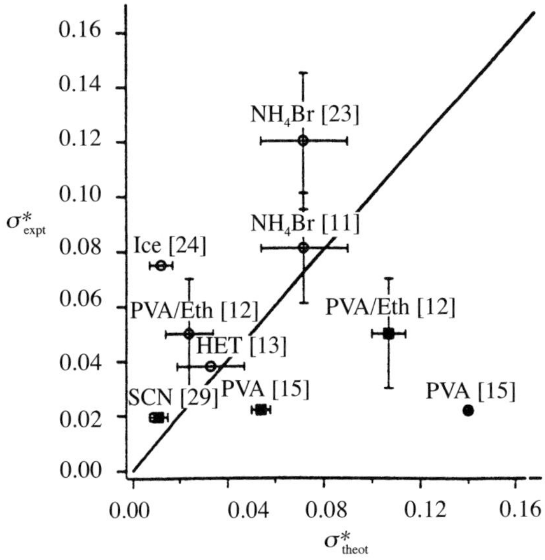

Figure 8.22 Experimental values of σ for different materials. The straight line indicates the ideal fit with the theory of the factor of anisotropy (Muschol, Liu and Cummins 1992).

Muschol and coworkers (1992) report the measured values of the quantity $\sigma = 2d_0 D/\rho^2 V$ in this figure for materials of different crystalline anisotropy, where ρ is the radius of curvature of the tip of the dendrite. Note some discrepancy in the results between different materials and for the same material between different experiments as well. The domain of values of σ thus spreads from 0.02 to 0.12 for the whole set of considered materials. The measurements of velocity are performed for the growth from a pure liquid and for the growth in solution as well. The straight line indicates the locus of the ideal agreement with the theory of solvability presented in the preceding paragraphs.

Both experimental and theoretical reasons can account for this disagreement. The experiments indicate that different factors perturb the proposed ideal theoretical configuration. For instance Glicksman (1994) and coworkers show with experiments in microgravity how the dependence of the velocity of a succinonitrile dendrite is modified by the gravity field with respect to the supercooling. In figure 8.23, like Raz and coworkers (1989), Emsellem and Tabeling (1994) visualise the field of concentration of solute around a dendrite of NH_4Br by an interferometric method. The lines of the same concentration of the fields (a) and (b) seem more perturbed than those obtained during a growth inside a gel in field (c), where the convective effects are inhibited. By comparison the field (d) is the theoretical field of the Ivantsov paraboloid of the same radius of curvature. Moreover if the anomalous light scattering described in paragraph 3.7 is due to the presence of a crystalline mesophase going along the dendrite, it could be the origin of effects not yet taken into account in this model.

Another main reason accounting for the dispersion of the results is the fragility of the solutions obtained at high velocity for the Saffman–Taylor finger, or at a low factor of anisotropy for the dendrites. The theory asserts that in these two limiting cases the selection occurs to within exponentially small terms, far smaller in order of magnitude than any perturbation caused by phenomena not taken into account in the model. For instance the relative size of a Saffman–Taylor finger does not tend to $1/2$ at high velocity but reaches lower values, undoubtedly determined by other causes such as thickness effects. In a similar way the factors of anisotropy of the materials considered here are small and lead to large values of C (observed experimentally) and therefore to exponentially small differences in the growth forms. Perturbations, either steady-state ones caused by parasite flows, or dynamic ones such as those caused by the phenomenon of encapsulation of a dissolved gas discussed in paragraph 3.7, must clearly overcome these small differences. Contrary to the Saffman–Taylor finger, which is a fragile growth form only at high velocity, the needle-like crystal is always a fragile form for unclear physical reasons explained by the factor of anisotropy. Nevertheless, as discussed in chapter 10, the numerical simulation of Karma and Rappel (1996) shows that thermodynamic noise alone is not sufficient to modify this fragility with a factor of anisotropy as low as the factor for succinonitrile. This behaviour is all the more particular since this factor is the only parameter which leads to a selection of steady-state needle-like crystals. As mentioned in paragraphs 8.7 and 8.8, the presence of impurities or no axisymmetry do not change anything in the problem. We must pay careful attention to the influence of the kinetic effects described in paragraph 3.2, since they could explain the variation of $\rho^2 V$ with Δ observed for certain materials.

As for the growth forms, let us consider the tips of a succinonitrile dendrite and a xenon dendrite (figure 10.14 in paragraph 10.5). The two dendrites have close val-

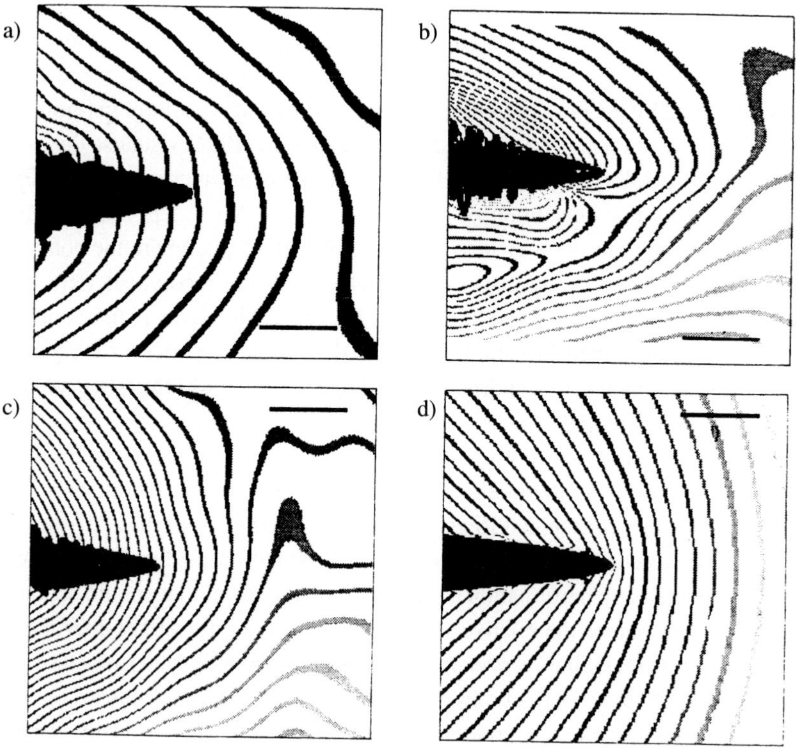

Figure 8.23 Isoconcentration lines for dendrites at $V = 8.4\ \mu\text{m} / \text{s}$. (a) and (b): experiments in solution, (c) experiment in a gel, (d) theoretical diffusion field (Emsellem and Tabeling 1994).

ues of C yet different shapes. The first one is close to a paraboloid. The associated value of C is large and is in good agreement with the axisymmetric solutions in paragraph 8.7. The second one is close to the four-arm shape predicted by Brener (1995). It does not depend on C but depends on the amplitude of a modulation of the shape of the fourfold-symmetry tip in the [100] plane, determined by a condition for solvability of the equations in the region of the tip discussed in paragraph 8.7. Other experiments are required to clarify this alternative approach.

The qualitative importance of the factor of crystalline anisotropy is undisputable for the dendritic morphology. In fact even if the value of C is fixed by other factors (arising from an unsteady source), equation (8.14) in paragraph 8.7 indicates that the velocity of the dendrite is inversely proportional to the capillary length, therefore at a maximum at the principal axis of symmetry of the crystal. These structures have therefore a tendency to grow along these axis since other structures growing in other directions would be inhibited by the screening effect caused by the faster structures.

Nevertheless its quantitative importance is maybe not as clear, as previously discussed.

8.12 CURVED FLAMES

A curved flame can be obtained in a vertical 5 cm diameter tube where the ignition of the fresh gases occurs at the bottom of the tube. In usual conditions the flame moves upwards and consumes the reactive gas in a relatively short time of a few seconds. To analyse the shape of the flame and its surrounding flow it is best to stabilise it in a tube by the method of laser tomography already exposed in paragraph 6.7 (Boyer 1980), where the fresh gases are injected into the upper part of the tube at the same velocity as the velocity of the propagation of the flame. In the same paragraph we consider a propane–air flame diluted with nitrogen to adjust the normal velocity of propagation. The velocity field of the fluid is measured by laser anemometry and the shapes of the flames by photography (Pelcé-Savornin *et al.* 1988). In figure 8.24 we report two shapes of the flame obtained in conditions of a very dilute reactive gas, in which the first one is about to go out whereas the second one goes out at the tip. The luminous head is the reaction zone where the reaction heat is produced. Behind, the flame goes out because of the proximity of the reaction front to the walls of the tube kept at room temperature.

a) b)

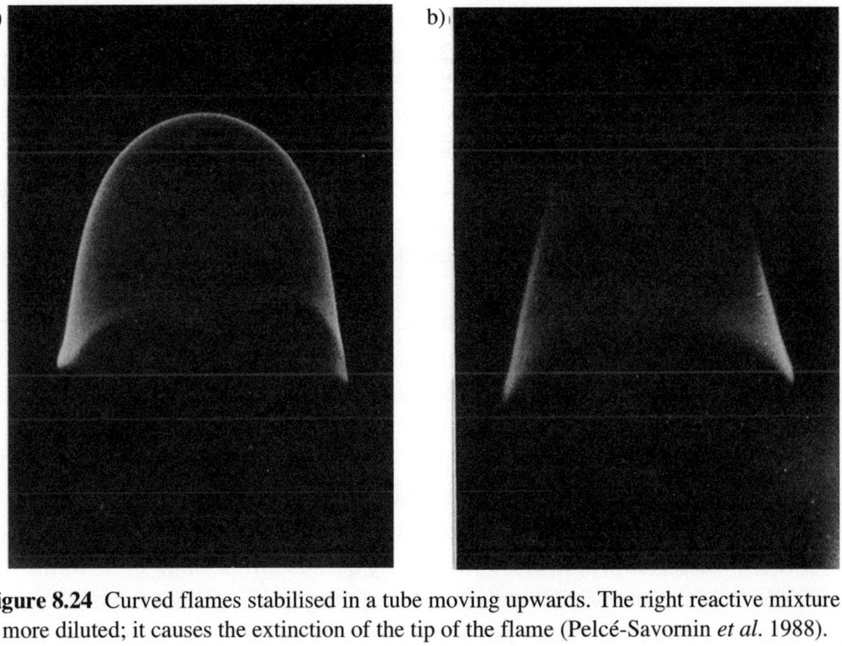

Figure 8.24 Curved flames stabilised in a tube moving upwards. The right reactive mixture is more diluted; it causes the extinction of the tip of the flame (Pelcé-Savornin *et al.* 1988).

In figure 8.25 we report the normal velocity of the flame measured at different points of the front from the centre to the walls. It increases by a factor of two, then suddenly decreases because of the proximity of the reaction front to the walls of the tube. The normal velocity is obviously not constant as assumed in paragraph 7.8, and the theory developed in the same paragraph can only be approximative.

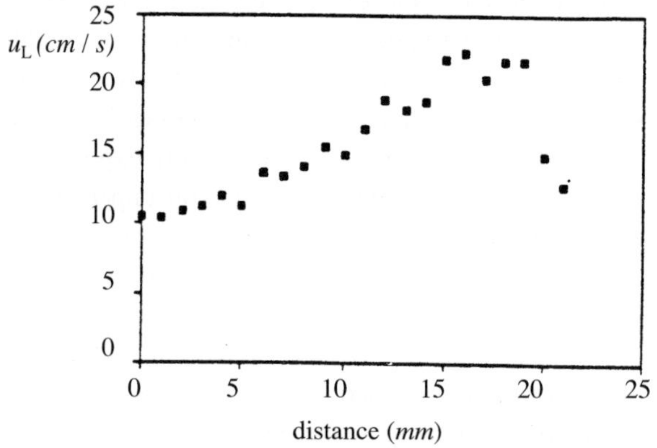

Figure 8.25 Normal velocity of the flame as a function of the radial distance in the tube (Pelcé-Savornin *et al.* 1988).

To interpret this change in normal velocity it is useful to show the data of the longitudinal velocity of the flow of fresh gases as a function of the distance from the tip (figure 8.26). The latter (black squares) is decomposed into a decreasing component near the tip of the flame (+), and a component which is non-zero only at the proximity of the flame and is associated to the effects of gas expansion. The first component of the velocity field of the fresh gases is therefore in good agreement with the one calculated in paragraph 7.8, since the flame is assumed to be infinitely thin in the model.

The decreasing field of axial velocity is the sign of a flow diverging from the tip to the walls. The convex flame is therefore in a stagnant flow, which, as we have analysed in paragraph 5.6, makes the effects of curvature stronger on the normal velocity of the flame propagation. More precisely the velocity gradient measured at the tip of the flame is of the order of 0.3 s^{-1} and the curvature of the front is of the order of 0.8 cm^{-1}. The change in the measured normal velocity can therefore be attributed to the effects of gradient and of curvature by equation (5.13) in paragraph 5.6, if the Markstein length L is of the order of 0.05 cm, a value close to the one obtained from the analysis of linear stability of the plane front. When the dilution of the reactive medium increases the normal velocity of the flame decreases and the effects of the flow gradient increase with respect to advection. The calculation in

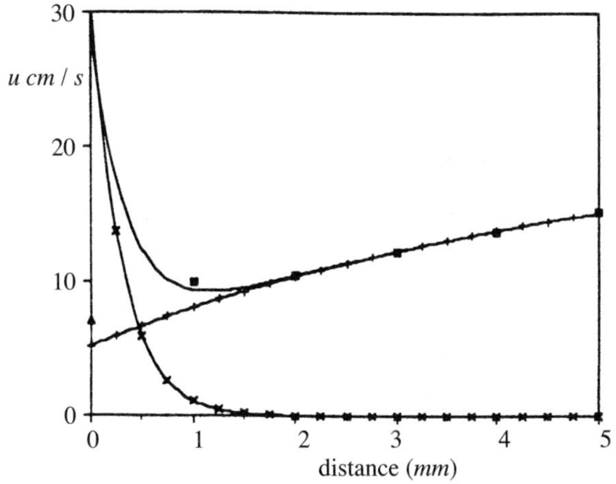

Figure 8.26 Longitudinal component of the velocity field as a function of the distance from the front of the flame (squares), decomposed into expansion effects in the thickness of the flame (crosses) and the deviation of the stream lines in paragraph 7.8 (+) (Pelcé-Savornin *et al.* 1988).

paragraph 5.6 then predicts a possible extinction of the flame, which might explain the partial extinction of the flame at the tip observed in figure 8.24.

BIBLIOGRAPHY

[1] A. Barbieri, D.C. Hong, J.S. Langer, 'Velocity Selection in the Symmetric Model of Dendritic Crystal Growth', *Phys. Rev.*, **A35**, p.1802, 1987.

[2] A. Barbieri, J.S. Langer, 'Predictions of Dendritic Growth Rates in the Linearized Solvability Theory', *Phys. Rev.*, **A39**, p. 5314, 1989.

[3] J. Bechhoefer, H. Guido, A. Libchaber, 'Solidification dans un Capillaire Rectan-gulaire', *C.R. Acad. Sci.*, **t. 306, II**, p. 619, 1988.

[4] M. Ben Amar, Y. Pomeau, 'Theory of Dendritic Growth in a Weakly Undercooled Melt', *Europhys. Lett.2*, **4**, p. 307, 1986.

[5] M. Ben Amar, B. Moussalam, 'Numerical Results on Two-Dimensional Dendritic Solid', *Physica*, **D25**, p. 155, 1986.

[6] M. Ben Amar, B. Moussalam, 'Absence of Selection in Directional Solidification', *Phys. Rev. Lett.*, **60, 4**, p. 317, 1988.

[7] M. Ben Amar, P. Pelcé, 'Impurity Effect on Dendritic Growth', *Phys. Rev.*, **A39**, pp. 4263-4269, 1989.

[8] M. Ben Amar, E. Brener, 'Theory of Pattern Selection in Three-Dimensional Nonaxisymmetric Dendritic Growth', *Phys. Rev. Lett.*, **71**, p. 589, 1993.

[9] M. Ben Amar, E. Brener, 'Parity Broken Dendrites', *Phys. Rev. Lett.*, **75**, p. 561, 1995.

[10] C.M. Bender, S.A. Orszag, *Advanced Mathematical Methods for Scientists and Engineers*, McGraw-Hill, New York, 1979.

[11] E. Ben-Jacob, R. Godbey, N.D. Goldenfeld, J. Koplik, H. Levine, T. Mueller, L.M. Sander, 'Experimental Demonstration of the Role of Anisotropy in Interfacial Pattern Formation', *Phys. Rev. Lett.*, **55**, p. 1315, 1985.

[12] B. Billia, H. Jamgotchian, L. Capella, 'Pattern Selection During Directional Solidification', *J. Crystal Growth*, **82**, p. 747, 1987.

[13] B. Billia, H. Jamgotchian, L. Capella, 'Pattern Selection During Solidification', *J. Crystal Growth*, **94**, p. 987, 1989.

[14] L. Boyer, 'Laser Tomographic Method for Flame Front Movement', *Combustion and Flame*, **39**, p. 321, 1980.

[15] E.A. Brener, 'Influence of kinetic effects on the growth of a two-dimensional dendrite', *Zh ETF*, **97** (7), p. 237, 1989.

[16] E.A. Brener, V.I. Mel'nikov, 'Pattern Selection in Two-Dimensional Dendritic Growth', *Advances in Physics*, **40, 1**, p. 53, 1991.

[17] M.H. Burden, J.D. Hunt, 'Cellular and Dendritic Growth', *J. Crystal Growth*, **22**, p. 109, 1974.

[18] E. Brener, M.B. Geilikman, D.E. Temkin, 'Growth of a Needle-Shaped Crystal in a Channel', *Sov. Phys.*, JETP **67** (5), 1988.

[19] B. Caroli, C. Caroli, B. Roulet, J.S. Langer, 'Solvability Condition for Needle Crystals at Large Undercoding in a non local Model of Solidification', *Phys. Rev.*, **A33**, p. 442, 1986.

[20] R. Combescot, T. Dombre, V. Hakim, Y. Pomeau, A. Pumir, 'Shape Selection of Saffman–Taylor Fingers', *Phys. Rev. Lett.*, **56**, p. 2036, 1986.

[21] Y. Couder, N. Gerard, M. Rabaud, 'Narrow Fingers in the Saffman–Taylor Instability', *Phys. Rev.*, **A34**, p. 5175, 1986.

[22] T. Dombre, V. Hakim, 'Saffman–Taylor Fingers and Directional Solidification at Low Velocity', *Phys. Rev.*, **A36**, p. 2811, 1987.

[23] T. Dombre, V. Hakim, Y. Pomeau, 'Sélection de la Largeur des Doigts dans l'Instabilité de Saffman–Taylor', *C.R. Acad. Sci.*, **302**, p. 803, 1986.

[24] A.T. Dorsey, O. Martin, 'Saffman–Taylor Fingers with Anisotropic Surface Tension', *Phys. Rev.*, **A35**, p. 3989, 1987.

[25] P.G. de Gennes, *The Physics of Liquid Crystals,* Oxford University Press, 1974.

[26] P.G. de Gennes, J. Prost, *The Physics of Liquid Crystals*, 2nd edition, Oxford University Press, 1992.

[27] V. Emsellem, P. Tabeling, 'Measurement of the Concentration Field around a Dendrite', *Europhys. Lett.*, **25 (4)**, p. 277, 1994.

[28] M. Georgelin, A. Pocheau, 'Onset of Sidebranching Instability in Directional Solidification', *Phys. Rev. E.*, **57**, p. 3189, 1998.

[29] M.E. Glicksman, R.J. Shaefer, J.D. Ayers, 'Dendritic Growth – A Test of Theory', *Metall. Trans.*, **A7**, p. 1747, 1976.

[30] M.E. Glicksman, M.B. Koss, E.A. Winsa, 'Dendritic Growth Velocities in Microgravity', *Phys. Rev. Lett.*, **73**, p. 573, 1994.

[31] D.C. Hong, J.S. Langer, 'Analytical Theory of the Selection Mecanism in the Saffman–Taylor Problem', *Phys. Rev. Lett.*, **56**, p. 2032, 1986.

[32] D.C. Hong, J.S. Langer, 'Pattern Selection and Tip Perturbations in the Saffman–Taylor Problem', *Phys. Rev.*, **A 36**, p. 2325, 1987.

[33] H. Honjo, Y. Sawada, 'Quantitative Measurements on the Morphology of a NH_4Br Dendritic Crystal Growth in a Capillary', *J. Crystal Growth*, **58**, p. 297, 1982.

[34] S.C. Huang, M.E. Glicksman, 'Fundamentals of Dendritic Solidification', *Acta Metall.*, **29**, p. 701, 1981.

[35] J.D. Hunt, D.G. McCartney, 'Numerical Finite Difference Model for Steady State Cellular Array Growth', *Acta metall*, **35**, p. 89, 1987.

[36] A. Karma, P. Pelcé, 'Oscillatory Instability of Deep Cells in Directional Solidification', *Phys. Rev.*, **A39**, p. 4162, 1988.

[37] A. Karma, W.J. Rappel, 'Numerical Simulation of Three-Dimensional Dendritic Growth', *Phys. Rev. Lett.*, **77**, p. 4050, 1996.

[38] K. Kassner, *Pattern Formation in Diffusion-Limited Crystal Growth,* World Scientific, 1996.

[39] D.A. Kessler, J. Koplik, H. Levine, 'Steady State Dendritic Crystal Growth', *Phys. Rev.*, **A 33**, p. 3352, 1986.

[40] D.A. Kessler, H. Levine, 'Growth Velocity of Three-Dimensional Dendritic Crystals', *Phys. Rev.*, **A36**, p. 4123, 1987.

[41] D.A. Kessler, H. Levine, 'Pattern selection in Three-Dimensional Dendritic Growth', *Acta Metall.*, **36**, p. 2693, 1988.

[42] D.A. Kessler, H. Levine, 'Steady-State Cellular Growth during Directional Solidifi-cation', *Phys. Rev.*, **A39**, p. 3041, 1989.

[43] M. Kruskal, H. Segur, 'Asymptotics beyond all Orders in a Model of Dendritic Crystals', *Aero. Res. Ass. of Princeton Tech. Memo*, 1985.

[44] P. Kurowski, C. Guthmann, S. de Cheveigne, 'Shapes, Wavelength Selection, and the Cellular-dendritic 'Transition' in Directional Solidification', *Phys. Rev.*, **A42**, p. 7368, 1990.

[45] L. Landau, *On the theory of Superconductivity*, collected papers ed. D. Ter. Haar Gordon and Breach, p. 217, 1937.

[46] J.S. Langer, 1986, in *Chance and Matter*, Lectures in the Theory of Pattern Formation, Les Houches Summer School, 1986, J. Souletie, J. Vannimenus, R. Stora, eds., North-Holland, New York, 1987, p. 629.

[47] J.W. McLean, P.G. Saffman, 'Effect of Surface Tension on the Shape of Fingers in a Hele–Shaw Cell', *J. Fluid. Mech.*, **102**, p. 455, 1981.

[48] J. Lipton, M.E. Glicksman, W. Kurz, 'Dendritic Growth into Undercooled Alloy Metals', *J. Mater. Sci. Eng.*, **65**, p. 57, 1984.

[49] G. Livescu, M.R. Srinivasan, H. Chou, O.N. Mesquita, H.Z. Cummins, 'Dynamic Light Scattering at the Nonequilibrium Crystal-melt Interface in Cyclohexanol', *Phys. Rev.*, **A36**, p. 2293, 1987.

[50] M. Mashaal, M. Ben Amar, V. Hakim, 'Directional Solidification at Low Velocity', *Phys. Rev.*, **A41**, p. 4421, 1990.

[51] D. Meiron, 'Selection of Steady States in the Two-Dimensional Symmetric Model of Dendritic Growth', *Phys. Rev.*, **A33**, p. 2704, 1986.

[52] P. Molho, A.J. Simon, A. Libchaber, 'Péclet Number and Crystal Growth in a Channel', *Phys. Rev.*, **A42**, p. 904, 1988.

[53] M. Muschol, D. Liu, H.Z. Cummins, 'Surface-Tension-Anisotropy Measurements of Succinonitrile and Pivalic acid: Comparison with microscopic solvability theory', *Phys. Rev.*, **A46**, p. 1038, 1992.

[54] G.E. Nash, M.E. Glicksman, 'Capillarity-Limited Steady-State Dendritic Growth', *Metall.*, **22**, p. 1283, 1974.

[55] P. Oswald, 'Dendritic Growth of a Discotic Liquid Crystal', *J. Phys. France*, **49**, p. 1083, 1988.

[56] P. Pelcé, 'Dynamique des fronts courbés', thesis held at the university of Provence, 1986.

[57] P. Pelcé, Y. Pomeau, 'Dendrites in the Small Undercooling Limit', *Studies in Applied Math.*, **74**, p. 245, 1986.

[58] C. Pelcé-Savornin, J. Quinard, G. Searby, 'The Flow Field of a Curved Flame Propagating Freely Upwards', *Comb. Sci. and Tech.*, **58**, p. 337, 1988.

[59] Y. Pomeau, M. Ben Amar, 'Dendritic Growth and Related Topics', in *Solids far from equilibrium*, C. Godrèche ed., Cambridge University Press, 1992.

[60] E. Raz, S.G. Lipson, E. Polturak, 'Dendritic Growth of Ammonium Chloride Crystals: Measurements of the Concentration Field and a Proposed Nucleation Model for Growth', *Phys. Rev.*, **A40**, p. 1088, 1989.

[61] D.A. Reinelt, P.G. Saffman, 'The Penetration of a Finger into a Viscous Fluid in a Channel and Tube', *Siam J. Sci. Stat. Comput.*, **6**, p. 542, 1985.

[62] P.G. Saffman, G.I. Taylor, 'The Penetration of a Fluid into a Porous Medium or Hele–Shaw Cell Containing a More Viscous Liquid', *Proc. Roy. Soc. Lond.*, **A245**, p. 155, 1958.

[63] E. Scheil, *Z. Metallk.*, 34, 70, 1942.

[64] B.I. Shraiman, 'Velocity Selection in the Saffman–Taylor Problem', *Phys. Rev. Lett.*, **56**, p. 2028, 1986.

[65] P. Tabeling, G. Zocchi, A. Libchaber, 'An Experimental Study of the Saffman–Taylor Instability', *J. Fluid. Mech.*, **177**, p. 67, 1987.

[66] H. Thome, M. Rabaud, V. Hakim, Y. Couder, 'The Saffman–Taylor Instability: from the Linear to the Circular Geometry', *Phys. Fluids*, **A 1 (2)**, p. 224, 1989.

[67] L.H. Ungar, R.A. Brown, 'Cellular Interface Morphologies in Directional Solidification. IV. The Formation of deep Cells', *Phys. Rev.*, **B31**, p. 5931, 1985.

[68] J.M. Vanden-Broeck, 'Fingers in a Hele–Shaw Cell with Surface Tension', *The Physics of Fluids*, **26**, p. 2033, 1983.

[69] J.A. Warren, J.S. Langer, 'Stability of Dendritic Arrays', *Phys. Rev.*, **A42**, p. 3518, 1990.

[70] J.D. Weeks, W. Van Saarloos, 'Stability of Cellular Patterns in Directional Solidification', *Phys. Rev.*, **A42**, p. 5056, 1989.

[71] G. Zocchi, B.E. Shaw, A. Libchaber, L.P. Kadanoff, 'Finger Narrowing under Local Perturbations in the Saffman–Taylor Problem', *Phys. Rev.*, **A36**, p. 1894, 1987.

CHAPTER 9

Stability of a curved shape

9.1 EXPERIMENTAL RESULTS

At high velocity of propagation the stationary shape of the Saffman–Taylor finger becomes unstable. We observe several modes or types of deformation of the curved shape. The first one, secondary branching, appears close to the tip and grows on one side of the finger (figure 9.1a). This deformation, which almost stays at rest in the reference frame of the laboratory, moves backwards with respect to the tip at the same velocity as the finger. The second mode occurs by the splitting of the tip into two equivalent parts (figure 9.1b). The two resulting parts compete for the growth, one is stopped and the process re-occurs.

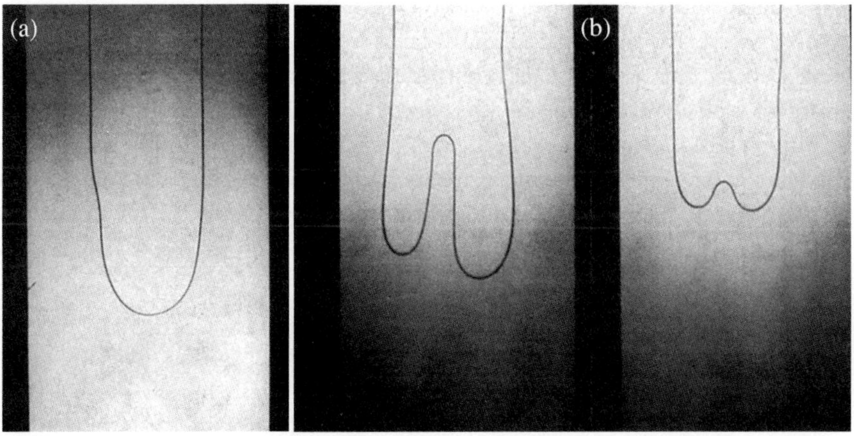

Figure 9.1 (a) Secondary branching; (b) tip-splitting (Tabeling *et al.* 1987).

The destabilisation of a finger does not occur at a well-defined velocity. Depending on the mode of preparation of the Hele–Shaw cell (whether the glass plates bordering the cell are perfectly clean and exactly parallel or not), the critical velocities of destabilisation of the finger can be very different (Tabeling, Zocchi and Libchaber 1987). Specifically, if the cell is perfect the finger can keep its stationary form up to high velocities of propagation, which justifies the property of anomalous stability. Indeed we have seen in the preceding chapter that when the velocity of the finger is high, its relative size, and hence its radius of curvature, saturates at a well-given value, whereas the wavelength λ_c of the destabilisation of the plane front becomes smaller and smaller. At the scale of λ_c the curved shape is almost planar and should therefore become destabilised into several cells.

Dendrites are different because they always have a non stationary form. From the tip secondary branches at rest in the reference frame of the laboratory are emitted more or less regularly. Figure 9.2 represents a series of snapshots of growth taken at successive intervals and superimposed. We note that when the tip moves at constant velocity the secondary branches grow along their axis. Seconday branching is usually rather irregular (Dougherty *et al.* 1987). In some cases, such as the growth of ^{3}He crystals above the roughening transition temperature (Rolley *et al.* 1986), dendrites featuring irregular and periodic secondary branching can be observed simultaneously. When they grow into a network, secondary branching of neighbouring dendrites have a higher correlation (Eshelman *et al.* 1988, Georgelin and Pocheau 1997).

It is worth noting in figure 9.3 that when the supercooling is small, the structure of these branches does not depend on the supercooling. Even if the radius of curvature changes a lot in the domain of supercooling, the relative distances between the secondary branches are the same. The distance from the tip from which the first secondary branch appears varies according to the material. For instance, this distance is larger for pivalic acid with a larger factor of crystalline anisotropy than for succinonitrile. The wavelength of the secondary branching is of the order of the radius of curvature of the tip, which is large compared to the most unstable wavelength of the planar interface, since the square of their ratio, C, is of the order of 100. Contrary to the Saffman–Taylor finger and except under special conditions (Koo *et al.* 1991) the mode which splits the tip is not observed.

A curved flame moving in a tube is also very robust. The flame can remain stable up to large Reynolds numbers, thus delaying the self-turbulence of combustion. This property of anomalous stability has remained paradoxical for quite a long time since the discovery of a fundamental instability of plane propagation by Darrieus and Landau.

To understand the nature of the modes of destabilisation we must establish an analysis of stability of the curved shape. As in chapter 6 the problem consists of determining how an initial perturbation of small amplitude of the curved shape changes with time. If all perturbations decay with time, the curved front is stable. From a mathematical point of view we linearise the equations of motion of the interface in the proximity of the stationary curved shape, and we determine the normal modes: we are looking for solutions of the form $f(l)\exp(\sigma t)$ where f is a function

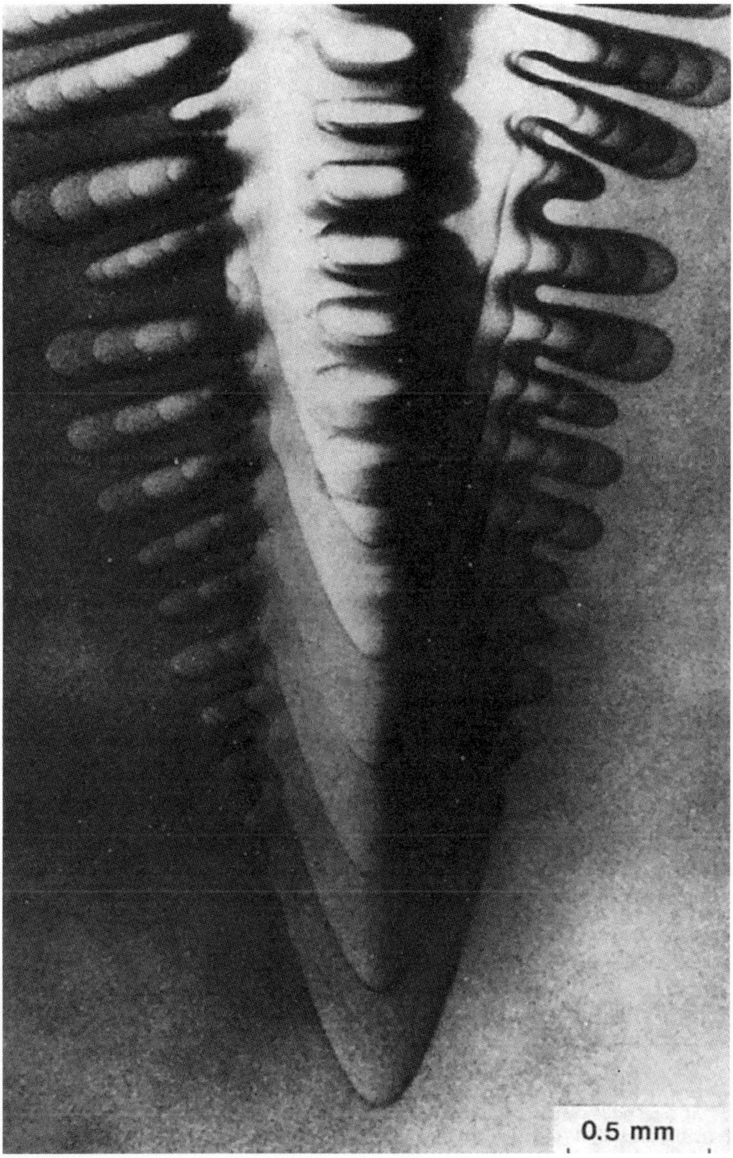

Figure 9.2 Superposition at successive intervals of snapshots of a dendrite, showing that the secondary branching is stationary in the reference frame of the laboratory (Huang and Glicksman 1981).

of the curvilinear coordinate of the non-perturbed form. The boundary conditions to apply to this function are not obvious. Indeed, far behind the tip, the finger or the dendrite is usually deformed by the extremity where the fluid is injected into the cell

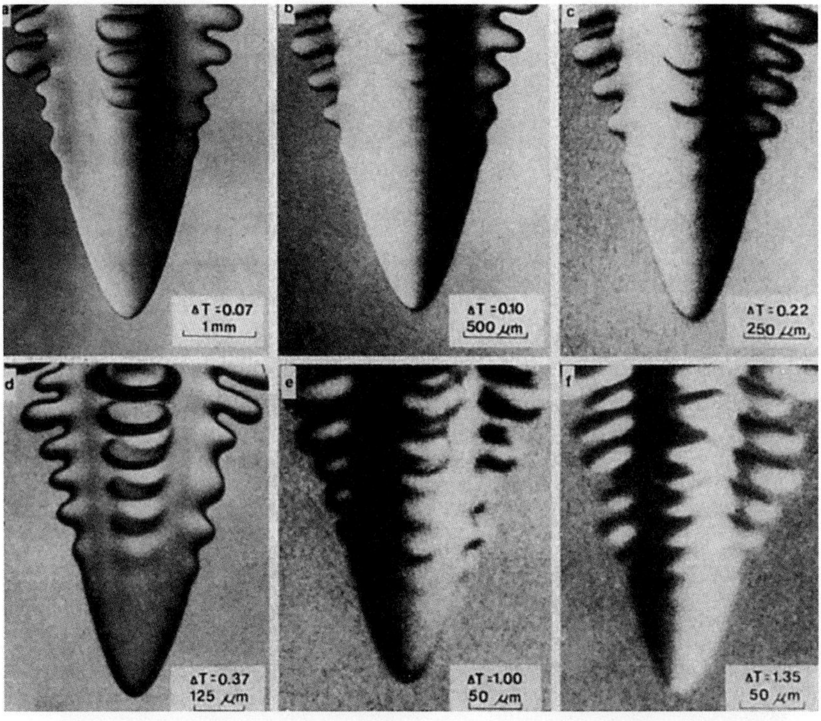

Figure 9.3 Snapshots of dendrites at different supercooling (Huang and Glicksman 1981).

or by the neighbouring dendrites. Its form changes noticeably from its ideal stationary shape. We will neglect the presence of these parasite deformations, and assume that far behind the tip these perturbations cancel out. The condition of stability is that all growth rates σ have a negative real part. An extra difficulty of the problem arises from the multiplicity of the stationary solutions (indexed by an integer n) determined in the preceding chapter. Do all the solutions have a physical meaning? We show that only one of these solutions is stable, the one for $n = 1$.

9.2 INSTABILITY OF EXACT SOLUTIONS WITHOUT SURFACE TENSION

The analysis of stability of exact solutions of fingers without surface tension was done by Taylor and Saffman (1959). Instead of the spatial variable $z = x + iy$ we use the variable $\zeta = \xi + i\eta$, in which the domain of the flow is the rectangle $\xi > 0$, $-1 < \eta < 1$. We must look for a conformal transformation $z = f(\zeta)$ corresponding to the motion of the finger. We start in the moving frame where the steady-state finger is at rest. We thus introduce the potential $\phi = \Phi - Ux$ and its conjugated stream function $\psi = \Psi - Uy$, where U is the velocity of the finger. The conformal transformation associated to the solutions without surface tension is, according to relation (7.4) in paragraph 7.5,

$$z_0(\zeta) = \zeta + \frac{2}{\pi}(1-\lambda)\mathrm{Log}\left(\frac{1+\exp(-\pi\zeta)}{2}\right). \tag{9.1}$$

For the analysis of stability we expand $z(\zeta, t) = z_0(\zeta) + z_1(\zeta, t)$ where z_1 is small compared to z_0. The kinematic condition of equality between the normal velocity of the interface and the normal velocity of the fluid is written as

$$\frac{\partial\psi}{\partial l} = \frac{\partial X}{\partial t}\cos\theta + \frac{\partial Y}{\partial t}\sin\theta \tag{9.2}$$

where θ is the angle between the direction normal to the interface and the direction of propagation, and X, Y have the values of x and y at the point $\xi = 0$. By using the relations $\cos\theta = \partial Y/\partial l$ and $\sin\theta = -\partial X/\partial l$ we obtain at the interface,

$$\frac{\partial\psi}{\partial\eta} = \frac{\partial X}{\partial t}\frac{\partial Y}{\partial\eta} - \frac{\partial Y}{\partial t}\frac{\partial X}{\partial\eta}. \tag{9.3}$$

By looking for solutions proportional to $\exp\sigma t$ we thus obtain at the first order of the expansion of relation (9.3),

$$\frac{\partial\psi_1}{\partial\eta} = \sigma\left(X_1\frac{\partial Y_0}{\partial\eta} - Y_1\frac{\partial X_0}{\partial\eta}\right). \tag{9.4}$$

By using $\phi_1 = -UX_1$, $\psi_1 = -UY_1$, and, considering the real part of relation (9.1),

$$X_0 = \frac{2}{\pi}(1-\lambda)\mathrm{Log}\left(\cos\left(\frac{\pi\eta}{2}\right)\right),\ Y_0 = \lambda\eta \tag{9.5}$$

relation (9.4) becomes

$$-\frac{\partial Y_1}{\partial\eta} = \frac{\sigma}{U}\left(\lambda X_1 + (1-\lambda)\tan\left(\frac{\pi\eta}{2}\right)Y_1\right). \tag{9.6}$$

Let us first consider symmetric solutions of equation (9.6) in the form

$$X_1 = \sum_{n=0}^{\infty} a_n\cos(n\pi\eta),\ Y_1 = -\sum_{n=1}^{\infty} a_n\sin(n\pi\eta). \tag{9.7}$$

We then obtain the following relations between the coefficients a_n

$$\begin{aligned}(\sigma(2\lambda-1) - U\pi)a_1 &= -2\lambda\sigma a_0\\ (\sigma(2\lambda-1) - (n+1)U\pi)a_{n+1} &= (nU\pi - \sigma)a_n \qquad n > 1.\end{aligned} \tag{9.8}$$

The trivial solution $\sigma = 0$, $a_n = 0$, for $n > 1$ corresponds to the uniform displacement of the finger, parallel to its axis. It is worth noticing that the possible growth rates are of the form $\sigma_n = n\pi U$, and therefore are the same as for the planar interface. According to paragraph 6.1 the growth rate of a perturbation of the planar interface is actually written as $\sigma = Vk$ and, for a symmetric perturbation compatible with the condition of null transverse velocity at the wall, $k = n\pi$ where $n > 1$. Thus, when the effects of surface tension are neglected, the geometric effect of curvature does not modify its natural instability.

Asymmetric modes are also solutions of equation (9.6). If we actually consider, instead of equation (9.7)

$$X_I = \sum_{n=0} a_n \sin\left(\left(n + \frac{1}{2}\right)\pi\eta\right), \quad Y_I = \sum_{n=0} a_n \cos\left(\left(n + \frac{1}{2}\right)\pi\eta\right) \tag{9.9}$$

we obtain, for $n > 0$

$$\pi\left(n + \frac{1}{2}\right)a_n + \pi\left(n + \frac{3}{2}\right)a_{n+1} = \frac{\sigma}{U}(a_n - (1 - 2\lambda)a_{n+1}) \tag{9.10}$$

the solutions of which are in the form $\sigma_n = \pi(n + 1/2)U$.

Langer and Müller-Krumbhaar (1978) have also developed an analysis of stability for the Ivantsov parabolas and paraboloids when the effects of surface tension are neglected. The conclusion of the calculation is similar since it leads to the instability of these curved shapes. Thus, the geometric effect alone is not responsible for their stabilisation.

9.3 MODES OF INSTABILITY

When the effects of surface tension are taken into account, the analysis of linear stability of these curved shapes cannot be carried out in a general way (that is for arbitrary values of the control parameters), but only by numerical means (Kessler, Koplik and Levine 1986). The method consists of determining as a first step integral equations for the shape of the considered interfaces, but also taking into account the unsteady terms. Then we determine perturbations proportional to $\exp \sigma t$, by discretising these equations. The set of the possible values of σ constitutes the spectrum of stability of the curved shapes. The analysis of stability concerns all the branches of solutions which, as we have seen in the preceding chapter, are indexed by an integer. Again the results obtained are similar for the Saffman–Taylor fingers and the needle-like crystals. Two kinds of modes are determined: a continuum of modes with a negative real part and a non-zero imaginary part. A series of discrete modes is organised in such a way that for a branch of index n, $n - 1$ real, positive and therefore unstable modes are associated. Thus, only the branch $n = 1$ is stable. The problem of selection, which was not fully solved with steady-state solutions, is solved by the analysis of stability. Quite curiously this branch of solutions is also the one for which the curved shapes have the highest velocity, which is in agreement with the principle stating that the quickest forms are naturally selected. These results, not easily accessible to the experimentalist, call for a simplified analytic representation.

9.4 EQUATION OF EVOLUTION FOR PERTURBATIONS OF THE STEADY-STATE CURVED SHAPE

We derive a simpler equation for perturbations, with a wavelength smaller compared to the characteristic size of the curved shape (Bensimon, Pelcé and Shraiman 1987). This approximation, called WKB, allows us to get closer to the more intuitive picture of the plane local instability. The equation is valid for both the Saffman–Taylor fingers and the needle-like crystals, and is very similar to the equation of Michelson and Sivashinsky derived in paragraph 10.6 for cellular flames. When the wavelength is small, diffusion dominates the effects of advection and the external field satisfies the Laplace equation. Thus, we can always associate perturbations of the potential $\delta\Phi$ and of the stream function $\delta\Psi$ to a perturbation of the interface shape $\gamma(l, t)$ (figure 9.4). As in paragraphs 8.4 and 8.6, the lengths are made dimensionless by the half-length of the cell for the Saffman–Taylor finger, and by the radius of curvature of the parabola for the needle-like crystal, whereas the times are made dimensionless by this length, divided by the velocity of propagation.

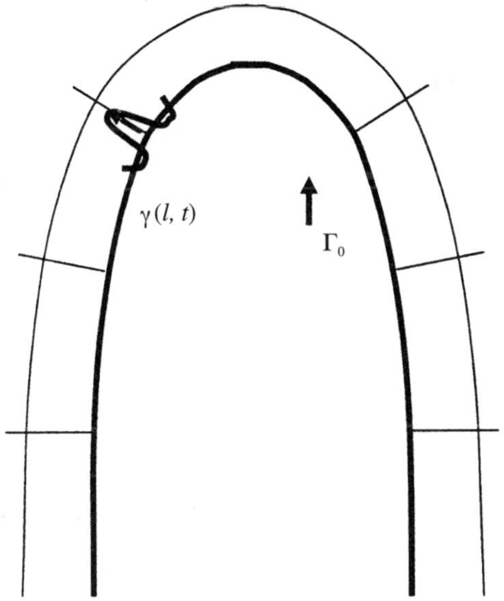

Figure 9.4 Schematic of the interface and of the coordinates used. We have sketched the interface (plain line) and the tangential and normal coordinates (thin lines).

Let $f(z, t) = \delta\Phi + i\delta\Psi$ be the analytic function. For a point $z(l)$ of the interface Γ_0, we can write

$$f(z(l)) = \frac{1}{\pi i}\int_{\Gamma_0} dz' \frac{f(z')}{z'-z}$$
$$= \frac{1}{\pi i}\int_{\Gamma_0} dl' f(z(l'))\frac{d}{dl'}\mathrm{Log}(z(l') - z(l)) \tag{9.1}$$

or even, by setting $z(l') - z(l) = \rho(l', l)\exp i\theta(l', l)$

$$\delta\Phi(l, t) = \frac{1}{\pi}\mathrm{P}\int_{-\infty}^{+\infty}\left(\delta\Psi(l', t)\frac{d\mathrm{Log}(\rho(l', l))}{dl'} + i\delta\Phi(l', t)\frac{d\mathrm{Log}(\theta(l', l))}{dl'}\right)dl' . \tag{9.2}$$

Since, as a first approximation, the interface can be considered planar, $d\mathrm{Log}(\rho(l, l'))/dl' \approx l/(l'-l)$, $d\mathrm{Log}(\theta(l', l))/dl' \approx 0$, and equation (9.2) can be approximated by

$$\delta\Phi(l, t) = \frac{1}{\pi}P\int_{-\infty}^{+\infty}\delta\Psi(l', t)\frac{1}{l'-l}dl' . \tag{9.3}$$

By using the Cauchy relation $\partial(\delta\Psi)/\partial n = \partial(\delta\Phi)/\partial l$ where the variables n and l indicate the derivations in the normal and tangential directions respectively, we obtain

$$\frac{\partial(\delta\Phi(l, t))}{\partial n} = \frac{1}{\pi}P\int_{-\infty}^{+\infty}\frac{\partial(\delta\Phi(l', t))}{\partial l'}\frac{1}{l'-l}dl' . \tag{9.4}$$

We apply equation (9.4) to the potential $\phi = (\Phi - Ux)/U$, where Φ is the potential in the reference frame of the laboratory (the definition of ϕ used here is slightly different from the definition used in paragraph 7.5). Thus, we have at the interface

$$\phi = -x + \frac{\nu}{R} \tag{9.5a}$$

$$\frac{\partial\phi}{\partial n} = \frac{\partial\gamma}{\partial t} \tag{9.5b}$$

where $\nu = (\gamma/12\eta U)(b/a)^2$ for the Saffman–Taylor finger, and $\nu = 1/C$ for the needle-like crystal. To calculate the total variations of ϕ at the interface, we write $n = n_0 - (\partial\gamma/\partial l)\tau_0$, $\phi \approx \phi_0 + \delta\phi$, where ϕ_0, τ_0 and n_0 are the steady-state potential, the vector tangent and the vector normal to the non-perturbed interface respectively.

We then have $\delta(\phi(l, t)) = \delta\phi(l, t) + (\partial\phi_0/\partial n_0)\gamma$, that is, according to relation (9.5a), since $(\partial\phi_0/\partial n_0)' = 0$,

$$\delta\phi(l, t) = -\left(\cos\theta + \nu\left(\left(\frac{1}{R_0}\right)^2 + \frac{\partial^2}{\partial l^2}\right)\right)\gamma . \tag{9.6}$$

In the same way,

$$\delta(\partial\phi(l,t)/\partial n) = \partial(\delta\phi + (\partial\phi_0/\partial n_0)\gamma)/\partial n_0 - (\partial\phi_0/\partial l)(\partial\gamma/\partial l).$$

In the system of coordinates at the interface,

$$\frac{\partial^2\phi_0}{\partial l^2} + \frac{\partial^2\phi_0}{\partial n_0^2} + \frac{1}{R_0}\frac{\partial\phi_0}{\partial n_0} = 0 \tag{9.7}$$

so that $\partial^2\phi_0/\partial n^2 = -\partial^2\phi_0/\partial l^2$. But, up to order $\sqrt{\nu}$, inclusive, according to equation (9.5a), $\partial\phi_0/\partial l = -\sin\theta$. We deduce by using equation (9.5b) and the expression of $\delta(\partial\phi(l,t))/\partial n$

$$\frac{\partial(\delta\phi(l,t))}{\partial n_0} = \frac{\partial\gamma}{\partial t} - \frac{\partial}{\partial l}(\sin\theta\gamma(l,t)). \tag{9.8}$$

Thus, equations (9.4), (9.6) and (9.8) allow us to obtain the equation for a perturbation of the shape of the interface

$$\frac{\partial\gamma}{\partial t} - \frac{\partial}{\partial l}(\sin\theta\gamma(l,t)) = -\frac{1}{\pi}P\int_{-\infty}^{+\infty}\frac{dl'}{l'-l}\left(\frac{\partial}{\partial l'}(\cos\theta\gamma) + \nu\frac{\partial^3\gamma}{\partial l'^3}\right) + O(\sqrt{\nu}). \tag{9.9}$$

In the case where the angle θ is assumed to be constant, the solutions of equation (9.9) are simply of the form $\gamma = \exp(\sigma t + ikl)$, where $\sigma = i\sin\theta k + |k|(\cos\theta - \nu k^2)$, and we find again the dispersion relation of the perturbations of the tilted plane front already mentioned in paragraphs 8.4 and 8.6. Thus, in the moving frame where the steady-state shape is at rest, a perturbation is at the same time advected along the interface by a tangential velocity of amplitude sin θ, and grows or decays by the instability of the initial plane front, depending whether the local wavenumber is in the band of the unstable wavenumbers or not. The width of this band decreases until it vanishes behind the curved shape, because of the modulation of the force of instability by the term cos θ. Thus, the dynamics of the perturbations seems quite universal. The unperturbed form, the analytic expression of which seems to be less relevant, only appears in the function $\theta(l)$. This equation is very similar to the Michelson–Sivashinsky equation determined in paragraph 9.6 for cellular flames, the $\cos\theta$ factor being absent since in this case the normal velocity is constant.

9.5 AMPLIFICATION OF THE THERMODYNAMIC FLUCTUATIONS

The continuum of modes can be understood in the following way in the framework of the curved flames (Zeldovich *et al.*1980). Let us consider an initial perturbation of the form of the interface, located in the proximity of the tip, which is of size λ_c, that is the most unstable wavelength of the plane front. This perturbation can be triggered by a thermodynamic fluctuation of the gas or by any other external cause. As we have already mentioned in the preceding chapter, the perturbation is amplified by the instability of the planar interface as it is pulled to the back of the curved shape. To analytically evaluate the growth rate of the amplitude of this perturbation we adopt the WKB limit already used before, that is when λ_c is far smaller than the radius of curvature of the tip, which here is of the order of the radius R of the tube. In the region I, the wavelength of the perturbation is far smaller than the local radius of curvature of the tip (figure 9.5). Its growth rate can therefore be written as

$$\Gamma(l) \approx \int_{\lambda_c}^{l} \frac{\sigma}{u_\tau} dl' \tag{9.1}$$

where $\sigma(k)$ is the local growth rate of the perturbations, determined by the analysis of linear stability of the plane front, and l the dimensional curvilinear coordinate of the non-perturbed steady-state front.

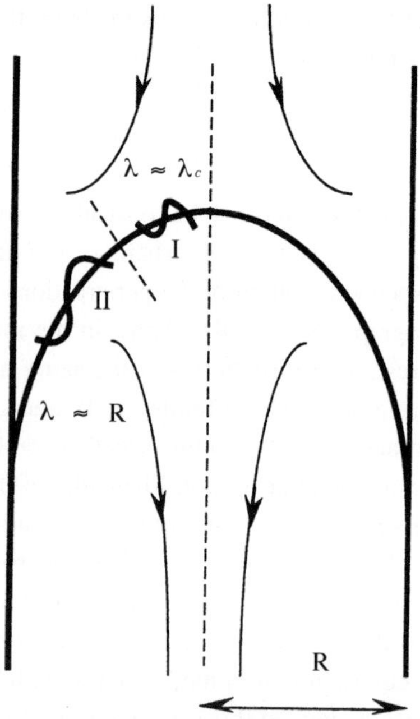

Figure 9.5 Evolution of a perturbation on a flame front.

Since the tangential flow is not uniform the wavelength of the perturbation spreads according to the relation

$$\frac{\delta\lambda}{\lambda} \approx \frac{\delta u_\tau}{u_\tau} \tag{9.2}$$

from which we can deduce, since at the initial time $l \approx \lambda_c / R << 1$, $\lambda \approx \lambda_c$

$$\lambda = \lambda_c \frac{u_\tau(l)}{u_\tau(\lambda_c)} \approx \frac{u_\tau(l)}{u'_\tau(0)} \tag{9.3}$$

where $u'_\tau(0)$ is the derivative with respect to the curvilinear coordinate of the tangential velocity at the tip. Around the tip the intensity of the tangential flow proportional to $\sin\theta$ linearly increases with the distance l. As mentioned by Zeldovich (1980) this effect resembles the stretching of the wavelength of radiation in the expanding universe according to the Hubble law, which states that two masses in the universe (stars, galaxies) move away from each other at a velocity proportional to their distance (Zeldovich and Novikov 1983). Since the wavelength of the perturbation of the front is large compared to the thickness of the flame, we can deduce from equation (6.12) in paragraph 6.7, by an expansion at small wavenumber, a local growth rate of the form

$$\sigma(k) = V_0 k(f(\gamma) - h(\gamma, \beta, Le)dk) \tag{9.4}$$

where γ, β and Le have been defined in paragraph 6.7. In particular, $\lambda_c = 4\pi(h/f)d$.

We thus obtain from equations (9.1), (9.3) and (9.4)

$$\Gamma(l) \approx \int_{\lambda_c}^{l} dl' \frac{2\pi V_0 f u'_\tau(0)\left(1 - \dfrac{\lambda_c u'_\tau(0)}{2u_\tau}\right)}{u_\tau^2} \tag{9.5}$$

from which we deduce the integrated growth rate of the perturbation in the region

$$\Gamma_{\mathrm{I}} \approx \frac{3\pi f}{2} \frac{V_0}{R u_\tau(0)} \frac{R}{\lambda_c} . \tag{9.6}$$

Since the first two factors are of the order of one, Γ_{I} is large compared to 1 by R/λ_c. We note that the net growth rate is almost determined by the proximity of the tip, where the tangential velocity is small. Actually in the region II $\lambda \approx R$ and $\Gamma_{\mathrm{II}} \approx \sigma\tau$, where $\sigma \approx U/\lambda$ is the growth rate of the Darrieus–Landau instability and $\tau \approx R/U$ the transient time of the perturbation. Thus, $\Gamma_{\mathrm{II}} \approx 1 \ll \Gamma_{\mathrm{I}}$.

The Saffman–Taylor finger and the dendrite are different since they have a very large longitudinal dimension. As observed by Tabeling and coworkers (1987) a perturbation of the shape of a Saffman–Taylor finger disappears far away from the tip. Indeed, since the interface becomes almost parallel to the walls of the cell, its normal velocity becomes very small and the instability of displacement disappears. In the case of the dendrite, the secondary branching develops, the local instability also disappears but further away than for a Saffman–Taylor finger, since a parabolic branch transversely spreads to infinity. Thus we must evaluate the growth rate of the perturbations in two parts (Bensimon *et al.* 1986, Pelcé 1986, Pelcé and Clavin 1987). The first one, as for the flame, is controlled by the region of the tip, where the perturbation notably increases because of a small drifting velocity. The second one integrates the local growth over the sides of the curved shape, far away from the tip (figure 9.6). To set the position l_f from which the perturbation begins to decrease we write that the local wavelength of a perturbation determined by equation (9.3) is

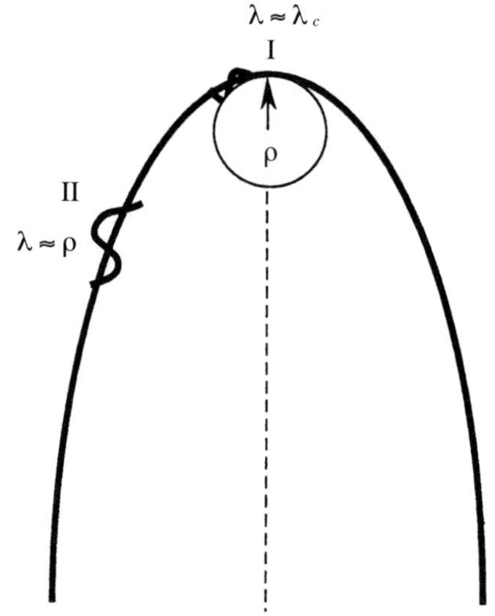

Figure 9.7 Evolution of a perturbation on a needle-like crystal.

Appendix: Evolution of a wave packet on a needle-like crystal (Caroli, Caroli and Roulet 1987, Langer 1987, Pieters 1988).

For a perturbation of constant frequency ω (Langer 1987) we must write, as in paragraph 8.4, at the zeroth order of the WKB method,

$$\xi(r, \omega) \approx \exp i\omega t \exp \int_0^l ikdl \approx \exp i\omega t \exp iW(r)$$

where $dW/dr = kdl/dr$, and k is the local wavenumber of the dispersion relation related to the frequency ω, that is which satisfies

$$i\omega = ik\sin\theta + k\left(\cos\theta - \frac{1}{C}k^2\right).$$

As we have noticed in this paragraph, far away from the tip frequencies and wavenumbers are small compared to the timescale and lengthscale imposed by the tip, such that

$$k \approx i\omega\exp(-i\theta) - i\omega^3 \frac{1}{C}\exp(-4i\theta).$$

Using $\cos\theta = 1/(1+r^2)^{1/2}$, $\sin\theta = r/(1+r^2)^{1/2}$ and $dl/dr = (1+r^2)^{1/2}$, we deduce far away from the tip,

$$\xi(r, \omega) \approx \exp\left(\sqrt{C}\left(\omega\left(r - i\left(\frac{r^2}{2} - t\right)\right) - \omega^3 \frac{r^2}{2}\right)\right)$$

where we have replaced the frequency scale with respect to the most stable growth rate of the planar interface by the change of ω into $\omega\sqrt{C}$. Thus, at fixed ω, the perturbation grows as it is dragged backwards (term proportional to ω), then decreases further away from the tip (term in ω^3). When the initial perturbation contains a relatively broad spectrum of frequencies (wave packet),

$$\xi(r,t) \approx \int d\omega S(\omega)\exp\left(\sqrt{C}\left(\omega\left(r - i\left(\frac{r^2}{2} - t\right)\right) - \omega^3\frac{r^2}{2}\right)\right)$$

$$\approx S(\omega_c)\left(\frac{C}{r}\right)^{1/4}\exp(\sqrt{C}\Phi(r,t))$$

where

$$\Phi(r,t) = \left(\frac{2}{3}\right)^{3/2} r^{1/2}\left(1 - \frac{i}{r}\left(\frac{r^2}{2} - t\right)\right) \approx \left(\frac{2}{3}\right)^{3/2} r^{1/2}\left(1 - \frac{3i}{2r}\left(\frac{r^2}{2} - t\right) - \frac{3}{8r^2}\left(\frac{r^2}{2} - t\right)^2\right).$$

We used the fact that, since C is large, the integral is dominated by the saddle point where the exponential factor is at an extremum at the frequency

$$\omega_c = \left(\frac{2}{3r}\right)^{1/2}\left(1 - \frac{i}{r}\left(\frac{r^2}{2} - t\right)\right)^{1/2} \approx \left(\frac{2}{3r}\right)^{1/2}$$

which decreases to 0 far away from the tip. Thus, the amplitude at the centre of the packet $(r^2 = 2t)$ increases according to $A(l) \approx C^{1/4} l^{-1/8}\exp(\sqrt{C}l^{1/4})$, where $l \approx r^2$ is the distance from the tip. If we go back to dimensional quantities, $\omega_c \approx \omega\rho/V\sqrt{C}$, from which we deduce $\omega \approx V\sqrt{C}/(\rho r)^{1/2}$ and $\lambda \approx V/\omega \approx (\rho r)^{1/2}/\sqrt{C} \approx \lambda_m$, the marginal wavelength determined by equation (9.7). Thus, instead of being damped at a given distance from the tip, far away, the wave packet perturbation becomes marginal. Triggering a perturbation of the interface by a pulsed laser rapidly applied to the tip, Qian and Cummins (1990) measure an amplification of the latter in agreement with the $\exp(\sqrt{C}l^{1/4})$ law. However, it is not clear at all whether the observed morphology of the perturbation is in agreement with the hypotheses of the calculation since the local wavelength rather seems to be of the order of the radius of curvature of the tip.

9.6 STABILITY CRITERION OF A CURVED SHAPE

Once we have determined Γ, the stability criterion of the curved flame is as follows: let A_i be the initial amplitude of the perturbation. By definition of the growth rate Γ, the amplitude of the perturbation at a distance l from the tip is written as

$$A(l) = A_i \exp\Gamma(l). \tag{9.1}$$

If we assume that the curved flame loses its stability when the amplitude of its deformation is of the order of its characteristic size R, the stability criterion of the curved flame is written as

$$\Gamma = -\mathrm{Log}(A_i/R) \tag{9.2}$$

where Γ, evaluated at a distance $l \approx R$, is determined by equation (9.6) in paragraph 9.5.

The remarkable property of the stability criterion defined by equation (9.2) is that it links microscopic and macroscopic properties. In fact, a relative amplitude of the perturbations as small as 10^{-n}, where n can be large like in the case of thermodynamic noise, creates a growth rate of the order n. However, from equation (9.6) mentioned above, Γ is, to within a numerical prefactor, the ratio R/λ_c between the radius of curvature of the tip and the most unstable wavelength of the plane front. The number n characterising the microscopic noise therefore sets the macroscopic morphology of the flame. This criterion also explains the property of anomalous stability of the flame. If the noise near the tip is small enough (n large), the curved flame stays stable, even if its radius of curvature exceeds the most unstable wavelength by an order of magnitude. More quantitatively let us assume, as in paragraph 2.5, that the interface noise is simply caused by thermodynamic fluctuations of the gas adjacent to the interface. We thus deduce from equation (2.22) in paragraph 2.5, $A_i \approx (1/\lambda\rho)^{1/2} \approx 10^{-11}\,\mathrm{m}$ where $\lambda \approx \lambda_c \approx 1$ mm and $\rho \approx 2 \cdot 10^{25}/\mathrm{m}^3$. If there was no parasite flow, the curved flame could remain stable up to a radius of the tube exceeding the most unstable wavelength of the planar interface by an order of magnitude.

In a similar way to equation (9.2), using equation (9.10) in paragraph 9.5, the stability criterion for the tip of a Saffman–Taylor finger is written as

$$\frac{a}{b}\sqrt{Ca} \approx -\frac{9\pi}{16}\mathrm{Log}\left(\frac{A_i}{a}\right). \tag{9.3}$$

If we assume that the glass plates confining the Hele–Shaw cell are clean, the origin of the perturbations can be attributed to the fluctuations of the interface only. Since the capillary energy is the only one involved, equation (2.20) in paragraph 2.5 can be applied, such that $A_i \approx (kT/\gamma)^{1/2}\mathrm{Log}(a/\lambda_c)^{1/2} \approx 10\,\text{Å}$. Thus, for $a \approx 5$ cm, $A_i/a \approx 10^{-7}$, and $a/\lambda_c \approx \sqrt{\kappa} \approx 10$. Again, the finger can remain stable even if its width is an order of magnitude larger than the most unstable wavelength of the planar interface.

The needle-like crystal is very particular, since the growth rate Γ at the tip, as determined by equation (9.12) in paragraph 9.5, depends on the constant C only. Thus, the stability criterion

$$\sqrt{C} \approx -\frac{9\sqrt{3}}{4}\mathrm{Log}\left(\frac{A_i}{\rho}\right) \tag{9.4}$$

is either always, or never met, depending on the nature of the crystal. As we have already discussed in paragraph 3.7, the origin of the perturbations of the liquid–solid interface is still unclear. In the simplest assumption we can consider an interface noise of capillary origin, as for the Saffman–Taylor finger, and obtain from equation (2.23) in paragraph 2.5, $A_i/\rho \approx 10^{-4}$, which leads to a value of C of the order of 100, precisely of the order of magnitude of the measured constants. This coincidence is probably related to the criterion of marginal stability suggested by Langer and Müller-Krumbhaar (1978) in the past, which indicated that the stability constant $\sigma \approx 1/C$ was determined by the marginal stability of the tip. In stationary state theory, the constant C is determined by the factor of anisotropy. From the general observation of the stability of the dendrites' tips, it seems that the factor of anisotropy of almost all the materials considered in paragraph 2.4 – though small – is sufficient for the right-hand side in equation (9.4) to be larger than the left-hand side. More precise experimental studies linking the different origins of noise, discussed in paragraph 3.7 according to criterion (9.4), seem necessary.

9.7 REFLECTION BY A POTENTIAL BARRIER OF A HIGH-ENERGY PARTICLE

Consider the problem of a particle of mass m moving in a one-dimensional potential with a wave function satisfying the stationary Schrödinger equation

$$-\frac{\bar{h}^2}{2m}\Delta\Psi + V(x)\Psi = E\Psi, \tag{9.1}$$

where $\bar{h} = \frac{h}{2\pi}$, h being the Planck constant. In Newtonian mechanics when the particle has an energy higher than the maximum of the potential barrier V_{max} (Pokrovskii and Khalatnikov 1961) it simply crosses the region of variable potential to be at $x = +\infty$ at long times. In quantum mechanics a part of the wave function is reflected by the potential barrier such that there exists a non-zero probability to find the particle at $x = -\infty$ at long times. To calculate in a general way the amplitude of the reflected wave we will assume that the scale of variation of Ψ is far faster than the scale of variation of the potential V. In this case, in the framework of the WKB approximation already used in paragraph 8.4, we are looking for Ψ in the form $\Psi(x) = \exp(iS/h)$ and we expand S in powers of h, $S = S_0 + (h/i)S_1 + \dots$. From equation (9.1) we deduce the equation for S,

$$\frac{1}{2m}\left(\frac{dS}{dx}\right)^2 - \frac{h}{2m}\frac{d^2S}{dx^2} = E - V. \tag{9.2}$$

Thus,

$$S_0(x) = \pm\int k(x)dx = \pm\int\sqrt{2m(E - V(x))}dx. \tag{9.3}$$

From equation (9.2) we deduce the equation for S_1:

$$\frac{dS_1}{dx}\frac{dS_0}{dx}+\frac{1}{2}\frac{d^2S_0}{dx^2} = 0 \tag{9.4}$$

from which we deduce that $S_1 = -(1/2)\mathrm{Log}k(x)$. The expansion in powers of h is only valid when the corrective term $(h/i)S_1$ does not exceed the dominating term S_0. This condition is met apart from the turning point x_0 where S_1 diverges, that is $k(x_0) = 0$. On both sides of the barrier, at $x = \pm\infty$, the particle is free. If we assume that at $x = +\infty$ the particle moves towards the right direction, that is $\Psi = a\exp(ip_\infty x) + b\exp(-ip_\infty x)$ where $p_\infty = h(2m\,E)^{1/2}$, then at $x = -\infty$, $\Psi = a\exp(ip_\infty x) + b\exp(-ip_\infty x)$ in general, corresponding to the superposition of an incident and a reflected wave.

In the WKB limit the general solution of equation (9.1) is written as $\Psi = a\Psi_+ + b\Psi$ where

$$\Psi_\pm(x) = \frac{1}{\sqrt{p}}\exp\left(\pm i\int^x pdx\right) \tag{9.5}$$

and $p = h(2m(E-V(x))^{1/2})$. It is clear that if we consider the superposition of waves on the real axis, $b = 0$ according to the condition at $x = +\infty$, and there is no reflected wave. This apparent paradox stems from the fact that the WKB solution is just an approximated solution and therefore the coefficients can only be defined to within $(h/pl)^2$ at best, where l is the typical lengthscale over which the potential changes.

To access the reflection coefficient we must consider the WKB solution on a line L, where the two waves $\Psi_\pm$ are of the same order of magnitude, that is a line $\mathrm{Im}\int^z pdz = \mathrm{cst}$, since, if we get away from this line, the ratio of the amplitudes of the two waves exponentially increases or decreases. This line must pass through the roots or the singularities of p^2, because, as we will show below, the two waves generated around them are of the same order of magnitude.

Let us consider the simplest case where z_0 is a simple root of p^2, that is $U(z_0) = E$. Around z_0, $p = C(z-z_0)^{1/2}$, and therefore

$$\int_{z_0}^{z} pdz = \frac{2}{3}C(z-z_0)^{3/2}. \tag{9.6}$$

From equation (9.6) it is clear that three lines $\text{Im}\int^{z} pdz = \text{cst}$ move away from z_0, with an angle $2\pi/3$ between them. Two of them – those which go along the x axis at $x = \pm\infty$ form the considered line L. Near z_0 the Schrödinger equation is of the form

$$\frac{d^2\Psi}{dz^2} + C^2(z - z_0)\Psi = 0. \tag{9.7}$$

The solution on the line L_l, on the left of z_0, corresponding to the transmitted wave, is determined from the Hankel functions

$$\Psi_{Ll} = C(z - z_0)^{1/2} H^{(1)}_{1/3}\left(\frac{2}{3}C(z - z_0)^{3/2}\right) \tag{9.8}$$

and behaves, for $C(z - z_0)^{3/2} >> 1$, as $\exp(i5\pi/12)\Psi_+$, where the lower limit of the integral in equation (9.5) is z_0. We determine the solution on the branch L_2 from the solution on L_l by a rotation of angle $-2\pi/3$, that is, if z is a point of the branch L_l,

$$\Psi_{L2} = C(\exp-(2i\pi/3)(z - z_0))^{1/2} H^{(1)}_{1/3}\left(\frac{2}{3}C(\exp-(2i\pi/3)(z - z_0))^{3/2}\right). \tag{9.9}$$

Using

$$H^{(1)}_{1/3}(\exp(-i\pi)z) = H^{(1)}_{1/3}(z) + \exp(-i\pi/3)H^{(2)}_{1/3}(z) \tag{9.10}$$

and the asymptotic behaviour of $H^{(1)}_{1/3}$ and $H^{(2)}_{1/3}$ for large z, we deduce, on L_2, $\Psi = \exp-(i5\pi/12)\Psi_- + \exp(-i\pi/3)\exp(i5\pi/12)\Psi_+$. Again, the lower limit of the integrals of equation (9.5) is z_0. We deduce the wave function at $x = -\infty$ with

$$\int_{z_0}^{z} pdz = p_\infty z - \alpha, \quad \alpha = \int_{-\infty}^{z_0} (p - p_\infty)dz + p_\infty z_0 \tag{9.11}$$

and therefore at $x = -\infty$,

$$\Psi \approx \frac{1}{\sqrt{p_\infty}}(\exp(ip_\infty x)\exp(-i(\alpha + 5\pi/12)) + \exp(-i\pi/3)\exp(-ip_\infty x)\exp(i(\alpha + 5\pi/12))) \tag{9.12}$$

where we can determine the reflection factor $f = b/a = i\exp(2i\alpha)$.

Thus, even if there is no real barrier, the non-uniformity of the potential generates a reflected wave. When the zero of the potential is in the upper half-plane, α is, from equation (9.11), a complex number with a positive imaginary part, since we assume that p is real here. The reflection factor is therefore an exponentially small number in the WKB limit.

$$\Psi_{L2}(z) = \bar{a_{11}}\exp\left(\frac{i}{\sqrt{\nu}}\int_{z_-}^{z_+} k_1 dz\right)\exp\left(\frac{i}{\sqrt{\nu}}\int_{z_+}^{z} k_1 dz\right) + \bar{a_{12}}\exp\left(\frac{i}{\sqrt{\nu}}\int_{z_-}^{z_+} k_2 dz\right)\exp\left(\frac{i}{\sqrt{\nu}}\int_{z_+}^{z} k_2 dz\right). \tag{9.10}$$

In the same way as for point z_- each wave Ψ_1 and Ψ_2 generates at the point z_+ a combination of the waves Ψ_1 and Ψ_2 on the line L_3 which moves away at $l = +\infty$ (figure 9.8). Thus, with obvious notations from equation (9.10) the coefficient of the increasing mode Ψ_2 is

$$b = \overset{+}{a_{21}}\bar{a_{11}}\exp\left(\frac{i}{\sqrt{y}}\int_{z_-}^{z_+} k_1 dz\right) + \bar{a_{21}}\overset{+}{a_{22}}\exp\left(\frac{i}{\sqrt{\nu}}\int_{z_-}^{z_+} k_2 dz\right). \tag{9.11}$$

This coefficient must cancel out since only the decreasing mode is allowed on the line L_3. A Bohr–Sommerfeld-type quantification condition results and is written as

$$\mathrm{Re}\left(\frac{1}{\sqrt{\nu}}\int_{z_-}^{z_+}(k_1 - k_2)dz\right) = \mathrm{Cst} + 2n\pi \tag{9.12}$$

since, on the line L_2, $\mathrm{Im}\int_{z_-}^{z_+}(k_1 - k_2)dz = 0$.

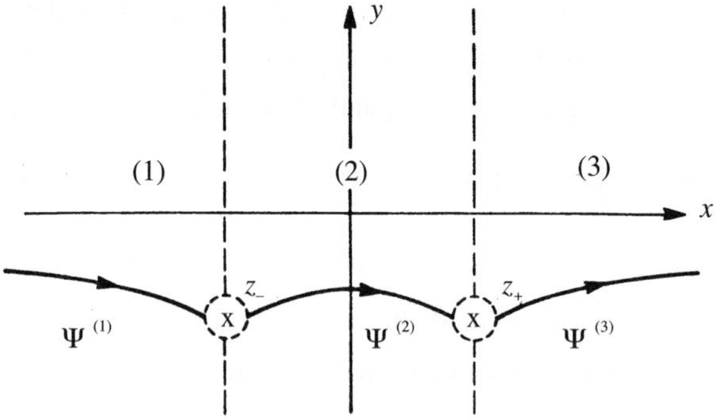

Figure 9.8 L lines and turning points (Bensimon *et al.* 1987).

If in general integral (9.12) must be numerically evaluated, an analytic calculation is possible when $\sigma\sqrt{\varepsilon}$ is a small positive number. We obtain from equation (9.5)

$$k_1(z) = \sigma\sqrt{\nu}\exp(-i\theta(z)) \tag{9.13a}$$

$$k_2(z) = \exp(i\theta(z)/2) - \frac{\sigma\sqrt{\nu}}{2}\exp(-i\theta(z)). \tag{9.13b}$$

For the Saffman–Taylor finger we consider the expression of the curvilinear axis as a function of the angle θ, given by equation (8.15) in paragraph 8.4. Since we assume that $\sigma\sqrt{\nu}$ is small, the turning points are close to $\eta = i$ from equation (9.8) and $z_{\pm} \approx (\pi/2)(-i \pm \sqrt{\alpha})$ where α is supposed to be positive. The condition of quantification (9.12) thus leads to

$$\frac{\alpha^{3/4}}{2\sqrt{2\nu}} = n + \frac{3\sigma}{2}. \tag{9.14}$$

For the needle-like crystals we consider the relation $\theta(l)$ of the Ivantsov parabolas given by equation (8.3) in paragraph 8.6. When β is small, the condition of resonance must be calculated between the points $z_0 = -i\pi/4 + i(8\beta)^{3/4}/3$ and $z_+ = -i\pi/4 + (8\beta)^{3/4}/3$ and we obtain

$$\frac{\beta^{7/4}}{\nu} = 6.08\left(\left(n + \frac{1}{2}\right)^2 + 0.45\beta\sigma^2\right). \tag{9.15}$$

Setting $\sigma = 0$ in equations (9.14) and (9.15) we find again the relation of selection of the stationary fingers or of the needle-like crystals determined in paragraphs 8.4 and 8.6. In fact, in this calculation the neutral modes are by definition perturbations of the curved shape without surface tension, which remain stationary. They therefore determine the stationary fingers at small surface tension. If we plot σ as a function of $\alpha^{3/4}/\sqrt{\nu}$, for all *n*, we obtain straight lines of positive slopes crossing the *x* axis at a series of points corresponding to the neutral modes (figure 9.9). It is clear that the branch *n* has $n-1$ unstable modes, as numerically shown by Kessler and Levine (1986). Thus, only the first branch of the solutions of stationary states, the branch of McLean and Saffman in paragraph 6.4, is linearly stable. This calculation fully solves the selection problem, since a single stationary curved shape is finally stable.

Using a slightly different formalism, Xu (1991) obtains a relation of mode quantification perfectly analogous to equation (9.12). He shows that this equation also has oscillating neutral modes ($\mathrm{Re}\sigma = 0$, $\mathrm{Im}\sigma \neq 0$) as its solutions. The turning points are very far away from the point $\eta = i$, and are not subjected to the '$\lambda - 1/2$' effect, in the case of the Saffman–Taylor finger or to the 'small β' effect, in the case of the needle-like crystal. These neutral modes would be associated to curved shapess for which the tip oscillates at a well-defined frequency, periodically emitting secondary branches on the sides. Oscillating fingers seem to have been observed by Homsy, in an extremely clean Hele–Shaw cell, and oscillating dendrites are also sometimes observed (Rolley *et al.* 1986). For dendrites this mode only exists at a well-defined value of *C*, which is therefore independent of any anisotropy factor, and thus satisfies the hypothesis of marginal stability introduced by Langer and Müller-Krumbhaar in the past (1978). Nevertheless, such modes have not been detected yet in complete numerical simulations of these equations.

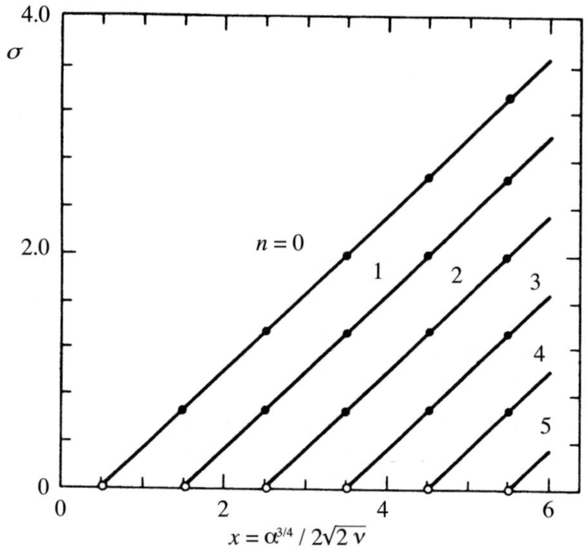

Figure 9.9 σ as a function of $\alpha^{3/4}/2\sqrt{2}\nu$ for different n (Bensimon *et al.* 1987).

9.9 THE CELL–DENDRITE TRANSITION

As mentioned in paragraph 8.10 a stationary cellular lattice is just a periodic juxtaposition of crystals in fictitious capillaries, the parallel walls of which are at the same distance from the cellular tips, and along which a temperature gradient is kept constant. Before dealing with the problem of the cell–dendrite transition, it is therefore useful to solve the problem of the stability of a crystal in a capillary.

The Saffman–Taylor branch of the stationary crystal in a capillary determined in paragraph 8.9 exhibits the unphysical property of having a higher velocity for a smaller oversaturation. This particularity is related to an instability of the kind of growth (Pelcé 1988). Since the Saffman–Taylor finger is linearly unstable this instability only comes from a characteristic which distinguishes the dynamics of viscous fingering and of the crystalline growth, that is the non-steadiness of the diffusion. When the crystal gets thicker, it releases more impurities in the same space and its growth therefore must slow down. However, on the Saffman–Taylor branch a slower crystal is thicker and the process is amplified. As we determine in appendix 1 the growth rate of the instability σ determined by

$$2\sigma \frac{D_c}{V^2} = -\frac{C\lambda(C)\dfrac{d\lambda(C)}{dC}}{\dfrac{d}{dC}(C\lambda(C))^2} \qquad (9.1)$$

is positive, since the derivative $d\lambda/dC$ of the McLean–Saffman function is negative. Depending on the initial conditions the crystal growth becomes slower and the crystal thicker as it grows (figure 9.10), and its tip resembles a Saffman–Taylor finger, as in the simulation by Kessler and Levine (1988). The crystal can also grow faster and get thinner, acquiring a dendritic morphology (Molho *et al.* 1990).

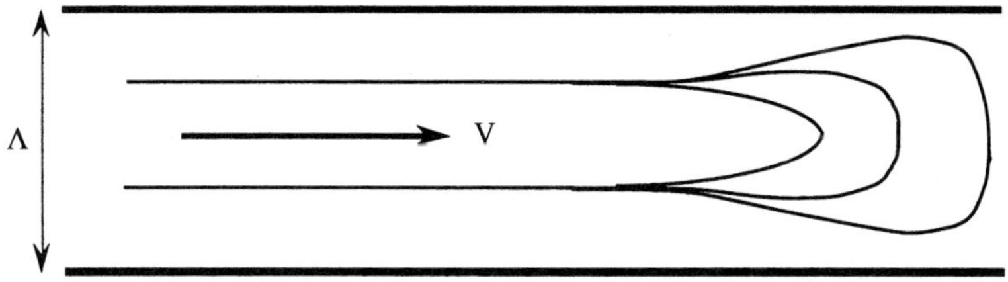

Figure 9.10 Crystal which gets thicker and slower in a capillary.

As shown in figure 9.11, and calculated from the results of appendix 1, the temperature gradient $(1/\nu \neq 0)$ inhibits the instability mentioned above since stronger and stronger gradients (decreasing values of ν) lead to $Re\sigma < 0$. The Saffman–Taylor branch is thus stabilised and the observed cellular forms can be associated to this branch. The gradient also introduces an imaginary part to the growth rate, which leads to an oscillation of the tips. In general, this oscillation is damped because the values of C actually achieved are large. As we show in appendix 2 the oversaturation of the cells in paragraph 8.10 decreases when the wavelength increases, and the wavelength of cells closer to the liquid has the tendency to increase and thus to eliminate the cells left behind. The value C of the cellular lattice can therefore only increase, making the value of $d\lambda/dC$ decrease below the instability threshold of the cells of the Saffman–Taylor branch.

For a given value of ν, when the cell reaches the plateau of the curve $\Delta(C)$ plotted in figure 8.17 in paragraph 8.10, the cell has a tip of radius of curvature of about the order of $\Lambda/2$, almost independent of the control parameters. Since the shape of the cell is close to a Saffman–Taylor finger, the growth rate of the Mullins–Sekerka instability is, as in paragraph 9.5, dominating in the region of the tip. A calculation analogous to that in paragraph 9.5 (Sarkar 1986) indicates that this growth rate is proportional to $\Lambda/2\lambda_c$, where

$$\lambda_c = 2\pi\sqrt{3}\left(\frac{d_{0c}l}{l - \frac{1}{l_T}}\right)^{1/2} \tag{9.2}$$

is the most unstable wavelength determined from equation (6.9) in paragraph 6.6.

Here d_{0c} and l_T are defined by the local concentration of the tip, that is, replacing c_∞/K by $c_\infty(1 + \Delta(1/K - 1))$ in the lengths used in paragraph 8.10. As in paragraph 9.6 the secondary branching can then occur when $\Lambda/2\lambda_c$ exceeds the logarithm of the amplitude of the interface fluctuations. Some coherence in the secondary branching from a cell to a neighbouring one can sometimes be observed (Eshelman *et al.* 1988, Georgelin and Pocheau 1998). The frequency of oscillation of the cells determined by the temperature gradient is too small to be the only cause responsible for this phenomenon. The interaction between the interface dynamics of neighbouring cells as well as the influence of the secondary branching on the motion of the tip must make an important contribution.

The modes of deformation of the cellular interface considered here are of either large or small wavelength, compared to the cellular wavelength. Other modes of instability involving wavelengths of the order of the cellular wavelength have been determined by Kopczynski and coworkers (1996) by the method of Floquet-Bloch, and observed by Georgelin and Pocheau (1997).

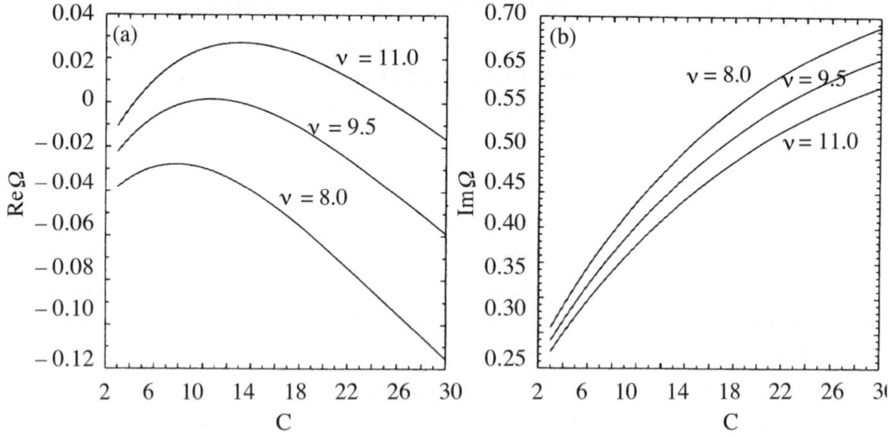

Figure 9.11 (a) Re Ω as a function of C for $K = 0.2$ and three values of ν. (b) Im Ω for the same parameter values. The instability appears at $\nu = 9.5$ and $C = 11$.

Appendix 1: Instability of a crystal in a capillary (Pelcé 1988).

From paragraph 8.9, since the Péclet number $P = \Lambda V/D_c$ is small, the region of the tip can be assimilated by a planar interface at $z = z_0(t)$ which releases in the one-dimensional diffusion layer a flux of impurities of effective concentration $\Phi \approx c - c_0$, modulated by the relative size of the finger

$$-D_c\frac{d\Phi}{dz} = \lambda c_0(1-K)\frac{dz_0}{dt}$$

where λ depends on $C = (\Lambda^2 dz_0/dt)/D_c d_{0c}$ by the McLean and Saffman curve, and $d_{0c} = \gamma T_0/|m|Qc_0(1-K)$. Ahead of the tip

$$\frac{\partial\Phi}{\partial t} = D_c\Delta\Phi$$

at $z = z_0(t)$, $\Phi \approx 0$, and, far in front of the finger, $\Phi = c_\infty - c_0$. Setting $dz_0/dt = V$ we find again the stationary form of the aforementioned paragraph, since then $\Phi = (c_\infty - c_0)(1-\exp(-(z-z_0)V/D_c))$ in the liquid. When the crystal slightly deviates from this stationary dynamics $\delta\Phi \approx A\exp(\sigma t - qz)$, $\delta z_0 \approx B\exp\sigma t$, such that, using the equations written above

$$\Delta\sqrt{1+4\sigma D_c/V^2} = \frac{d}{dC}(C\lambda(C))2\sigma D_c/V^2 + \Delta$$

the non-null solution of which is

$$2\sigma D_c/V^2 = -\frac{C\lambda(C)\dfrac{d\lambda(C)}{dC}}{\dfrac{d}{dC}(C\lambda(C))^2}$$

which is positive, when λ decreases with C.

When a temperature gradient is present (Karma and Pelcé 1989), using equation (8.3) in paragraph 8.10, we replace the first equation of this appendix by

$$-D_c\frac{d\Phi}{dz} = \lambda c_0(1-\mathrm{K})\frac{dz_0}{dt} - (1-\lambda)\frac{D_c G}{m}$$

where $\Phi \approx c - c_\infty/K - Gz/m$ is the new effective concentration. As in the same paragraph λ depends on C according to the McLean–Saffman relation, which states that the instantaneous value of C is

$$C = \frac{\Lambda^2\dfrac{dz_0}{dt}}{D_c d_{0c}}\left(1-(1-K)\frac{z_0}{l_\mathrm{T}} - \frac{D_c}{l_\mathrm{T}\dfrac{dz_0}{dt}}\right)$$

where $d_{0c} = \gamma T_0 K/|m|Qc_\infty(1-K)$ is the capillary length, and $l_T = |m|c_\infty(1-K)/KG$ the thermal length. At the tip of the cell, $c = c_\infty/K + Gz_0/m$, such that $\Phi \approx 0$. Setting $dz_0/dt = V$ we find the stationary solutions characterised by equations (8.6) and (8.7) in paragraph 8.10. Again, when the crystal slightly gets away from this stationary dynamics, $\delta\Phi \approx A\exp(\sigma t - qz)$, $\delta z_0 \approx B\exp\sigma t$, such that, using the equations mentioned above

$$\left(\Delta - \frac{1}{2\nu}\right)\sqrt{1+2\Omega} = a\Omega + b$$

where $\Omega = 2\sigma D_c/V^2$, Δ being defined by equation (8.6) in paragraph 8.10, a and b by the relations

$$a = \frac{K + \frac{1}{2\nu}(1-K)(1-\lambda)}{1-\lambda(1-K)} \frac{d}{dC}(C\lambda(C))$$

$$b = \frac{\frac{(1-\lambda)}{\nu} + K\lambda\left(1+\frac{1}{2\nu}\right)}{1-\lambda(1-K)} - \frac{1-K}{\nu}\frac{d}{dC}(C\lambda(C))$$

where $1/2\nu = D_c/Vl_T$.

Appendix 2: Derivative of the cellular wavelength at small Péclet number.

As already noticed by Jackson and Hunt (1966) regarding the eutectic interfaces when a cellular lattice wrinkles the effective walls move away from each other at the tips of smaller oversaturation, where the curvature $1/R$ of the line joining the ends of the tips is positive and get closer at the tips of stronger oversaturation (figure 9.12).

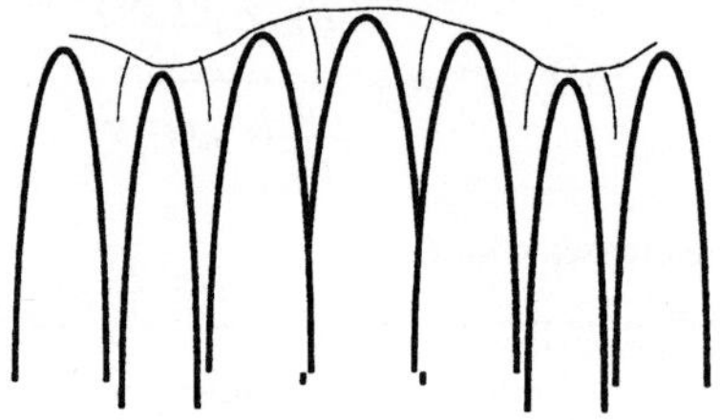

Figure 9.12 Effective wrinkled interface.

From an analysis analogous to that of Langer (1980) in the eutectic interfaces we write

$$\frac{1}{\Lambda}\frac{d\Lambda}{dt} = \vec{v}\cdot\vec{n}\frac{1}{R}$$

from which we deduce

$$\frac{d(\delta\Lambda)}{dt} = -\Lambda V\frac{\partial^2 z_0}{\partial y^2}.$$

Furthermore from the definition of the oversaturation Δ of paragraph 8.10, $(d\Delta/d\Lambda)\delta\Lambda = -\delta z_0/l_T$, such that

$$\sigma = -\Lambda V l_T\frac{d\Delta}{d\Lambda}k^2$$

where σ and k are the growth rate and the wavenumber of the effective interface respectively.

BIBLIOGRAPHY

[1] D. Bensimon, 'Stability of Viscous Fingering', *Phys. Rev.*, **A33**, p. 1302, 1986.

[2] D. Bensimon, L.P. Kadanoff, S. Liang, B.I. Shraiman, C. Tang, 'Viscous Flow in Two Dimensions', *Rev. Mod. Phys.,* **58**, p. 977, 1986.

[3] D. Bensimon, P. Pelcé, B. Shraiman, 'Dynamics of Curved Fronts and Pattern Selection', *J. Physique*, **48**, p. 2081, 1987.

[4] Ph. Bouissou, A. Chiffaudel, B. Perrin, P. Tabeling, 'Dendritic Side-Branching Forced by an External Flow', *Europhys. Letters*, **13**, p. 89, 1990.

[5] B. Caroli, C. Caroli, B. Roulet, 'On the Linear Stability of Needle Crystals: Evolution of a Zeldovich Localized Front Deformation', *J. Physique*, **48**, p. 1423, 1987.

[6] A.J. Degregoria, L.W. Schwartz, 'A Boundary-Integral Method for Two-Phase Displacement in Hele–Shaw Cells', *J. Fluid. Mech.*, **164**, p. 383, 1986.

[7] A. Dougherty, P.D. Kaplan, J.P. Gollub, 'Development of Side Branching in Dendritic Crystal Growth', *Phys. Rev. Lett.*, **58**, **16**, p. 341, 1987.

[8] M.A. Eshelman, V. Seetharaman, R. Trivedi, 'Cellular Spacings', *Acta Metall.*, **36**, p. 1165, 1988.

[9] M. Georgelin, A. Pocheau, 'Onset of Sidebranching in Directional Solidification', *Phys. Rev. E*, **57**, p. 3189, 1998.

[10] M. Georgelin, A. Pocheau, 'Oscillatory Instability, Limit Cycle, and Transition to Doublets in Directional Solidification', *Phys. Rev. Lett.*, **79**, p. 2698, 1997.

[11] K.A. Jackson, J.D. Hunt, 'Lamellar and Rod Eutectic Growth', *Transactions of the Metallurgical society of AIME*, **236**, p. 1129, 1966.

[12] A. Karma, P. Pelcé, 'Oscillatory Instability of Deep Cells in Directional Solidification', *Phys. Rev.*, **A39**, p. 4162, 1989.

[13] D. Kessler, H. Levine, 'Theory of the Saffman–Taylor 'Finger' Pattern', *Phys. Rev.*, **A33**, p. 2621, 1986.

[14] D. Kessler, H. Levine, 'Stability of Dendritic Crystals', *Phys. Rev. Lett.*, **57**, p. 3069, 1986.

[15] D. Kessler, H. Levine, *Mod. Phys. Lett.*, **2**, p. 945, 1988.

[16] K.K. Koo, R. Ananth, W.N. Gill, 'Tip-Splitting in Dendritic Growth of Ice Crystals', *Phys. Rev.*, **A44**, p. 3782, 1991.

[17] P. Kopczynski, W.J. Rappel, A. Karma, 'Critical Role of Cristalline Anisotropy in the Stability of Cellular Array Structures in Directional Solidification', *Phys. Rev. Lett.*, **77**, p. 3387, 1996.

[18] J.S. Langer, 'Eutectic Solidification and Marginal Stability', *Phys. Rev. Lett.*, **44**, p. 1023, 1980.

[19] J.S. Langer, 'Dendritic Sidebranching in the Three-Dimensional Symmetric Model in the Presence of Noise', *Phys. Rev.*, **A36**, p. 3350, 1987.

[20] J.S. Langer, H. Müller-Krumbhaar, 'Theory of Dendritic Growth - I. Elements of Stability Analysis; II. Instabilities in the Limit of Vanishing Surface Tension; III. Effect of Surface Tension', *Acta Metallurgica*, **26**, p. 1681, 1978.

[21] P. Molho, A.J. Simon, A. Libchaber, 'Péclet Number and Crystal Growth in a Channel', *Phys. Rev.*, **A42**, p. 904, 1988.

[22] P. Pelcé, 1986, 'Dynamique des Fronts Courbés', thèse d'État, université de Provence and, 'Introduction', *in Dynamics of Curved Fronts*, Academic Press, New York (1988).

[23] P. Pelcé, 'Unsteady Crystal Growth in a Channel in the Small Péclet Number Limit', *Europhys. Letters*, **3**, p. 907, 1988.

[24] P. Pelcé, P. Clavin, 'The Stability of Curved Fronts', *Europhys. Letters*, **3**, p. 907, 1987.

[25] R. Pieters, J.S. Langer, 'Noise-Driven Sidebranching in the Boundary-Layer Model of Dentritic Solidification', *Phys. Rev. Lett.*, **56**, p. 1948, 1986.

[26] R. Pieters, 'Noise-Induced Sidebranching in the Boundary-Layer Model of Dendritic Solidification', *Phys. Rev.*, **A37**, p. 3126, 1988.

[27] V.L. Pokrovskii, I.M. Khalatnikov, 'On the Problem of Above-Barrier Reflexion of High-Energy Particles', *Soviet Physics JETP*, **13**, p. 1207-1210, 1961.

[28] X.W. Qian, H.Z. Cummins, 'Dendritic Sidebranching Initiation by a Localized Heat Pulse', *Phys. Rev. Lett.*, **64**, p. 3038, 1990.

[29] E. Rolley, S. Balibar, F. Gallet, 'The First Roughening Transition of ^{3}He Crystals', *Europhys. Letters*, **2 (3)**, p. 247, 1986.

[30] P.G. Saffman, 'Viscous Fingering in Hele–Shaw Cells', *J. Fluid. Mech.*, **173**, p. 73, 1986.

[31] S. Sarkar, 'A WKB Type Analysis of Marginal Stability in Directional Solidification', *Physics Letters*, **A117**, p. 137, 1986.

[32] P. Tabeling, G. Zocchi, A. Libchaber, 'An Experimental Study of the Saffman–Taylor Instability', *J. Fluid. Mech.*, **177**, p. 67, 1987.

[33] J.J. Xu, 'Interfacial Wave Theory of Solidification: Dendritic Pattern Formation and Selection of Growth Velocity', *Phys. Rev.*, **A43**, p. 930, 1991.

[34] J.J. Xu, 'Global Instability of Viscous Fingering in a Hele–Shaw Cell: Formation of Oscillatory Fingers', *Euro. Jnl. of Applied Mathematics*, **2**, p. 105, 1991.

[35] Ya.B. Zeldovich, 'Structure and Stability of Steady Laminar Flame at Moderatly Large Reynolds Number', *Combustion and Flame,* **40**, p. 225, 1980.

[36] Ya.B. Zeldovich, A.G. Istratov, N.I. Kidin, V.B. Librovich, 1980, 'Flame Propagation in Tubes: Hydrodynamics and Stability', in *Selected Works of Ya.B. Zeldovich*, vol. 1, p. 459, Princeton University Press, 1992.

[37] Ya.B. Zeldovich, I.D. Novikov, 'The Structure and Evolution of the Universe', vol. 2 *Relativistic Astrophysics*, the university of Chicago Press, 1983.

CHAPTER 10

Evolution of growth forms

10.1 GEOMETRICAL MODELS

Consider a deformable curve C, described by an ensemble of mobile points $M(x(t), y(t))$ and defined by $\Phi(x, y, t) = 0$. During the deformation of curve C,

$$\frac{d\Phi}{dt} = \vec{v} \cdot \vec{\nabla}\Phi + \frac{\partial \Phi}{\partial t} = 0 \tag{10.1}$$

where $\vec{v} = U\vec{n} + V\vec{t}$ is the velocity of points M. U and V are the normal and tangential velocities, respectively, and $\vec{n}$ and $\vec{t}$ are vectors normal and tangential to the curve. Since $\vec{\nabla}\Phi$ is normal to the curve, the equation for evolution (10.1) is reduced to

$$(\vec{v} \cdot \vec{n})\left(\vec{n} \cdot \vec{\nabla}\Phi\right) + \frac{\partial \Phi}{\partial t} = 0\,. \tag{10.2}$$

Thus, the motion of the curve is only described by the normal velocity of these points. The tangential velocity appears as a degree of freedom corresponding to the invariance of the gauge or the parametrisation. The choice of parametrisation depends on the problem in question. The simplest case is clearly $V = 0$, but certain constraints may impose other possibilities. The relative velocity of two neighbouring points is written as

$$\frac{d}{dt}(\vec{X}(l+\delta l) - \vec{X}(l)) = \frac{d}{dt}(\exp(i\theta)\delta l) = \left(i\frac{d\theta}{dt}\delta l + \frac{d(\delta l)}{dt}\right)\exp(i\theta) \tag{10.3}$$

where l is the curvilinear coordinate, $\vec{X}(l)$ is a point on the curve, $\tau = d\vec{X}/dl = \exp(i\theta)$ is the unit vector tangential to the curve, and θ is the angle between the tangent and a given direction.

We can write the same quantity

$$\left(\frac{d}{dt}\vec{X}\right)\delta l = \frac{d}{dl}(-iU\exp(i\theta) + V\exp(i\theta))\delta l\,. \tag{10.4}$$

Identifying the real and imaginary parts of equations (10.3) and (10.4) leads to

$$\frac{1}{\delta l}\frac{d(\delta l)}{dt} = U\kappa + \frac{\partial V}{\partial l} \tag{10.5a}$$

$$\frac{d\theta}{dt} = -\frac{\partial U}{\partial l} + V\kappa . \tag{10.5b}$$

Thus, using the relation

$$\frac{d}{dt}\left(\frac{\partial\theta}{\partial l}\right) = -\frac{1}{\delta l}\frac{d(\delta l)}{dt}\frac{\partial\theta}{\partial l} + \frac{\partial}{\partial l}\left(\frac{d\theta}{dt}\right) \tag{10.6}$$

we deduce the equation of the evolution of the curve

$$\frac{d\kappa}{dt} = -\left(\kappa^2 + \frac{\partial^2}{\partial l^2}\right)U + V\frac{\partial\kappa}{\partial l} \tag{10.7}$$

or using

$$\frac{dl}{dt} = \int_0^l U\kappa dl + V(l) - V(0) \tag{10.8}$$

we get

$$\frac{\partial\kappa}{\partial t} = -\left(\kappa^2 + \frac{\partial^2}{\partial l^2}\right)U + \left(V(0) - \int_0^l \kappa U dl\right)\frac{\partial\kappa}{\partial l} . \tag{10.9}$$

A simplified model of crystal growth consists of approximating the interface to a curve, the normal velocity of which only depends on its curvature and its derivatives with respect to the curvilinear coordinate (Brower, Kessler, Koplik and Levine 1983)

$$U = F(\kappa) + \gamma\frac{\partial^2\kappa}{\partial l^2} . \tag{10.10}$$

$F(\kappa)$ must increase proportionally to κ for large radii, since, in this limit, from paragraph 5.4, the velocity is limited by the latent heat or the solute given out. It must first decrease and then become negative for smaller radii, because they reach or surpass the radius of nucleation. The term proportional to $\partial^2\kappa/\partial l^2$ represents the stabilising effects of the surface tension. Consequently, by the effects of crystalline anisotropy the coefficient γ can possibly become dependent on the angle θ between the normal and a given direction, as can the coefficients of $F(\kappa)$ by the effects of kinetic anisotropy. Since the points defining the interface do not play any particular physical role the simplest parametrisation can be chosen: it is the one in which the tangential velocity of the displacement is zero. We therefore numerically integrate equation (10.9) with the condition that $V(0) = 0$. Taking $F(\kappa) = \kappa + \alpha\kappa^2 - \beta\kappa^3$, $\gamma = 1$, and multiplying the second term of equation (10.10) by $(1 + \varepsilon\cos(4\theta))$, we find the two different evolutions, depending on the value of ε (Kessler *et al.* 1984). As shown in figure 10.1a, when the surface tension is isotropic ($\varepsilon = 0$), the interface forms lobes which are periodically split in two. When it is anisotropic ($\varepsilon \neq 0$), tips are formed and grow at a constant velocity, while producing secondary branching on the sides (figure 10.1b). This model, although very simplified, illustrates the important role of the anisotropy of the surface tension in the dynamics of dendrites.

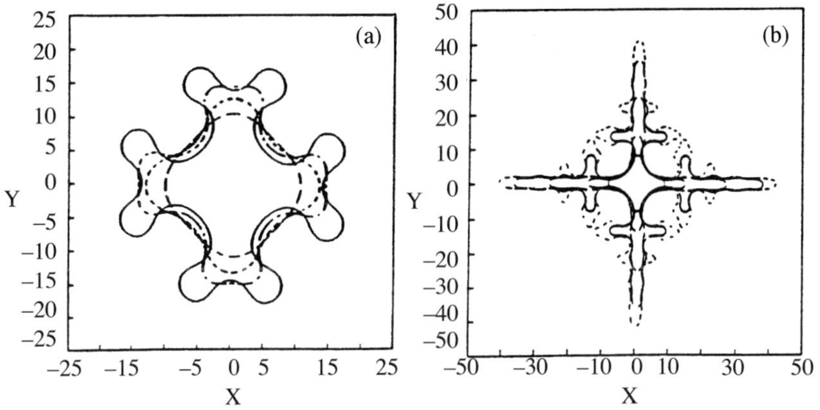

Figure 10.1 (a) $\alpha = 1$, $\beta = 0.25$ and $\varepsilon = 0$ (Kessler *et al.* 1984); (b) $\alpha = 4$, $\beta = 1.5$ and $\varepsilon = 0.15$ (Kessler *et al.* 1984).

A more elaborate model of the 'boundary layer', intermediary between the extremely simplified model described above and the physical model of crystalline growth described in paragraph 5.3, was studied by Ben-Jacob and coworkers (1983, 1984). He coupled the dynamics of an interface to a concentration field localised in a layer which lies along the interface. He therefore reduced the non-local dynamics of crystalline growth to a local dynamics, yet conserving its physical properties. Again, depending on whether the anisotropy factor is zero or not, the interface develops lobes which are periodically cut in two or dendrites which grow at a constant velocity.

10.2 DYNAMICS OF POLES

Determining the evolution of the shape of an interface with time starting from initial conditions is without a doubt the most intriguing problem, but also the most difficult. In general, numerical simulations of the equations remain the only possible way to tackle the problem. However, in the case where surface tension effects are neglected, exact solutions can be determined from the dynamics. As in chapter 7, the method consists of determining a conformal transformation which transfers the interface into a simple form, which is a circle of unit radius in that case. In this paragraph, the considered shapes of the interface evolve with time, thus the conformal transformations will also vary with time.

The inside of the disc of unit radius $|\xi| = 1$ is thus transformed into the fluid domain outside the interface by the conformal transformation $z = x + iy = f(\xi, t)$. The shape of the interface can therefore be written as $\gamma(s, t) = f(\exp is, t)$, where s is a real number between 0 and 2π. We thus deduce the vector normal to the interface $n = i(\partial f/\partial s)/|\partial f/\partial s| = \xi(\partial f/\partial \xi)/|\partial f/\partial \xi|$, as well as the normal velocity of the interface

$$\vec{v}_i \cdot \vec{n} = -\operatorname{Re}\left(\frac{\xi \dfrac{\partial f}{\partial \xi}\dfrac{\overline{\partial f}}{\partial t}}{\left|\dfrac{\partial f}{\partial \xi}\right|}\right). \tag{10.1}$$

In the plane of the variable ξ, since the interface is the circle of unit radius, the complex potential associated with the flow is simply

$$W = -\frac{Q}{2\pi}\operatorname{Log}\xi \tag{10.2}$$

where, as in paragraph 6.2, Qb is the injected volume of fluid per unit time. It follows that $dW/dz = -(Q/2\pi)(1/\xi)/(\partial f/\partial \xi)$, and therefore $\partial\phi/\partial n = \operatorname{Re}(ndW/dz) = (Q/2\pi)|\partial f/\partial \xi|$. Thus, applying boundary conditions (5.5) in paragraph 5.2, the conformal transformation f must satisfy equation

$$\operatorname{Re}\left(\xi \frac{\partial f}{\partial \xi}\frac{\overline{\partial f}}{\partial t}\right) = -\frac{Q}{2\pi} \tag{10.3}$$

on the unit circle $|\xi| = 1$. Consider first a round bubble for which $f(\xi) = a_0(t)/\xi$.

Equation (10.3) then determines the law of evolution for $a_0(t)$, that is, $a_0 da_0/dt = Q/2\pi$, giving again, as in paragraph 6.2, $a_0(t) = (Q(t-t_0)/\pi)^{1/2}$.

Now consider the shape of a finger evolving towards a Saffman–Taylor finger in a Hele–Shaw cell of unit width (Saffman 1959)

$$f(\xi, t) = d(t) - \operatorname{Log}\xi + 2(1-\lambda)\operatorname{Log}((1 + a(t)\xi)/2). \tag{10.4}$$

Coefficients $a(t)$ and $d(t)$ are determined by inserting transform (10.4) into equation (10.3). We thus obtain the two relations

$$\frac{d}{dt}(\lambda d(t) + (1-\lambda)\operatorname{Log}a(t)) = V \tag{10.5a}$$

$$d(t) - \operatorname{Log}a(t) + 2(1-\lambda)\operatorname{Log}(1 - a^2(t)) = \text{Cst} \tag{10.5b}$$

where λ is the relative thickness of the final stationary finger and $V = Q$ is the uniform fluid velocity far upstream from the finger. If $0 < a(0) << 1$, equation (10.4) determines a weak amplitude perturbation of the planar interface $x = 2a(0)(1-\lambda)\cos y$. At long times, $a(t)$ tends towards 1 so that from equation (10.5a), $d \approx Vt/\lambda$. We thus find one of the stationary Saffman-Taylor fingers

$$f(\xi, t) = \frac{Vt}{\lambda} - \mathrm{Log}\xi + 2(1-\lambda)\mathrm{Log}((1+\xi)/2)\,. \tag{10.6}$$

It is remarkable that this kind of transformation can be generalised to more complex forms, working in circular geometries where the splitting of the lobes can be arbitrarily generated, as well as in a Hele–Shaw cell.

To describe the forms in a circular geometry, let us consider for example the transformations of the form (Bensimon and Pelcé 1986, Howinson 1986)

$$f(\xi, t) = \frac{a_0(t)}{\xi} + \sum_{m=1}^{M} \alpha_m \sum_{l=0}^{N-1} \exp\left(-\frac{2\pi i l}{N}\right) \mathrm{Log}\left(a_m(t)\exp\left(\frac{2\pi i l}{N}\right) - \xi\right) \tag{10.7}$$

where M and N are integers, α_m are complex numbers and $a_m(t)$ complex numbers moving with time, of modulus greater than 1. The law of evolution for the coefficients a_m is obtained by inserting equation (10.7) into equation (10.3). For instance, when $N = 2$, we get

$$\begin{aligned} \frac{d}{dt}\left(a_0^2 - 2a_0 \sum_{m=1}^{M} \mathrm{Re}\,(\alpha_m a_m)\right) &= \frac{Q}{\pi} \\ \alpha_n a_0 a_n - \sum_{m=1}^{M} \alpha_n \overline{\alpha}_m \mathrm{Log}\left(\frac{a_n \overline{a}_m + 1}{a_n \overline{a}_m - 1}\right) &= \mathrm{Cst} \end{aligned} \tag{10.8}$$

for $n = 1, 2 \ldots, M$. Each logarithm of the sum (10.7) is associated with a cut, which spreads from the singularity $a_m(t)\exp(2i\pi l/N)$, located outside the circle of unit radius, towards infinity. As we can see from the structure of equations (10.8), a_0 increases towards infinity with time, proportionally to $\sqrt{t}$. This is the same as in the first example, though each singularity is moving towards the border of the unit circle. As soon as one singularity gets very close to the circle, it creates a hole in the real shape and in the end forms a channel where the fluid is almost at rest.

For example, the superposition of four pairs of singularities are presented in figure 10.2, $M = 4$, $N = 2$ (Bensimon and Pelcé 1986). Note $a_m(0) = r_m \exp(i\theta_m)$ and $\alpha_m = r_m^\alpha \exp(i\theta_m^\alpha)$. On the left in figure 10.2, $a_0 = 3$; $r = 3, 5, 8, 8$; $\theta = 0, \pi/2, \pi/4, 3\pi/4$; $r^\alpha = 1, 1, 1, 1$; $\theta^\alpha = \theta$. The first pair of singularities $(r = 3)$ gets close to the unit circle and creates two holes at $\theta = 0$, π. Then follows the second pair $(r = 5)$ at $\theta = \pi/2, 3\pi/2$, and finally the last two $(r = 8)$ at $\theta = \pi/4, 3\pi/4, 5\pi/4, 7\pi/4$. The choice of the initial positions of the singularities determines the final repartitioning of the tip-splittings. Other choices of initial constants model other shapes (on the right).

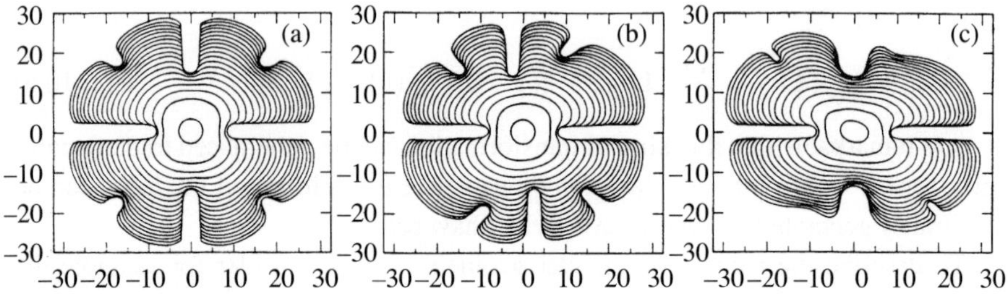

Figure 10.2 Evolution of interfaces by the pole motion (Bensimon and Pelcé 1986).

Instead of considering pairs of singularities one can also consider sextets ($N = 6$), and determine analogous equations to equations (10.8) (Howinson 1986). We obtain structures with an hexagonal symmetry, the lobes of which can eventually be split in two, which are analogous to those observed in experiments (Paterson 1981).

In the case of a linear geometry, consider transformations of the form

$$f(\xi, t) = d(t) - \mathrm{Log}\xi + \frac{1}{N}\sum_{n=1}^{N} 2(1-\lambda_n)\mathrm{Log}(1 + a_n(t)\xi)/2 \qquad (10.9)$$

where $0 < a_n < 1$ and $(1/N)\sum_{1}^{N}\lambda_n = \lambda$ are between 0 and 1. This transformation is a solution of equation (10.3) if the time-dependent coefficients satisfy the equations

$$\frac{d}{dt}\left(\lambda d + \frac{1}{N}\sum_{m=1}^{N}(1-\lambda_n)\mathrm{Log}a_n\right) = V \qquad (10.10a)$$

$$1 - \mathrm{Log}a_n + \frac{1}{N}\sum_{m=1}^{N} 2(1-\lambda_m)\mathrm{Log}(1 - a_m a_n) = \mathrm{Cst}\ . \qquad (10.10b)$$

At long times coefficients a_n tend towards 1 and $d \approx Vt/\lambda$. The asymptotic form is therefore that of a Saffman–Taylor finger of relative thickness λ determined from equation (10.6).

Consider for example, in figure 10.3a, the case $N = 2$, $a_1(0) = 0.2$, $a_2(0) = 0.6$, $\lambda_1 = 1.8$, $\lambda_2 = -0.2$. A relatively large finger $\lambda = 0.8$ develops, leaving behind a typical necking of the forms observed in experiments, absent in the previously studied first form $(N = 1)$.

Now consider $N = 3$, $a_1(0) = 0.2$, $a_2(0) = 0.04$, $a_3(0) = 0.4$, $\lambda_1 = 2.65$, $\lambda_2 = -0.05$, $\lambda_3 = -0.8$, $\lambda = 0.6$ (figure 10.3b). If the evolution of the tip is analogous to the preceding cases, that of the back of the finger is different. A secondary finger, analogous to those which are observed when the stationary finger becomes unstable, appears several radii of curvature away from the tip.

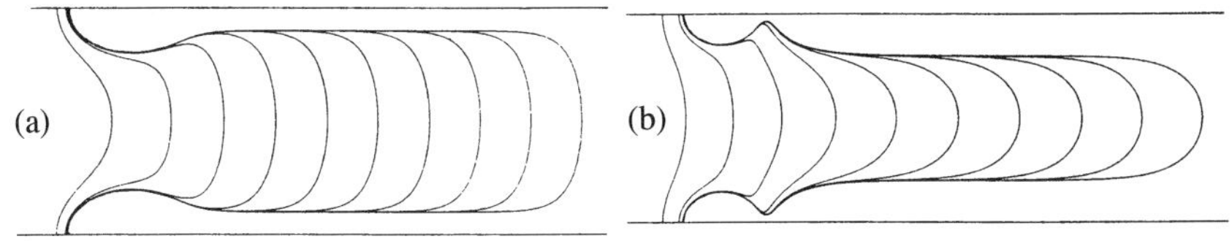

Figure 10.3 (a) Necking (Howinson 1986); (b) Secondary finger (Howinson 1986).

Certain solutions without surface tension are only valid for a finite time. Consider for example (Shraiman and Bensimon 1984) the conformal transformation

$$f(\xi, t) = a(t)\xi\left(1 + \frac{b(t)}{n-1}\xi^{-n}\right). \tag{10.11}$$

When b is small, the shape of the contour is a perturbation at n lobes of the circle of radius a. Using equation (10.3), we then get

$$\frac{da}{dt} = \frac{1}{a(1-b^2)}; \quad \frac{db}{dt} = (n-2)\frac{b}{a^2(1-b^2)} \tag{10.12}$$

from which we deduce $b \approx a^{n-2}$. Thus from equation (10.12), when $n > 2$, b tends towards 1 at a finite time, which is manifested by a singularity of the contour slope. Indeed, when $b = 1$, $d\gamma/ds = ia((1-\cos(ns)) + i\sin(ns))$. In particular, close to $s = 0$, $dx/ds \approx -s$, that is $x \approx -s^2$ and $dy/ds \approx s^2$, that is $y \approx s^3$. The singularity is therefore a cusp $x \approx -|y|^{2/3}$. If most of those growth forms resemble those observed in experiments, the mathematical intrigues still remain. Indeed, their evolution does not take into account the effects of surface tension, which, as we have discussed in chapter 8, are essential to the problems of selection and stability. Thus, all the solutions determined in this paragraph are unstable and are therefore not physically relevant. Nevertheless, with the problem still unsolved, they could appear as external solutions (in the sense of paragraph 5.2), that is, valid within a time interval where the surface tension effects are negligible, and therefore linked to the solutions where these effects are taken into account.

$$\frac{1}{R} = -\,\mathrm{Im}\left(\frac{\dfrac{\partial^2 f}{\partial s^2}}{\dfrac{\partial f}{\partial s}\left|\dfrac{\partial f}{\partial s}\right|}\right). \tag{10.5}$$

In the z variable, the potential W does not have the simple form of equation (10.2) in paragraph 10.2, since the real part is no longer constant over the unit circle. We must find a complex function defined inside the unit disc, whose real part is proportional to expression (10.5). This can be achieved using the integral Poisson formula. This relation indicates that an analytic function defined inside the unit disc $f(\xi) = a(\xi) + ib(\xi)$, whose real part over the unit disc is given, $a(s) = a_0 + \Sigma(a_n \exp(ins) + a_n^* \exp(-ins))$, can be written as $f(\xi) = \mathrm{A}\{f\}(\xi) = a_0 + 2(\Sigma a_n \xi^n)$. We can thus write the expression for the complex potential W in the disc of unit radius,

$$W = -\frac{Q}{2\pi}\mathrm{Log}\xi + \frac{\sigma b^2}{12\eta}A\left\{\frac{1}{R}\right\}(\xi). \tag{10.6}$$

The evolution equation for the conformal transformation is then deduced in the same way as in paragraph 10.2:

$$\frac{\partial f}{\partial t} = n\frac{\partial \Phi}{\partial n} = n\mathrm{Re}\left(n\frac{dW}{dz}\right) = \xi\frac{\partial f}{\partial \xi}\left(\frac{\mathrm{Re}\left(\xi\dfrac{dW}{d\xi}\right)}{\left|\xi\dfrac{\partial f}{\partial \xi}\right|^2}\right) \tag{10.7}$$

at the border of the unit disc, if it is assumed that the points determining the parametrisation of the interface are moving along the normal direction. However, equation (10.7) does not keep the analytic nature of the conformal transformation, since the

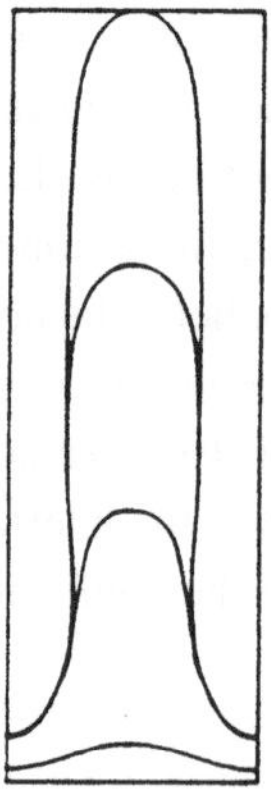

Figure 10.4 Stationary finger (Degregoria and Schwartz 1986).

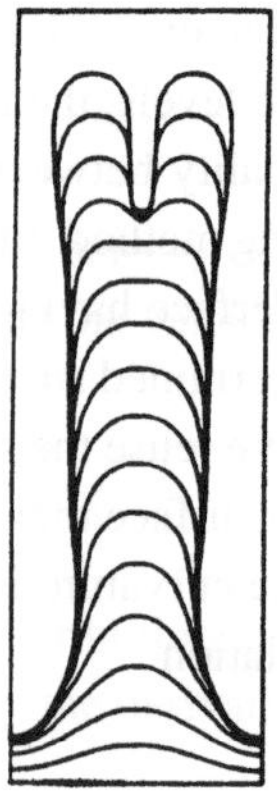

Figure 10.5 Tip splitting (Degregoria and Schwartz 1986).

right-hand term is not an analytic function. In order to keep the analytic nature of function f, we must make the right-hand member analytic, or in other words, add a tangential component C to the displacement of the parametrisation points, where C is an adequate real constant. It is therefore necessary to determine a constant C such that

$$\frac{\partial f}{\partial t} = \xi \frac{\partial f}{\partial \xi}\left(\frac{\mathrm{Re}\left(\xi \frac{\partial \Phi}{\partial \xi}\right)}{\left|\xi \frac{\partial f}{\partial \xi}\right|^2} + iC\right) \tag{10.8}$$

is analytic. For that, we use the Poisson formula again. Noting that the first term in the brackets of the right-hand side of equation (10.8) is the real part of the function needed in the unit circle, we can write

$$\frac{\partial f}{\partial t} = \xi \frac{\partial f}{\partial \xi} A\left\{\frac{\mathrm{Re}\left(\xi \frac{\partial \Phi}{\partial \xi}\right)}{\left|\xi \frac{\partial f}{\partial \xi}\right|^2}\right\} \tag{10.9}$$

then, using equation (10.6),

$$\frac{\partial f}{\partial t} = -\frac{Q}{2\pi} \xi \frac{\partial f}{\partial \xi} \mathrm{A}\left\{\frac{1 - \frac{\sigma b^2}{12\eta}\mathrm{Re}\left(\xi \frac{\partial (A\{1/R\})}{\partial \xi}\right)}{\left|\xi \frac{\partial f}{\partial \xi}\right|^2}\right\} \tag{10.10}$$

for all the points of the unit circle.

This equation, the structure of which resembles that of equation (9.9) in paragraph 9.4, can only be integrated numerically. We discretise the interface into N points, and we keep N coefficients of the Fourier series mentioned at the beginning of the paragraph for the evolution of quantity $A\{\ldots\}$. In the absence of surface tension the simulation reproduces the exact solutions determined in paragraph 7.5. When the effects of surface tension are taken into account, we reproduce the McLean and Saffman family of stationary fingers. At very low surface tension, an asymmetric mode is observed (figure 10.6). In figure 10.7, we note the logarithmic dependence of the threshold of instability as a function of the amplitude of the noise, theoretically evaluated in paragraph 9.6.

A final fourth method comes from statistical physics (Kadanoff 1985) and will be presented in more detail in the following chapter. It uses the fact that the stationary probability density of a random walk satisfies the Laplace equation. Thus by releasing walkers moving randomly on an array of points and waiting for them to attach near the already attached walkers, according to certain rules, we obtain an aggregate of points, the envelope of which has exactly the shape of a viscous finger. Thus, contrary to the other methods, there is no equation to be solved here.

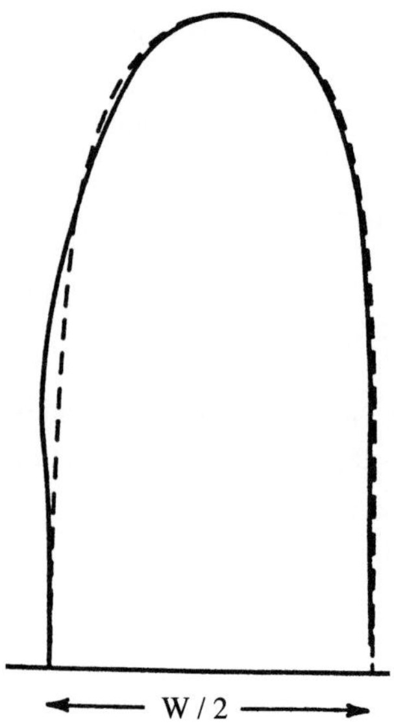

Figure 10.6 Asymmetric mode (Bensimon *et al.* 1986).

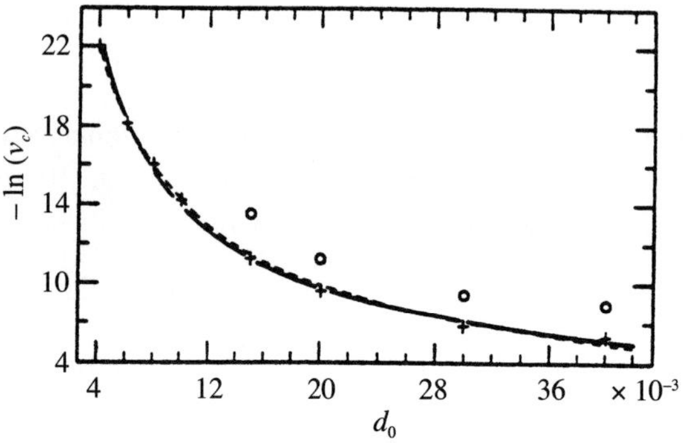

Figure 10.7 Threshold of instability (Bensimon *et al.* 1986).

10.4 ANISOTROPIC HELE–SHAW FLOWS

To underline the importance of crystalline anisotropy in the structure of dendrites, Ben-Jacob and coworkers (1985) had the idea to introduce an artificial anisotropy, by engraving a well-defined symmetric array on the plates of a circular Hele–Shaw cell. The engravings effectively introduce a crystalline anisotropy factor. In fact, in appendix Appendix: of paragraph 2.2, we have shown that the interface is curved on a large scale compared to the gap between the plates, so the interface can be considered as a curve with an effective surface tension $(\pi/4)\gamma$. The engravings induce a rapid change in the thickness, which must modulate the correction factor of the surface tension depending on the symmetry of the lattice. The more viscous fluid is coloured glycerine and the less viscous fluid is air. As shown in figure 10.8a, at very low velocities the interface forms facets, as in faceted crystals described in paragraph 2.4. At higher velocities (figure 10.8b), the interface forms lobes which are cut in two, as if there was no anisotropy. At still higher velocities (figure 10.8c), needles and eventually dendritic structures are formed.

In a conceptually related experiment, Couder and coworkers (1986) place a small bubble at the tip of one of the viscous fingers which grow in a Hele–Shaw cell of circular geometry. As shown in figure 10.9, the corresponding finger accelerates

Figure 10.8 Fingering in an engraved Hele–Shaw cell : (a) faceted growth, (b) tip-splitting, (c) needles, (d) dendrites (Ben-Jacob *et al.* 1985).

and rapidly acquires the autonomy of a dendrite. From paragraphs 8.4 and 8.11, bubbles and crystalline anisotropy have a common property of giving fingers and crystals relative thicknesses smaller than $1/2$, which allows the dendritic patterns to develop.

To analyse this effect more quantitatively, Almgren *et al.* (1993) perform a numerical simulation of viscous fingering in a Hele–Shaw cell of a circular geometry, with the surface tension being anisotropic, and its symmetry of order 4. The method of numerical integration is analogous to that of Degregoria and Schwartz (1986) described in the preceding paragraph. As shown in figure 10.10a, for a relatively large factor of anisotropy $\varepsilon = 0.5$, the interface develops four branches, at first very fast, then progressively slower. The length and width of a branch then evolve as $t^{3/5}$ and $t^{2/5}$, respectively (figure 10.10b). For a lower level of anisotropy, secondary branching develops and scaling factors are less accurate.

Since the normal velocity of the interface is proportional to the normal gradient of the laplacian field, the area of the growth form grows proportionally to t. Thus, when the surface tension is isotropic the two directions in space are equivalent, and both lengths and widths grow as $\sqrt{t}$. The factor of anisotropy distinguishes the two directions and if we assume that the length varies as t^{α}, the width of the finger would vary as $t^{1-\alpha}$. First note that the observed scaling factors, $\alpha = 3/5$, lead to $\rho^2 V = Cst$, where ρ is the radius of curvature of the tip and V its velocity. In fact, $\rho = (d^2y/dx^2)^{-1} \approx t^{2-3\alpha}$, $V = dx/dt \approx t^{\alpha-1}$ and so $\rho^2 V \approx t^{3-5\alpha}$.

The fact that $\rho^2 V$ is also constant here can be deduced from the same dimensional argument as that in paragraph 8.1. It is less clear that the dependence ε is the same as for a crystalline dendrite. In fact, as we have shown in paragraph 8.6, the

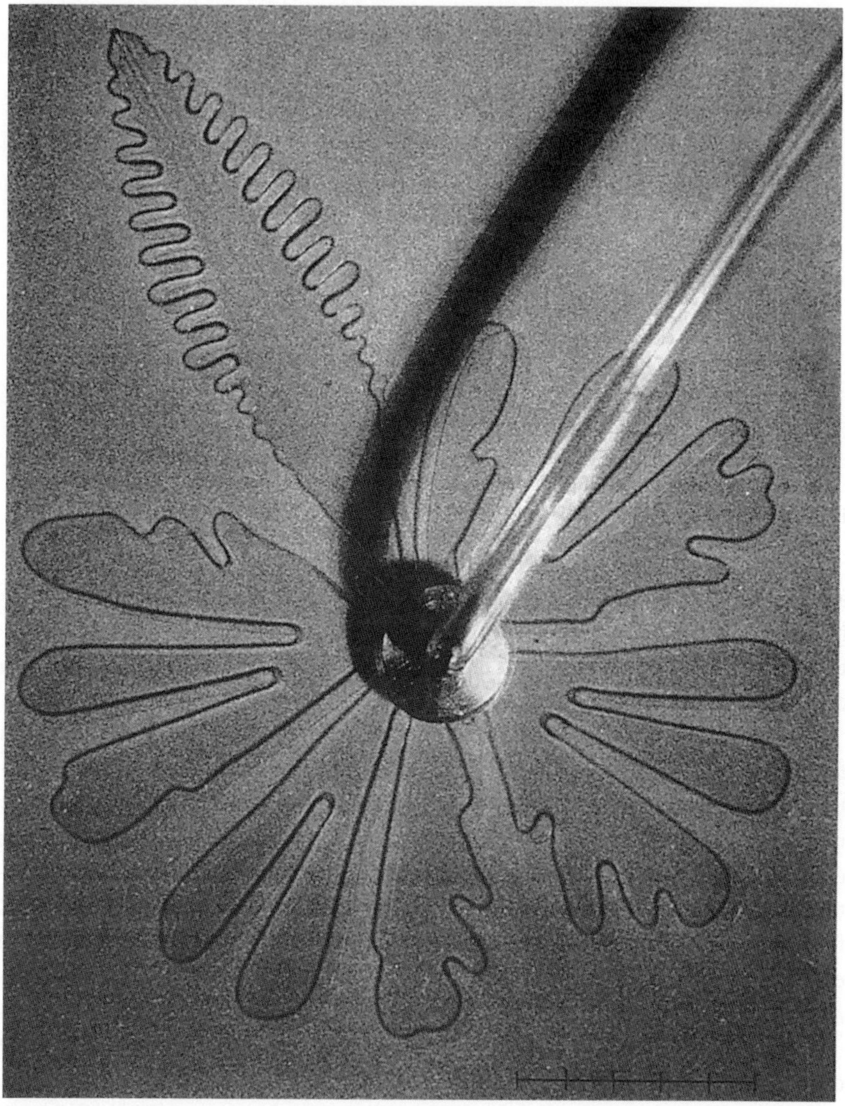

Figure 10.9 Dendritic finger (Couder *et al.* 1986).

constant $\rho^2 V$ is determined by both the marginal mode of the planar local front $k_m(l)$, determined from the dispersion relation, and the neighbouring form without surface tension, which is in this case a parabola. If at small Péclet number, the marginal modes of a viscous finger and a liquid–solid interface have the same expression, the form without surface tension which we can consider here, can only be approximated by a parabola at sufficiently long times. In fact, when the width of a

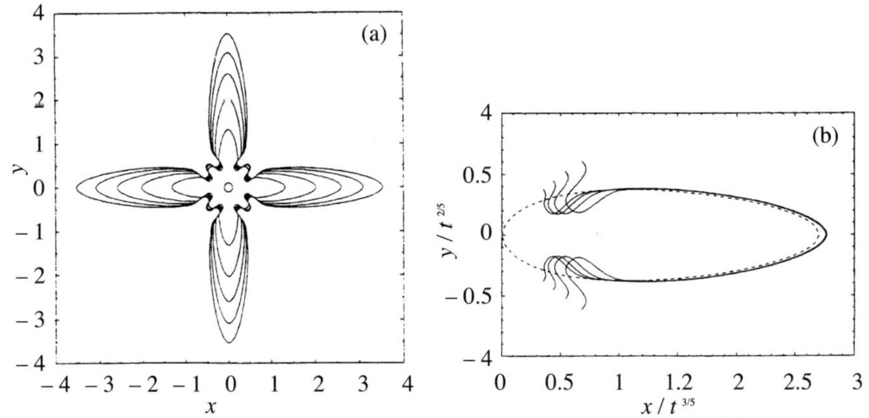

Figure 10.10 (a) Anisotropic fingering $\varepsilon = 0.5$ (Almgren *et al.* 1993); (b) self-similar form $\varepsilon = 0.5$ (Almgren *et al.* 1993).

branch grows much slower than its length, the growth form can be approximated to a cross of size $R(t) = At^{\alpha}$ at long times. The potential Φ, which satisfies $\Phi = 0$ at the cross, is determined by the conformal transformation on the circle of the appendix of paragraph 7.2

$$W = \Phi + i\Psi = \frac{1}{2}\cosh^{-1}\left(\frac{z}{R}\right)^2 ; \quad \frac{dW}{dz} = \frac{z}{\sqrt{z^4 - R^4}} . \tag{10.1}$$

We thus deduce the vertical velocity of the upper edge of the horizontal branch on the right of the cross

$$\frac{dy}{dt} = \frac{x}{\sqrt{x_{\text{tip}}^4 - x^4}} ; \quad 0 < x < x_{\text{tip}}(t) = At^{3/5} \tag{10.2}$$

and so the shape of the branch

$$y(x, t) = t^{2/5}\frac{5}{3A}\left(\frac{x}{x_{\text{tip}}}\right)^{2/3}\int_{x/x_{\text{tip}}}^{1}\frac{ds}{s^{2/3}\sqrt{1 - s^4}} . \tag{10.3}$$

As mentioned above, $\rho \approx t^{1/5}$. In coordinates which are made dimensionless by the radius of curvature, $x_{\text{tip}} - x = t^{1/5}\xi$, $y = t^{1/5}\eta$, we expand the form determined by equation (10.3)

$$\xi = \left(\frac{3}{5}\right)^2 A\eta^2 + B\frac{\eta^4}{t^{2/5}} + \ldots \tag{10.4}$$

Thus the shape of the branch becomes increasingly parabolic. The constant $\rho^2 V$ is therefore to a good approximation given by the value of the crystalline dendrites in paragraph 8.6. The approximation is even better if the factor of anisotropy is small, since in that case, the condition for solvability is satisfied up to exponentially small terms, which can be neglected compared to those in the derivation of equation (10.4).

10.5 DENDRITIC GROWTH

We first consider the growth of a liquid–solid interface in two dimensions, in a supercooled liquid confined by a circular contour of radius R over which the temperature is maintained constant at T_∞ (Kessler, Koplik and Levine 1984). As can be deduced from paragraph 5.3, if the growth is quasi-stationary,

$$\Delta u = 0 \tag{10.1}$$

in the two media where the dimensionless temperature $u = (T - T_\infty)/(T_0 - T_\infty)$ is chosen in an unusual fashion to cancel out at the external border. At the interface,

$$\begin{gathered}\left(\vec{\nabla} u_s - \vec{\nabla} u_l\right) \cdot \vec{n} = 4\pi \vec{v} \cdot \vec{n} \\ u = 1 - (1 - \varepsilon \cos(4\theta)) \frac{1}{R}\end{gathered} \tag{10.2}$$

and $u(R) = 0$. To write simpler mathematical terms, the lengths are made dimensionless by $d = (\gamma T/Q)(T_0 - T_\infty)$, and the time by $Qd^2/(4\pi c_p(T_0 - T_\infty)D)$. In the same way as in paragraph 7.6, we consider the contour confined by C, which lies along the external circle and the interface, on the inside and the outside, by making a little hole of negligible size at the interface. Thus, since the two diffusion coefficients are the same, we avoid having to consider the liquid and the solid as two different media. As in paragraph 7.6, the value of u can be expressed at one point in the volume, as a function of its value and its normal gradient on the above-mentioned contour

$$u(p_0) = \int_C dl \left(G(p_0|p) \vec{\nabla} u(p) - u(p) \vec{\nabla} G(p_0|p) \right) \cdot \vec{n} \,. \tag{10.3}$$

Since $u(R) = 0$, it is convenient to cancel out the contribution of the external contour, by choosing a Green function which is cancelled on that contour. Since this function satisfies

$$\Delta G(p_0|p) = -\delta(p_0 - p) \tag{10.4}$$

we deduce

$$G(p_0|p) = \mathrm{Log}(\vec{r}_0 - \vec{r})^2 - \mathrm{Log}\left(\frac{R^2 \vec{r}}{|\vec{r}|^2} - \vec{r}_0\right)^2 + \mathrm{Log}\frac{R^2}{|\vec{r}|^2} \,. \tag{10.5}$$

We can thus write, for all the points on the interface,

$$1-(1+\varepsilon\cos(m\theta))\frac{1}{R} = \int_{C_0} dl' G(p_0|p)U(p) \tag{10.6}$$

where U is the normal velocity of the contour $\vec{v}\cdot\vec{n}$. We parametrise the curve by $\theta(u)$, θ being the angle between the normal and a given direction, and $u = l/L$, the relative curvilinear coordinate, $0 < u < 1$, where L is the total length. From equations (10.5) in paragraph 10.1, we thus deduce

$$\frac{\partial\theta}{\partial t} = -\frac{1}{L}\frac{\partial U}{\partial u} - L\kappa(u)\left(\int_0^u\left(du'\kappa(u')U(u') - u\int_0^1 du'\kappa(u')U(u')\right)\right) \tag{10.7a}$$

$$\frac{dL}{dt} = L\int_0^1 du'\kappa(u')U(u'). \tag{10.7b}$$

At a given time t, we assume $\theta(u_i)$, $i = 1, \ldots, N$ and L are known and we invert equation (10.6) to determine the normal velocities $U(u_i)$; followed by the new $\theta(u_i)$ and L by equations (10.7). The initial condition is a perturbed circle of radius $r_0 = 10^3$, by a perturbation of four-fold symmetry and $R = 10^6$. Remarkably, this model has the same properties as the geometric model discussed in paragraph 10.1. As shown in figure 10.11a, when $\varepsilon = 0$ the lobes grow and are then split in two provided that they are of a certain length. When $\varepsilon = 0.2$ the lobes give rise to dendrites which develop in a stationary way (figure 10.11b). Even though it is an order of magnitude larger than the real anisotropy factors, this simulation qualitatively shows the important effect of a factor of anisotropy.

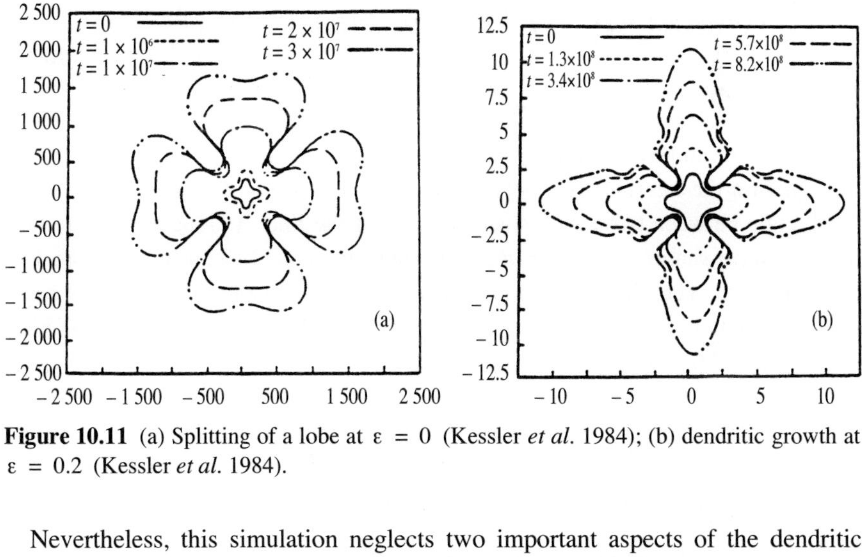

Figure 10.11 (a) Splitting of a lobe at $\varepsilon = 0$ (Kessler *et al.* 1984); (b) dendritic growth at $\varepsilon = 0.2$ (Kessler *et al.* 1984).

Nevertheless, this simulation neglects two important aspects of the dendritic growth: the effects of advection caused by the movement of the tip and the possible difference between the two diffusion coefficients in the liquid and the solid phases. The effects of advection are introduced by the term $\partial c/\partial t$ in the diffusion equation. It is possible to write an integral equation for the interface which is analogous to that in paragraph 8.5, but where the corresponding Green function explicitly depends on time. Thus, the concentration field at a point depends on the forms of the interface at all preceding instants, which considerably increases the calculation time. An approximation of Saito, Goldbeck-Wood and Müller-Krumbhaar (1988) consists of assuming the concentration field stationary in the moving frame of the tip. We thus write

$$\Delta u + \frac{2}{l}\frac{\partial u}{\partial \xi} = 0 \tag{10.8}$$

where $u = (c - c_\infty)/(c_0 - c_\infty)$, $l = 2D/V$, V is the tip velocity at time t, and $\xi = z - Vt$. At the interface, we still have the local relation

$$\vec{v} \cdot \vec{n} = -D\vec{\nabla} u \cdot \vec{n} \tag{10.9}$$

where $\vec{v} \cdot \vec{n}$ is the normal velocity of the interface in the laboratory coordinates and the Gibbs–Thomson equation

$$u = \Delta - \frac{d_0}{R}(1 - \varepsilon\cos(4\theta)). \tag{10.10}$$

We revisit the integral equation for a liquid–solid interface, in which the movement is limited by the diffusion of a solute in the liquid $(D_{cs}) = 0$, determined in appendix 1 of paragraph 7.6.

$$c(p_0)u(p_0) = \int_{\Gamma_c} \Big(u(p)\vec{\nabla} G(p_0|p) - G(p_0|p)\vec{\nabla} u(p)\Big) \cdot \vec{n} \\ -\frac{2}{l}n_z u(p)G(p_0|p))dx \tag{10.11}$$

where $c(p_0)$ can possibly be different to $1/2$. Indeed, during the discretisation of the equation the interface is a polygonal line and $c(p_0) = \phi/2\pi$, where ϕ is the local angle between two successive segments of the line. As long as the interface does not have singularities, $\phi = \pi$ at all points and $c = 1/2$, as in appendix 1 of paragraph 7.6. The Green function is determined by equation (7.14) in the same paragraph in two dimensions, the solution of which, already used in equation (7.22) in paragraph 7.6, is written as

$$G(p_0|p) = \frac{1}{2\pi}\exp-\frac{(z_0 - z)}{l}K_0\left(\frac{\left|\vec{r}_0 - \vec{r}\right|}{l}\right). \tag{10.12}$$

We can rewrite equation (10.11) in the form

$$\int_{\Gamma_c} dlG(p_0|p)\frac{\partial u}{\partial n}(p) = \int_{\Gamma_c} dlH(p_0|p)u(p) \tag{10.13}$$

where

$$H(p_0, p) = \frac{\partial G}{\partial n}(p_0, p) - \frac{2}{l} n_z G(p_0, p) - c(p)\delta(p_0|p) \ . \tag{10.14}$$

Since the form of the interface is known at time t, we calculate $u(p)$ by equation (10.10) then $\partial u / \partial n$ by equation (10.13).

Its form at instant $t + dt$ is then deduced from equation (10.9). For example, when the state is stationary, propagating at velocity V in the z direction, $\delta z = V \cos\theta \ \delta t$. Here the state is generally non-stationary and the approximation consists of setting the new velocity of the tip to the velocity V from the reference frame. In fact, due to a problem of numerical stability, we introduce a relaxation equation,

$$\frac{dV}{dt} = -\frac{1}{\tau}(V(t) - v_0(t)) \tag{10.15}$$

where τ is a much faster time than the time scale of the considered physical problem. As in the previous simulation, a simulation with a zero anisotropy factor always leads to a splitting of the tip into several parts. Figure 10.12 shows three evolutions of dendritic forms for $\Delta = 0.5$ and three values of the anisotropy factor, 0.048 (a), 0.091 (b), 0.167 (c), respectively.

We determine the values of $\sigma = d_0 V / 2Dp^2 = 2Dd_0 / \rho^2 V$ as a function of β for values of $\Delta = 0.25$ and $\Delta = 0.5$, where $p(\Delta)$ is determined from the Ivantsov relation in two dimensions in paragraph 7.6. The chosen anisotropy factors are $\beta = 0.05$; 0.10; 0.15 and 0.20 for $\Delta = 0.25$ and $\beta = 0.048$; 0.091; 0.13 and 0.167 for $\Delta = 0.5$. For those parameter values, the obtained dendritic forms have tips and a relatively stationary secondary branching. Two points with similar anisotropy factors and different Δ have similar values of σ, showing that σ only depends on the anisotropy factor. The points obtained numerically are in good agreement with the theoretical predictions of stationary solutions of Misbah (1987) from paragraph 7.8 since the model under consideration there is for a single diffusion coefficient.

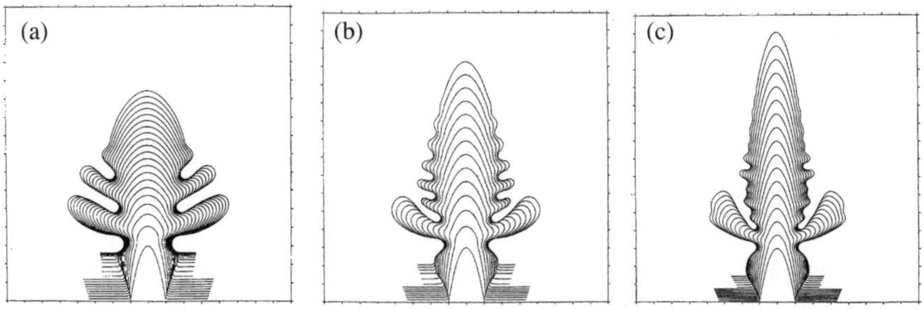

Figure 10.12 (a) $\Delta = 0.5$, $\varepsilon = 0.048$ (Saito *et al.* 1988); (b) $\Delta = 0.5$, $\varepsilon = 0.091$; (c) $\Delta = 0.5$, $\varepsilon = 0.167$.

When the effect of the temperature gradient is taken into account, we come back to the classical configuration of directional growth, already discussed in paragraphs 8.11 and 9.9 (Saito *et al.* 1989), with the only difference being that the cellular wavelength is imposed here. We come back to cellular and dendritic forms at small and large wavelengths of the lattice, respectively (figure 10.13a). As shown in figure 10.13b, the temperature gradient practically does not affect the condition for solvability, whereas it considerably modifies the Ivantsov relation.

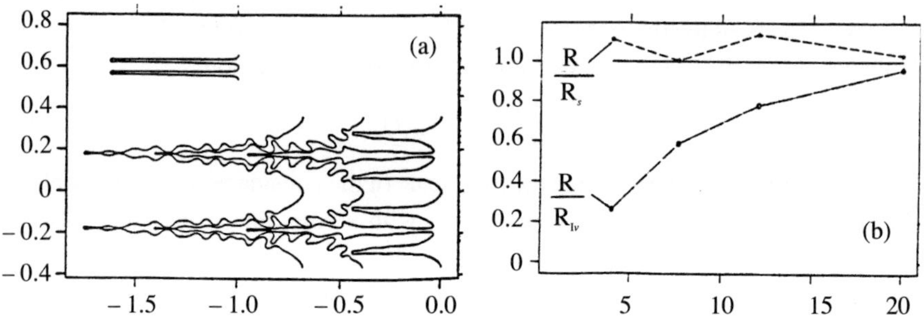

Figure 10.13 (a) Crystalline forms in directional growth. At the top is a cell with a small Péclet number. At the bottom is the evolution of a dendritic form with a larger wavelength of the lattice (Saito *et al.* 1989); (b) radius of curvature as a function of velocity made dimensionless by the Ivantsov radius R_{Iv} and by the radius determined by the condition for solvability R_s (Saito *et al.* 1989).

One of the important specificities of a three-dimensional dendrite is that the crystalline anisotropy modulates the crystal growth in a plane perpendicular to the principal growth axis. Thus, by sectioning the dendrite by such a plane, at some distance from the tip, if the symmetry is cubic we observe the shape of four branches, the length of which grows as the plane under consideration moves away from the tip. At large distances from the tip, the shape of the crystal is very far from an Ivantsov paraboloid and, as we have mentioned in paragraph 8.7, the perturbation methods are not applicable. Noting that the length of the branches close to the tip is even shorter than the diffusion length, and ignoring the diffusion in the direction parallel to the principal growth axis, we come back to a perfectly analogous problem to that considered in paragraph 10.4 (Brener 1993). As in that same paragraph, let x and y be the longitudinal and transversal coordinates associated with such a branch, and z the coordinate of the principal growth axis. In such a regime of intermediate growth the self-similar pattern is of length proportional to $t^{3/5}$, and of width proportional to $t^{2/5}$. To establish the proportionality coefficients, note that the area of the crystal is proportional to time, as for a circle whose radius grows according to relation $R^2 = \mathrm{P}(\Delta)D_T t$ (see appendix 1 of paragraph 5.4). By setting $t = -z/V$, we come back to the case of the stationary parabola, since then $z = -R^2/\rho$ and $P(\Delta) = \rho V/D_T$ is the Péclet number of stationary growth. Thus, if $x = At^{3/5}$, $y = (PD_T/A)t^{2/5}$. As in paragraph 10.4, the growth constant of the secondary branch $C_2 = \rho_2^2 V_2/D_T d_{0T}$ allows us to determine $A \approx P_2^{4/5} D_T^{3/5}/C_2^{1/5} d_{0T}^{1/5}$ using $V_2 \approx At^{-2/5}$ and $\rho_2 \approx y^2/x$, where the index 2 characterises the quantities which are associated to it. The form of a branch of a dendrite is therefore determined by replacing time t by $|z|/V$ in equations (10.2) and (10.3) in paragraph 10.4. In other words, by making the lengths dimensionless by ρ, the principal radius of curvature of the dendrite becomes

$$y(x, z) = \left(\frac{5}{3}|z|^{2/5}\right)\left(\frac{C_2}{C}\right)^{1/5}\left(\frac{x}{x_{\text{tip}}}\right)^{2/3}\int_{x/x_{\text{tip}}}^{1}\frac{ds}{s^{2/3}\sqrt{1-s^4}} \tag{10.16}$$

the altitude of the branch being

$$x_{\text{tip}}(z) = \left(\frac{5}{3}|z|\right)^{3/5}\left(\frac{C_2}{C}\right)^{1/5} \tag{10.17}$$

where C is the growth constant of the principal dendrite in good agreement with the experiments of Bisang and Bilgram (1995) on xenon dendrites (figure 10.14).

In figure 10.15, we show a numerical simulation of three-dimensional dendritic growth performed by Karma and Rappel (1996), which at the same time resolves the structure of the interface and the diffusion field. The equilibrium structure of the interface is the same as that of Cahn and Hilliard in paragraph 2.3, and the movement of such an interface satisfies the dissipative dynamics, described in paragraph 5.4 (Langer 1986, Caginalp 1989). The advantage of that method is, as for flame fronts (Denet and Haldenwang 1995), to avoid the complications coming from the geometry of the interface and only simulating reaction-diffusion type equations. The dendrites grow along the <100> axis for a factor of anisotropy $\varepsilon = 0.047$, defined in appendix 1 of paragraph 2.4. As shown in figure 10.15b, the values of $\sigma^* = 1/C$ determined by the simulation are in agreement with those of the numeric solution of the integral equation in paragraph 8.7, and larger than those from the linear theory of the same paragraph.

By comparison, the experimental values for succinonitrile (SCN) and for pivalic acid (PVA) are indicated at the bottom of the figure.

10.6 CELLULAR FLAMES

In paragraph 7.8, we have already noted that the structure of a stationary cellular flame is complex, not only because of the physio-chemical processes which take place inside the flame, but also because of a more complicated flow than that calculated in a Hele–Shaw cell, and notably because of the presence of vorticity in burnt

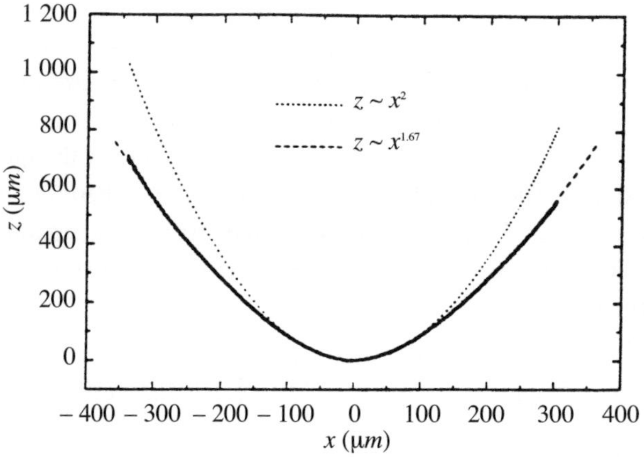

Figure 10.14 Contour of a xenon dendrite in the region of the tip (Bisang and Bilgram 1995).

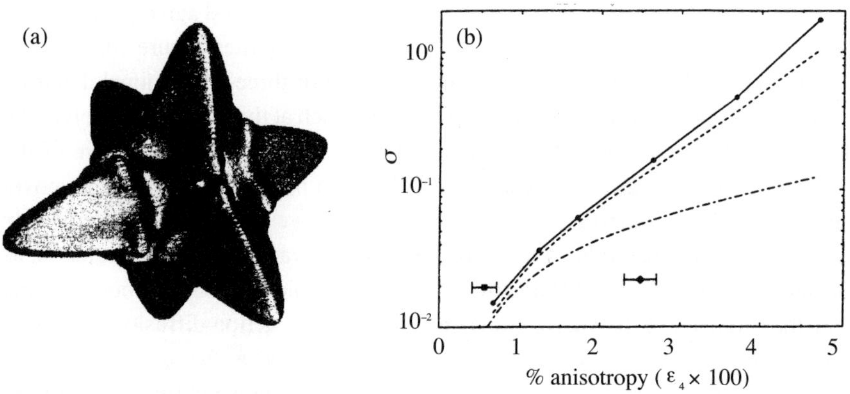

Figure 10.15 (a) Growing dendrite in the <100> axis for $\varepsilon = 0.047$ (Karma and Rappel 1996); (b) values of $\sigma^* = 1/C$ as a function of 100 ε. Simulation (dots); numerical analysis (dashes); linear solvability (dot–dash) (Karma and Rappel 1996).

gases. It is therefore beyond the scope of this discussion to describe the whole dynamics of cellular flames. A simplified approach was proposed by Sivashinsky (1977, 1983) if the gas expansion coefficient $\gamma = (\rho_1 - \rho_2)/\rho_1$ is small. When $\gamma = 0$, the flow stays at rest and the dynamics of the front is exactly that considered in paragraph 3.5, that is, $\vec{v} \cdot \vec{n} = V_0 - LD/R$, where L is the Markstein number, and

$$\frac{\frac{\partial \xi}{\partial t}}{\sqrt{1+\left(\vec{\nabla}\xi\right)^2}} = V_0 + LD\Delta\xi . \tag{10.1}$$

If $L > 0$, which is generally the case, the shape of the slightly perturbed planar flame returns to its equilibrium position by a diffusion equation $\partial\xi'/\partial t = LD\Delta\xi'$, where $\xi' = \xi - V_0 t$. In paragraph 3.6 we had envisaged the case in which the interface would be perturbed from its initial shape by noise of sufficiently large amplitude such that the term $\left(\vec{\nabla}\xi\right)^2$ is conserved. The situation is the same here except that the perturbation is caused by the flow itself, which is of a weak amplitude as γ is assumed to be small, but sufficient to retain the non-linear term. Thus, since the considered normal velocity is the relative velocity between the flame and the flow, equation (10.1) becomes

$$\frac{\frac{\partial \xi}{\partial t} - u - v\ \frac{\partial \xi}{\partial x}}{\sqrt{1+\left(\vec{\nabla}\xi\right)^2}} = V_0 + LD\ \frac{\partial^2 \xi}{\partial x^2} . \tag{10.2}$$

If we consider simply a perturbation of a two-dimensional form $\xi(x, t)$ and (u, v) the perturbation of the flow, or even, since the amplitude of the perturbation is small

$$\frac{\partial \xi'}{\partial t} = u(\xi') + v(\xi')\ \frac{\partial \xi'}{\partial x} + LD\ \frac{\partial^2 \xi'}{\partial x^2} + \frac{1}{2}\left(\vec{\nabla}\xi'\right)^2 . \tag{10.3}$$

From equations (6.4) and (6.7) in paragraph 6.7, u and v are of the same order of magnitude. Thus the term $v\partial\xi'/dx$ is negligible compared to u, and equation (10.3) becomes

$$\frac{\partial \xi'}{\partial t} = u(\xi') + LD\ \frac{\partial^2 \xi'}{\partial x^2} + \frac{1}{2}\left(\vec{\nabla}\xi'\right)^2 . \tag{10.4}$$

To determine $u(\xi')$, note that if we omit the non-linear term in equation (10.3) we should get the dispersion relation (6.12) in paragraph 6.7 as long as γ is small. In that approximation the coefficients of that relation are written as

$$A = 2;\ \ B - 2V_0k(1 + Ldk)\ ;\ \ C = -\gamma V_0^2k^2 + 2LV_0^2dk^3 . \tag{10.5}$$

The two terms of the coefficient C must be of the same order of magnitude, such that $k \approx \gamma/Ld$. Then $B \approx 2V_0k$, such that the σ term dominates that of σ^2. We thus deduce the simplified dispersion relation

$$\sigma = \frac{\gamma}{2} V_0 |k| - L d k^2 . \tag{10.6}$$

Comparing with the linear part of equation (10.4) we deduce that the Fourier component of u *is* $\hat{u}(k) = (\gamma/2) V_0 |k| \xi'(k)$, or from paragraph 8.4

$$u(x, t) = \frac{\gamma V_0}{2} \Lambda \xi' = \frac{\gamma V_0}{2\pi} P \int_{-\infty}^{+\infty} dx' \frac{\frac{\partial}{\partial x'} \xi'(x', t)}{x - x'} . \tag{10.7}$$

As in paragraph 3.6 we can get closer to an equation of a form which is more familiar to hydrodynamicists by considering the local gradient of the interface $\Psi = -d\xi'/dx$ which satisfies

$$\frac{\partial \Psi}{\partial t} + \Psi \frac{\partial \Psi}{\partial x} = \Lambda \Psi + \nu \frac{\partial^2 \Psi}{\partial x^2} . \tag{10.8}$$

Here, lengths and times are made dimensionless by $l = 2LD/\gamma V_0$, $\tau = 2l/\gamma V_0$ and $\nu = LD$. Different numerical studies of equation (10.8) (Michelson and Sivashinsky 1977, Pumir 1985, Thual *et al.* 1985) show that, starting from arbitrary initial conditions the flame front in the end acquires a typical stationary cellular form. As shown in figure 10.16a, the cells are curved towards the cold gas and are linked by the thin regions in which their curvature varies rapidly. We can also see the rapid change in the slope in figure 10.16b.

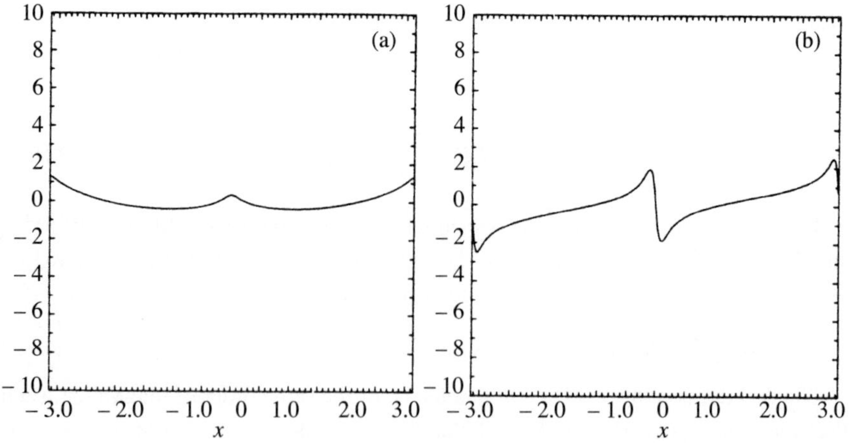

Figure 10.16 (a) Flame front (Thual *et al.* 1985); (b) slope of the front (Thual *et al.* 1985).

Appendix: Polar decomposition

Lee and Chen (1982), then Thual and his coworkers (1985), have shown that equation (10.8) has exact solutions obtained by polar decomposition in the way described in paragraph 10.2. Indeed,

$$\Psi(x,t) = -2\nu\sum_{i=1}^{2N}\frac{1}{x-z_i(t)}$$

where $z_i(t)$ are the poles in the complex plane, is the solution of equation (10.8) if the poles satisfy the equations

$$\frac{dz_i}{dt} = -2\nu\sum_{i\neq j}\frac{1}{z_i-z_j} - i\operatorname{sign}(\operatorname{Im}(z_i)) .$$

First note that if $p_z(x) = 1/(x-z)$, then $\Lambda p_z(x) = i\ \operatorname{sign}\ (\operatorname{Im}z)\partial/\partial x(p_z(x))$. Indeed, for example when z is in the upper half-plane, the integration contour of integral (10.7) must pass through the lower half-plane and travel around point x clockwise. The previous equation is deduced using the identity

$$\sum_{i\neq j}\frac{1}{x-z_i}\ \frac{1}{x-z_j} = \sum_i\frac{1}{(x-z_i)^2} + \sum_{i\neq j}\frac{2}{z_i-z_j}\ \frac{1}{x-z_i} .$$

In the simplest configuration let us consider two complex conjugate poles $z(t) = a(t) \pm ib(t)$ with $b > 0$. We therefore deduce from the pole equations,

$$\frac{dz}{dt} = -\frac{2\nu}{z-\bar{z}} - i$$

such that $a(t) = a(0)$ and

$$\frac{db}{dt} = \frac{\nu}{b} - 1 .$$

Thus the two effects oppose one another. The Darrieus–Landau instability, here represented by an operator Λ and by -1 in the last equation, tends to bring the poles back towards the real axis. The stabilising term of the instability, represented by the coefficient ν, tends to push them back. An equilibrium position results at $b_{eq} = \nu$. The corresponding solution of equation (10.8) is written as

$$\Psi(x) = -\frac{4\nu x}{x^2+\nu^2}$$

and thus the shape of the flame front

$$\xi(x) = -2\nu\operatorname{Log}(x^2+\nu^2)$$

is a single cell moving in an infinite medium. For a form of the flame front moving inside a tube, which is a more realistic experimental situation, we must consider a periodic distribution of poles. We can then constrain the domain of their evolution to the band of the real part comprised between 0 and 2π. Thus

$$\Psi(x,t) = -\nu \sum_{i=1}^{2N} \cot\left(\frac{x-z_i}{2}\right)$$

is the solution of equation (10.8), if

$$\frac{dz_i}{dt} = -\nu \sum_{i \neq j} \cot\left(\frac{z_i - z_j}{2}\right) - i\,\mathrm{sign}(\mathrm{Im}(z_i)) .$$

As before, considering only a pole and its complex conjugate we deduce $b_{eq} = \tanh^{-1}\nu$ and

$$\Psi(x) = -\frac{2\nu \sin x}{(1-\nu^2)^{-1/2} - \cos x}$$

shown in figure 10.17, similarly to figure 10.6b.

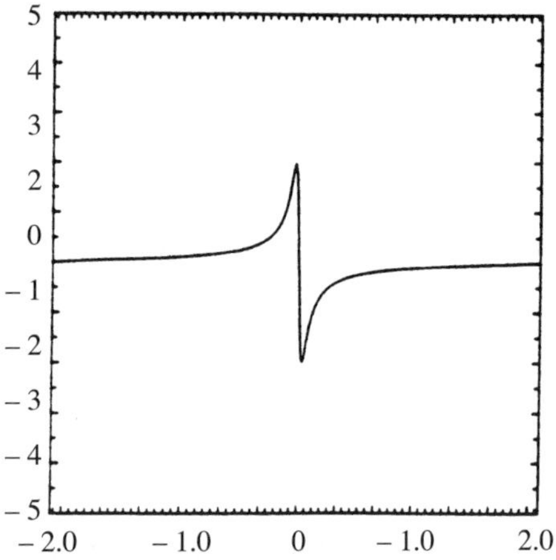

Figure 10.17 Slope of the front for a periodic stationary solution of two complex conjugate poles (Thual *et al.* 1985).

10.7 FLOWS INDUCED BY ELECTRO-DEPOSITION

Consider an electro-deposition experiment in a Hele–Shaw cell (figure 10.18). The cell is filled with a solution of $ZnSO_4$ or $CuSO_4$, enriched in metal ions by a dissolving anode and depleted of the same ions at the cathode by its growth. An electric field of the order of 20 V/cm, maintained between the two electrodes and directed towards the cathode, induces the process of electro-deposition during which the metal surface develops several branches at the bottom of the Hele–Shaw cell, as can be seen in figure 10.18.

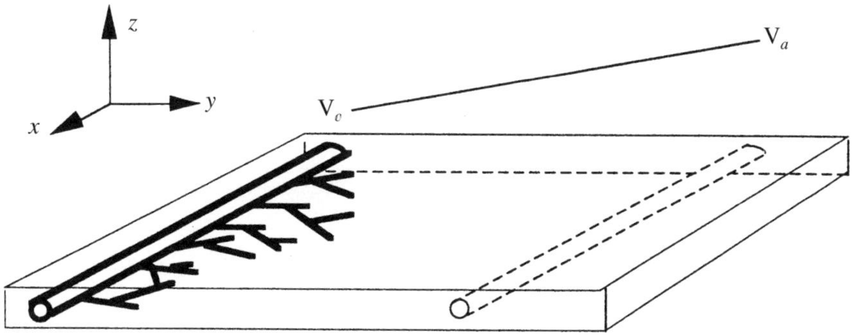

Figure 10.18 Schematic of the experiment.

It appears that the dynamics of the branches is strongly influenced by the presence of flows which accompany the electro-deposition process (Fleury *et al.* 1992, Fleury *et al.* 1994). In fact if the solution is filled with fine particles with trajectories which highlight the stream lines, a periodic succession of vortices is observed, the resulting stream leaving from the middle of two branches and entering at the tip of the branches (figure 10.19a). The vortices obtained in this way eliminate the ions by advection from the regions between branches, concentrating them along the arches joining the the branch peaks (figure 10.19b). The concentration field of the ions is therefore significantly modified compared to that which would result from diffusion alone. Thus the inter-branch regions, having lost their ions, block the lateral growth of the branches. However, the tip which is well supplied grows much more quickly.

The flows can originate from two sources: the convective one is associated with the gravity field and the differences in density generated by the concentration gradient of ions (Huth *et al.* 1995), while the electro-osmotic one, where the fluid is set in motion by the intermediary space charges, is due to the electric field (Alexander and Bruinsma 1990, Fleury *et al.* 1992). As we have seen in paragraph 5.8, the space charges are localised near the walls, therefore at the surface of the aggregate and the cell walls in this case. Since the aggregate is metallic, its surface is equipotential and the tangential electric field is zero. We could therefore hastily deduce from equation (5.14) in the cited paragraph, that there is no electro-osmotic flow generated. Indeed, the tangential electric field is not constant in the Debye layer, but grows linearly from 0 up to an external value imposed by the geometry of the conductors (Levich 1962). The velocity of the osmotic flow generated by such a field only represents a fraction of that given in equation (5.14) in the same paragraph, where the tangential electric field is that which is set at the external limit of the Debye layer. We can simplify the geometry shown in figure 10.19a by a succession of rectilinear

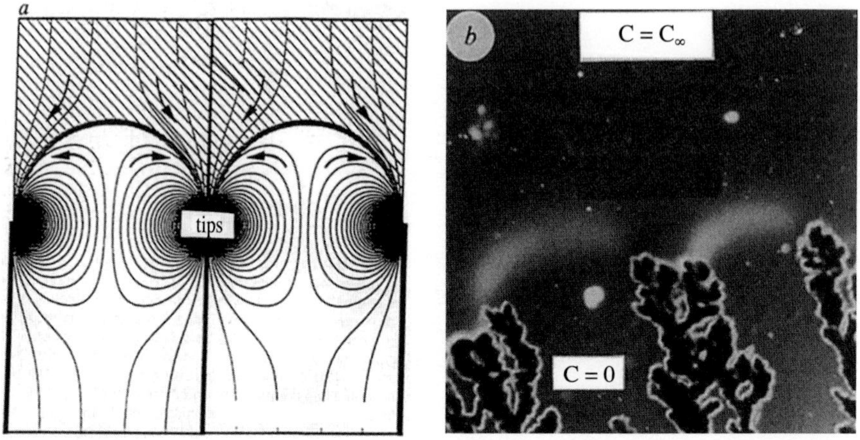

Figure 10.19 (a) Flow structure (Fleury *et al.* 1994); (b) arches and branches (Fleury *et al.* 1994).

branches in the axis of the electric field, with a regular spacing of λ. Since the tangential electric field essentially only exists near the tips we can summarise the electro-osmotic effect by a force $\vec{f}$, of intensity f and in the direction of the electric field, applied to the fluid at the tip. As we show in the appendix such a distribution of forces generates a velocity field $\vec{W} = \text{curl}\,\Psi$ in the Hele–Shaw cell, where

$$\Psi = \frac{b}{24\pi\eta} f \sum_{k=-\infty}^{k=+\infty} \frac{x - k\lambda}{r_k^2} \qquad (10.1)$$

Huth and coworkers (1995) show that the flow (10.1) is in good quantitative agreement with experiments, if $f \approx 5 \cdot 10^{-10}\,\text{N}$, $b \approx 100\ \mu\text{m}$ and $\lambda \approx 500\ \mu\text{m}$. If we compare the intensity of the flow to the intensity which would be produced by a charged species described in paragraph 5.8, we should write $v \approx bf/24\pi\eta\lambda^2 \approx \varepsilon\varepsilon_0 E^2 d/\eta$, where d is the thickness of the charged species and $f \approx 24\pi\varepsilon\varepsilon_0 E^2 d\lambda^2/b$. Taking $E \approx 100\ \text{V}/m$, $\varepsilon \approx 80$ and $\varepsilon_0 \approx 8{,}9 \cdot 10^{-12}\text{C}^2/\text{J} \cdot \text{m}$, we get $f \approx 10^{-6} d$ and $d = 500\ \mu\text{m}$, values which are several orders of magnitude larger than the theoretical predictions (Chazalviel 1990). For the control parameter values considered here, the effects of normal convection play an important role.

Appendix: Flow generated from a periodic distribution of point forces in a Hele–Shaw cell.

Starting with equation (5.3) in paragraph 5.2, taking into account an external force $\vec{f}$

$$\vec{w} = -\frac{b^2}{12\eta}\left(\vec{\nabla} p + \vec{f}\right)$$

so that by taking the curl,

$$\overrightarrow{\mathrm{curl}}\vec{w} = -\frac{b^2}{12\eta}\overrightarrow{\mathrm{curl}}\vec{f}.$$

Assuming $\vec{f} = (f/b)\vec{x}\delta(r - r_{tip})$, where the Dirac function is in two dimensions here, we then deduce

$$\Delta\psi = \frac{b}{12\eta}f\frac{\partial\delta(r)}{\partial x}$$

where $\vec{w} = \mathrm{curl}\ \psi$. The parameter f then has the dimensions of a force. The solution is

$$\psi = \frac{1}{2\pi}\ \frac{b}{12\eta}f\frac{x}{r^2}.$$

Since the equation for the velocity field is linear, the flow generated by a periodic distribution of point forces is the sum of the preceding terms, and equation (10.1) is then deduced.

BIBLIOGRAPHY

[1] R. Almgren, W.S. Dai, V. Hakim, 'Scaling Behavior in Anisotropic Hele–Shaw Flow', *Phys. Rev. Lett.*, **71**, p. 3461, 1993.

[2] H. Aref, G. Tryggvason, 'Numerical Experiments on Hele–Shaw Flow with a Sharp Interface', *J. Fluid. Mech.*, **136**, p. 1, 1983.

[3] E. Ben-Jacob, N. Goldenfeld, J.S. Langer, G. Schon, 'Dynamics of Interfacial Pattern Formation', *Phys. Rev. Lett.*, **51**, p. 1930, 1983.

[4] E. Ben-Jacob, N. Goldenfeld, J.S. Langer, G. Schon, 'Boundary-Layer Model of Pattern Formation in Solidification', *Phys. Rev.*, **A29**, p. 330, 1984.

[5] E. Ben-Jacob, N. Goldenfeld, B.G. Kotliar, J.S. Langer, 'Pattern Selection in Dendritic Solidification', *Phys. Rev. Lett.*, **53**, p.2110, 1984.

[6] E. Ben-Jacob, R. Godbey, N.D. Goldenfeld, J. Koplik, H. Levine, T. Mueller, L.M. Sander, 'Experimental Demonstration of the Role of Anisotropy in Interfacial Pattern Formation', *Phys. Rev. Lett.*, **55**, p. 1315, 1985.

[7] D. Bensimon, L.P. Kadanoff, S. Liang, B.I. Shraiman, C. Tang, 'Viscous Flow in Two Dimensions', *Rev. Mod. Phys.*, **58**, p. 977, 1986.

[8] D. Bensimon, P. Pelcé, 'Tip-Splitting Solution to a Stefan Problem', *Phys. Rev.*, **A33**, p. 4477, 1986.

[9] U. Bisang, J.H. Bilgram, 'Shape of the Tip and the Formation of Sidebranches of Xenon Dendrites', *Phys. Rev. Lett.*, **75**, p. 3898, 1995.

[10] E. Brener, 'Needle-Crystal Solution in Three-Dimensional Dendritic Growth', *Phys. Rev. Lett.*, **71**, p. 3653, 1993.

[11] R. Brower, D. Kessler, J. Koplik, H. Levine, 'Geometrical Approach to Moving-Interface Dynamics', *Phys. Rev. Lett.*, **51**, p. 1111, 1983.

[12] S. Alexander, R. Bruinsma, 'Theory of Electrohydrodynamic Instabilities in Electrolytic Cells', *J. Chem. Phys.,* **92**, p. 3074, 1990.

[13] G. Caginalp, 'Stefan and Hele–Shaw Types Models as Asymptotic Limits of the Phase-Fields Equations', *Phys. Rev.*, **A39**, p. 5887, 1989.

[14] J.N. Chazalviel, 'Electrochemical Aspects of the Generation of Ramified Metallic Electrodeposits', *Phys. Rev.*, **A42**, p. 7355, 1990.

[15] Y. Couder, O. Cardoso, D. Dupuy, P. Tavernier, 'Dendritic Growth in the Saffman–Taylor Experiment', *Europhysics letters*, **2**, pp. 437, 1986.

[16] A.J. Degregoria, L.W. Schwartz, 'A Boundary-Integral Method for Two-Phase Displacement in Hele–Shaw Cells', *J. Fluid. Mech.*, **164**, p. 383, 1986.

[17] B. Denet, P. Haldenwang, 'A Numerical Study of Premixed Flames Darrieus–Landau Instability', *Combust. Sci. and Tech.*, **104**, p. 143, 1995.

[18] V. Fleury, J.N. Chazalviel, M. Rosso, 'Theory and Experimental Evidence of Electroconvection Around Electrochemical Deposits', *Phys. Rev.*, **E68**, p. 2492, 1992.

[19] V. Fleury, J.H. Kaufman, D.B. Hibbert, 'Mechanism of a Morphology Transition in Ramified Electrochemical Growth', *Nature*, **367**, p. 435, 1994.

[20] S.D. Howinson, 'Fingering in Hele–Shaw Cells', *J. Fluid. Mech.,* **167**, p. 439, 1986.

[21] J.M. Huth, H.L Swinney, W.D. McCormick, A.Kuhn, F. Argoul, 'Role of Convection in Thin-Layer Electrodeposition', *Phys. Rev.,* **E51**, p. 3444, 1995.

[22] A. Karma, W.J. Rappel, 'Numerical Simulation of Three-Dimensional Dendritic Growth', *Phys. Rev. Lett.*, **77**, p. 4050, 1996.

[23] D.A. Kessler, J. Koplik, H. Levine, 'Numerical Simulation of Two-Dimensional Snowflake Growth', *Phys. Rev.*, **A30**, p. 2820, 1984.

[24] D.A. Kessler, J. Koplik, H. Levine, 'Geometrical Models of Interface Evolution. II. Numerical Simulations', *Phys. Rev.,* **A30**, p. 3161, 1984.

[25] J.S. Langer, in *Directions in Condensed Matter,* World Scientific, Singapore, 1986, p.164.

[26] Y.C. Lee, H.H. Chen, *Phys. Scr.*, **T2**, p. 41, 1982.

[27] V.G. Levich, *Physiochemical Hydrodynamics*, Prentice Hall, 1962.

[28] D.M. Michelson, G.I. Sivashinsky, 'Nonlinear Analysis of Hydrodynamic Instability in Laminar Flames – II. Numerical Experiments', *Acta Astronautica,* 1977.

[29] A. Pumir, 'Equation Describing Wrinkled Flame Fronts', *Phys. Rev.*, **A31**, p. 543, 1985.

[30] P.G. Saffman, 'Exact Solutions for the Growth of Fingers from a Flat Interface between Two Fluids in a Porous Medium or Hele–Shaw Cell', *Q.J. Mech Appl. Math.*, **12**, p. 146, 1959.

[31] Y. Saito, G. Goldbeck-Wood, H. Müller-Krumbhaar, 'Numerical Simulation of Dendritic Growth', *Phys. Rev.*, **A38**, p. 2148, 1988.

[32] Y. Saito, C. Misbah, H. Müller-Krumbhaar, 'Directional Solidification: Transition from Cells to Dendrites', *Phys. Rev. Lett.*, **63**, 21, p. 2377, 1989.

[33] B. Shraiman, D. Bensimon, 'Singularities in Nonlocal Interface Dynamics', *Phys. Rev.*, **A30**, p. 2840, 1984.

[34] G.I. Sivashinsky, 'Nonlinear Analysis of Hydrodynamic Instability in Laminar Flames – I. Derivation of Basic Equations', *Acta Astronautica*, **4**, p. 1177, 1977.

[35] G.I. Sivashinsky, 'Instabilities, Pattern Formation, and Turbulence in Flames', *Ann. Rev. Fluid. Mech.*, **15**, p. 179, 1983.

[36] O. Thual, U. Frisch, M. Henon, 'Application of Pole Decomposition to an Equation Governing the Dynamics of Wrinkled Flame Front', *J. Physique,* **46**, p. 1485, 1985.

CHAPTER 11

Stochastic growth forms

11.1 DIFFUSION-LIMITED AGGREGATION (DLA)

Consider a square lattice of points in the centre of which we have placed a particle at rest. From an arbitrary point at the edge of the lattice we introduce a second particle which leaps randomly from one site to a neighbouring one in an equiprobable way. As soon as the particle reaches a site next to the central particle, it is attached and thus nucleates an aggregate of two particles. We repeat the same process several times; as soon as the new particle reaches a site next to the aggregate it is attached in the same way. After many operations the aggregate (DLA cluster) has a complex and very jagged shape, such as that obtained by Witten and Sander (1981) sketched on figure 11.1.

Figure 11.1 Aggregate of 3000 particles on a square lattice (Witten and Sander 1981).

The aggregate is self-similar. Upon magnification it has the same structure as on a larger scale. Apart from the smallest scale – the size of a particle – and the largest scale – the size of the lattice – all the remaining lengthscales are equivalent. More precisely let us introduce the correlation function

$$c(\vec{r}) = \frac{1}{N}\sum_{\vec{r}'}\rho(\vec{r}+\vec{r}')\rho(\vec{r}') \tag{11.1}$$

where $\rho(\vec{r})$ is the local density which is equal to 1 or 0 depending on whether the site $\vec{r}$ is occupied or not. If this function is averaged over several aggregates the anisotropy effects from the lattice disappear and the correlation function depends on r only. As we show in figure 11.2,

$$c(r) = r^{-\alpha} \tag{11.2}$$

with $\alpha \approx 0.34$ over the considered range of scales. This power law reflects the self-similarity of the aggregate; if the lengths are stretched by a scaling factor of b, the correlation function is just modified by the constant multiplication factor $b^{-\alpha}$.

The aggregate occupies only a small part of the space. Large voids are present between the two neighbouring branches. The branches are very thin – their thickness is at most two particles – but they cover a notable part of the space because of their

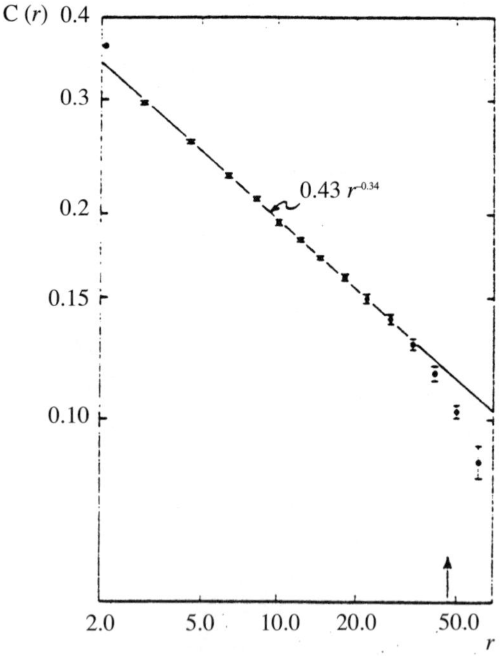

Figure 11.2 Density correlation function, as a function of the distance averaged over six aggregates (Witten and Sander 1981).

multiple ramifications. More precisely, if $N(R)$ is the number of particles in a circle of radius R,

$$N(R) \approx R^D \tag{11.3}$$

where $D \approx 1.66$. The aggregate is said to be fractal, of dimension *D*, intermediate between $D = 1$, which characterises an aggregate formed by a small number of straight branches, and $D = 2$, a dense aggregate (Mandelbrot 1982). *D* and α are simply related by $D = 2 - \alpha$, since

$$N(R) \approx \int_0^R c(r)dr \approx R^{2-\alpha}. \tag{11.4}$$

A classic example of a fractal object is the Cantor set. From a segment of unit length which is divided into three equal parts, the central part is removed. The same process is repeated, assuming that each solid segment is a new segment of unit length. This results in a very fragmented set of intermediate dimension between a point and a curve (figure 11.3). At the *k*th generation we thus generate $N_k = 2^k$ segments of length $l_k = 3^{-k}$.

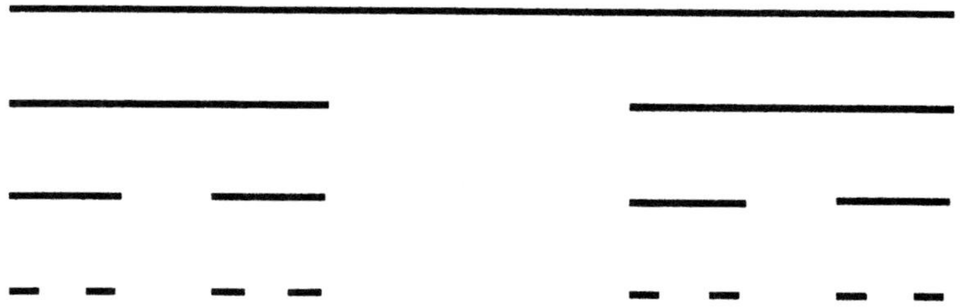

Figure 11.3 Successive generations of a Cantor set.

Rather than counting the number of points $N(l)$ in a segment of size *l*, it is more convenient here to determine the minimum number of segments of size *l*, $N(l)$ needed to cover the set. Choosing $l \approx l_k$, $n(l) \approx 2^k$, for each *l*,

$$N \approx N(l)n(l) \approx l^D n(l). \tag{11.5}$$

Thus, $n(l) \approx l^{-D}$ and $D = \text{Log}2/\text{Log}3 \approx 0.63$. In general if each solid part occupies the fraction $1/p$ of the unit segment, $D(p) = \text{Log}2/\text{Log}p$. As noted by Arnéodo and coworkers (1992) we find this kind of structure at the intersection of the cluster of DLA and a circle of quite large radius. The two long segments of the second generation of the Cantor set drawn above correspond to two main branches; four segments correspond to four secondary branches, and so on. Nevertheless the fractal dimension of this set is slightly smaller than that of the Cantor set, 0.6. For a planar cluster of DLA the fractal nature of the structure would mainly be azimuthal since the dimension of the aggregate is close to $1 + 0.6$.

More precisely, as shown by Witten and Sander (1983), the structure of the aggregate is determined by the structure of its borders. As shown in figure 11.4, out of 3000 particles, the last 1500 ones to be attached are light shaded. The new particles are preferentially attached close to the tips of the aggregate. They very seldom penetrate the lacunary spaces. The inner structure just reflects the history of the growth process.

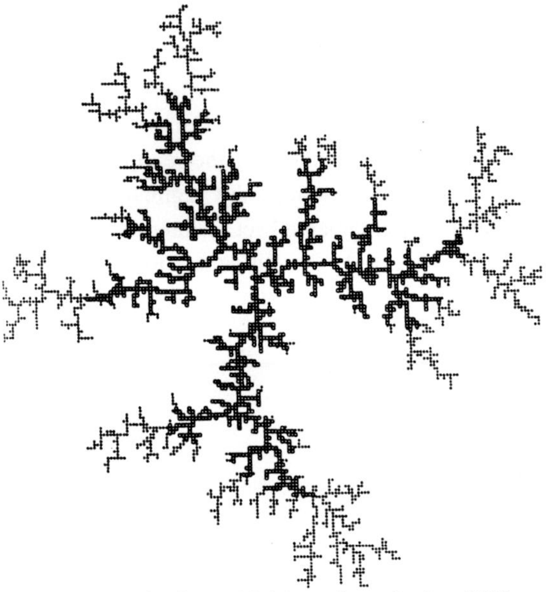

Figure 11.4 Same aggregate as in figure 11.1 but where the last 1500 particles are represented in light shading (Witten and Sander 1983).

To analyse this effect more quantitatively Plischke and Racz (1984) numerically determine for $d = 2$ the probability $p(r, N)dr$ that a N-particle deposits at a distance r of the centre of mass of the existing aggregate. For different values of N, $p(r, N)$ is a Gaussian (figure 11.5)

$$p(r, N) \approx \frac{1}{\sqrt{2\pi}\xi_N} \exp\left(- \frac{(r - r_N)^2}{2\xi_N^2} \right) \tag{11.6}$$

where r_N is the mean radius of deposition, and ξ_N the thickness of the active zone of deposition. r_N and ξ_N increase with N according to the power laws

$$r_N \approx r_0 N^{\nu}\,;\;\; \xi_N \approx \xi_0 N^{\nu'} \tag{11.7}$$

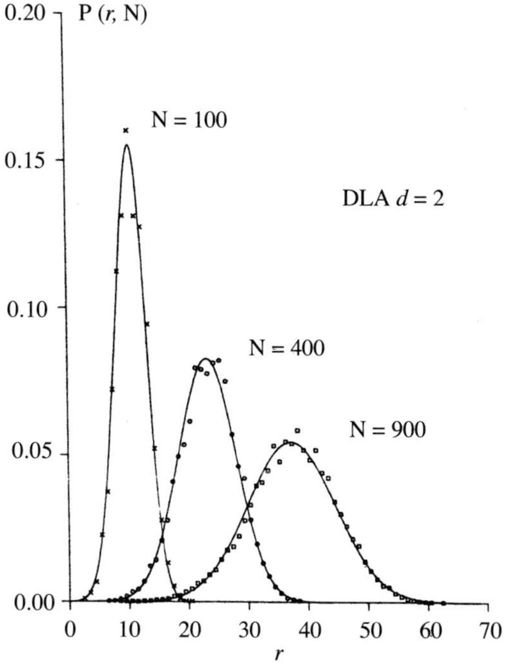

Figure 11.5 $p(r, N)$ as a function of r for different values of N. The solid curves are Gaussians approximating averages over 200 aggregates (Plischke and Racz 1984).

where $\nu \approx 0.58$ and $\nu' \approx 0.48$. In the domain of N considered – here less than 3000 – $\xi_N << r_N$, which quantitatively reflects the existence of the active border mentioned above.

Meakin (1983) carried out simulations of the growth of aggregates with the same rules, but on a hypercubic lattice with a dimension varying between 2 and 6. The aggregates obtained are fractal with thedimensions given in table 11.1

Table 11.1 Fractal dimension of diffusion-limited aggregates for different space dimensions.

d	D	D_m
2	1.70 ± 0.06	1.67
3	2.53 ± 0.06	2.50
4	3.31 ± 0.1	3.40
5	4.20 ± 0.16	4.33
6	5.3	5.29

These fractal dimensions are in remarkably good agreement with the relation of Muthukumar (1983): $D_m = (d^2+1)/(d+1)$. Different arguments allow us to obtain this relation (Muthukumar 1983, Tokuyama and Kawasaki 1984, and Hentschel 1984), but a rigorous determination is still to be done. Racz and Plischke (1985) determine the probability density $p(r, N)$ for $d = 3$. This probability density can be written in the same form as equation (11.6), ξ_N and r_N satisfying equation (11.7) with $\nu = 0.401 \pm 0.09$ and $\nu' = 0.32 \pm 0.05$.

The aggregates analysed above contain relatively few particles, approximately 3000, and one can question whether the corresponding scaling laws are still valid when N becomes larger. In particular, does the fractal dimension not vary with N? Over clusters of 10^4 to 10^5 particles Amitrano and coworkers (1991) observe a slow variation of the fractal dimension with the radial distance, which essentially decreases beyond half the radius of the aggregate. On aggregates of $3 \cdot 10^7$ particles, Mandelbrot (1992) observes a fractal dimension $D = 1.71$, well-defined over the whole of the domain of the aggregate even if the lacunary aspect of the internal regions decreases when the number of particles increases. As stated by Lam (1995) the possible changes in the observed scaling laws with respect to the number of particles can only be due to edge effects. In particular when N becomes large the exponents ν and ν' related to the probability density of the active border should become equal since the ratio ξ_N/r_N converges towards a finite value.

11.2 RELATION WITH VISCOUS FINGERING

To determine the link between this structure and the growth forms in the preceding chapters consider the probability $u(\vec{r}, k)$ that the random walker reaches the site $\vec{r}$ after the kth step. As in all random walks, u satisfies the equation

$$u(\vec{r}, k+1) = \frac{1}{z}\sum_{\vec{l}} u(\vec{r}+\vec{l}, k) \tag{11.1}$$

where the sum applies to the neighbouring sites of the considered point. This equation is just the discrete version of the diffusion equation

$$\frac{\partial u}{\partial t} = \Delta u . \tag{11.2}$$

At the border of the aggregate we must have $u = 0$. In fact, when a particle reaches a site next to the aggregate, it is attached straight away and therefore disappears from the domain where it can freely move. The probability that a site next to the aggregate gains a particle at the time $k+1$ is given by equation (11.1), where the sum is limited to the free sites, next to the point under consideration. Since $u = 0$ at the interface this equation can simply be written as

$$\vec{v}_i \cdot \vec{n} = \eta \vec{\nabla} u \cdot \vec{n} \tag{11.3}$$

where we have identified the normal velocity of the interface $\vec{v}_i \cdot \vec{n}$ to the probability of the aggregate to gain a particle at a given point of its perimeter. Under the conditions of the numerical experiment the growth of the cluster is very slow and we can neglect the time dependence of equation (11.2). Replacing u by $-p$, where p is the pressure, we are back to the Hele–Shaw-type problem of a free boundary, when the effects of surface tension are neglected.

We have already mentioned the existence of an instability of the uniform and planar displacement in chapter 6. This is also the case here, which explains the branching of the aggregate. Moreover, when the effects of surface tension are neglected the growth rate related to the perturbation of the wavenumber k is proportional to this wavenumber and does not involve any other lengthscale: the smaller the wavelengths the faster they grow. This property must be related to the self-similarity of the aggregate, up to a size of the order of the period of the lattice, where the continuum model breaks down.

To find the viscous fingering we must average the growth dynamics of a DLA cluster in a certain way. A first method designed by Tang (1985) consists of not attaching the random walker at the first collision with the aggregate. A counter is placed at each border site. Each time a particle reaches a given site the number of the counter is increased by one. When this number exceeds a number m set in advance, the particle is attached. Nevertheless, because of the absence of surface tension the averaged dynamics is unstable and as a consequence the Saffman–Taylor patterns cannot be found. However, the reverse dynamics, obtained by walking backwards in time, is stable. To model it we introduce diffusing 'anti-random walkers', i.e. particles which move as walkers. Nevertheless, when the counter of a given site exceeds m, the anti-particle, instead of attaching, annihilates the particle on the contact site and makes the interface move back. Thus, as shown in figure 11.6, if the initial aggregate is a finger with a given width, but with a tip of arbitrary shape, the dynamics of the 'anti-random walkers' allows us to obtain the Saffman–Taylor finger with the same relative width.

The relation between the growth of a DLA cluster and viscous fingering in a Hele–Shaw cell can also be obtained by the following operation (Arnéodo *et al.* 1989). We successively grow N clusters of DLA containing the same number of particles M in a rectangular cell. All these clusters are different from each other, but for a reason yet to be explained, their superposition (figure 11.7) allows to find one of the Saffman–Taylor finger stationary states determined in paragraph 7.5. After superposition of all these clusters we define $r(x, y)$, the number of particles on the site (x, y) divided by N. For a given abscissa x the $r(x, y)$ profile is maximum at the centre of the cell, with a value of r_{max}, and decreases when moving towards the

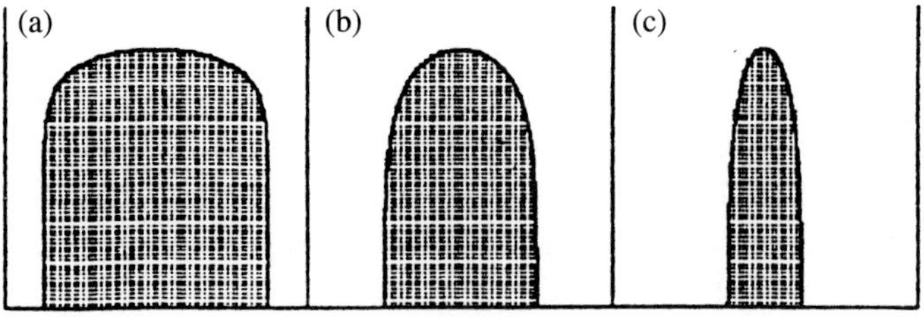

Figure 11.6 Saffman–Taylor fingers obtained by the 'anti-random walkers' dynamics (Tang 1985). The relative widths are (a) 0.75; (b) 0.5; (c) 0.25 respectively. The width of the channel is 64 sites and $m = 20$.

edges. Remarkably, the line $r = r_{max}/2$ is the Saffman–Taylor finger of relative width $\lambda = 1/2$ (figure 11.6). Similarly, the same operation, but in this case for cluster growth in a circular section with an angle $\pi/2$ at the centre, produces an analogous result. The line where $r = r_{max}/2$ is the stationary solution of petal shape in an angular section of relative occupancy $\lambda = 0.82$ from paragraph 7.5.

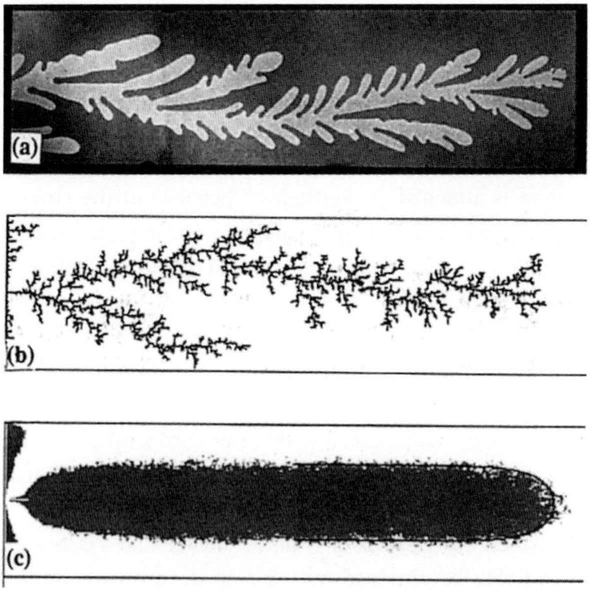

Figure 11.7 (a) Snapshot of a very unstable Saffman–Taylor finger. (b) Aggregate of 6000 particles in linear geometry. (c) Superposition of 510 (b)-type aggregates, where the region $r > r_{max}/2$ is filled in grey. The solid line is the Saffman–Taylor finger of $\lambda = 1/2$ (Arnéodo *et al.* 1989).

11.3 KINETIC EFFECT

The previous rule stated that a particle sticks to the cluster as it reaches a neighbouring site, therefore with a sticking probability $p_c = 1$. We now suppose that when it reaches a site next to the cluster it attaches itself with probability p_c in this case smaller or equal to 1, and therefore the probability that it stays free to move is clearly $1 - p_c$ (Witten and Sander 1983). This rule determines a new lengthscale, possibly far larger than the period of the lattice. In fact, let us simply consider one-dimensional dynamics with 0, l, $2l$ indexing the border and the two neighbouring sites, respectively. In the same way that we determined equation (11.1) in paragraph 11.2, we now obtain

$$\begin{aligned} u(k+1, l) &= (1-p_c)u(k,0) + \frac{1}{2}u(k, 2l) \\ u(k+1, 0) &= \frac{1}{2}u(k, l) \end{aligned} \tag{11.1}$$

since a particle has probability $1/2$ to come from a free neighbouring site and probability $(1-p_c)$ to be free at the border site. In the same stationary condition as in the aforementioned paragraph $u(k+1, l) = u(k, l)$, such that $p_c u(l) = u(2l) - u(l) = l\partial u/\partial x$. We also deduce the value of u at the border site

$$u(0) = \left(\frac{1}{p_c} - 1\right) l \frac{\partial u}{dx} \tag{11.2}$$

Thus, when $p_c = 1$, we find $u(0) = 0$. From equation (11.3) in the same paragraph the value of the probability to find a particle on a border site is proportional to the normal velocity of the advancing interface, that is

$$u(0) = \left(\frac{1}{p_c} - 1\right)\frac{1}{\eta}\vec{v}_i \cdot \vec{n}. \tag{11.3}$$

Thus, decreasing the sticking probability introduces a kinetic effect, not a surface tension one. This effect is nevertheless stabilising, since the parts of the front moving fastest have a value of u larger than the one at the interface. This leads to a decrease in the probability gradient and therefore in the velocity of the advancing front. A more quantitative analysis of this effect can be obtained by analysing the linear stability of the growth of the circular cluster of radius R. In a way similar to that developed in paragraph 6.2 we write the solution of the Laplace equation in the form

$$u = A\mathrm{Log}r + B + g(t)\cos(n\theta)r^{-n} \tag{11.4}$$

for $r > R$, where the last term of equation (11.4) is a small correction of the stationary solution. The application of the boundary conditions (11.3) in paragraphs 11.2 and 11.3 then leads to the evolution equation for g:

$$\left(1 + \frac{ln}{R}\left(\frac{1}{p_c} - 1\right)\right)\frac{1}{g}\frac{dg}{dt} = (n-1)\frac{1}{R}\frac{dR}{dt}. \qquad (11.5)$$

Thus, the modes $n \geq 1$ are unstable, slowed down by the kinetic effect occurring when $p_c \leq 1$. A simple method to stabilise the interface consists of making the sticking probability p_c depend either on the environment of a given point on the borderline (Witten and Sander 1983), or on its curvature (Vicsek 1985), the sticking probability being higher in the dips or in the concave parts than close to the convex tips. Assuming a dependency $p_c(\kappa)$ where κ is the curvature of the interface we obtain by making equation (11.3) linear and by using equation (6.8) in paragraph 6.2 the new evolution equation

$$\left(1 + \frac{ln}{R}\left(\frac{1}{p_c} - 1\right)\right)\frac{1}{g}\frac{dg}{dt} = (n-1)\frac{1}{R}\frac{dR}{dt}\left(1 - n(n+1)\frac{1}{R}\right) \qquad (11.6)$$

a relation very similar to equation (6.11) in the aforementioned paragraph. We can still notice an important difference: the relative growth rate $(dg/dt)/g$ cancels out with the growth rate of the cluster in equation (11.6) whereas it does not cancel out under the same conditions in equation (6.11) mentioned above. The $p_c(\kappa)$ effect does not correctly model the effect of surface tension.

11.4 SURFACE TENSION

To correctly model the effects of surface tension we must give the random walkers the possibility to leave the interface (Kadanoff 1985), as in the model of the so-called disaggregation of Botet and Jullien (1985). We define a Green function $g(r, s)$ over the lattice, where r belongs to the free space or to the borderline of this free space, and s, only to the borderline. g satisfies the discrete version of the Laplace equation, that is $\Delta_r g(r, s) = 0$, where $\Delta f(x, y) = 4f(x, y) - f(x+1, y) - f(x-1, y) - f(x, y+1) - f(x, y-1)$. At the borderline $g(r, s) = \delta(r, s)$, where δ is the Kronecker symbol. Note that this Green function is different from the one used in paragraph 7.6. This function has a simple probabilistic interpretation. We release N random walkers at point s, which disappear as they reach the borderline. Then, if $N(r)$ is the number of times that the random walker gets through point r, the ratio $N(r)/N$ tends towards g when N is large. In fact, the probability density of the random walkers satisfies the Laplace equation as g does, and by construction, $N(r)/N$ is either 1 or 0 at the borderline. The solution $p(r)$ of the Laplace equation with the given boundary condition $\psi(s)$ at the borderline is written more simply than in paragraph 7.6 in the form

$$p(r) = \sum_s g(r, s)\psi(s) \qquad (11.1)$$

where the sum acts on all the points of the borderline. From equation (11.3) in paragraph 11.2 the velocity of displacement of the interface is proportional to the gradient of the considered field, that is

$$\begin{aligned} v &= -C(\psi(s) - p(r)) \\ &= C\sum_{s_1} g(r, s_1)\psi(s_1) - C\sum_{s_1} g(s, s_1)\psi(s) \end{aligned} \tag{11.2}$$

where r is the point next to point s inside the fluid. Thus, at the interface $\psi(s) = d_0\kappa$, whereas at infinity $\psi = \psi_\infty$. We can them approximate the sum (11.2) by the expression

$$v \approx C\sum_{s_{1\infty}} g(r, s_1)\psi_\infty + C\sum_{s_{1i}} g(s, s_1)(\psi(s_1) - \psi(s)) \,. \tag{11.3}$$

In fact in the absence of surface tension $\psi(s) = 0$ at the interface, and the expression of v only consists of the first sum of equation (11.3). When the effects of surface tension are taken into account, but are of a small order of magnitude, we approximate $g(r, s_1)$ by $g(s, s_1)$ in the terms proportional to the curvature. Thus, the interface moves under the effect of two types of random walks. The first one consists of walkers coming from infinity and attaching at the interface, as in the case of the DLA cluster. The second one, represented by the second sum of equation (11.3), consists of walkers leaving the interface to attach somewhere else on it (figure 11.8). The probability of a walk joining two given points s and s_1 of the interface is weighted by the difference in curvature between these two points. To determine it we evaluate the curvature of the interface at the given point s as follows (Vicsek 1984). Let n be the number of particles belonging to the aggregate included in the circle of radius R and of centre s. Then, if the variations of surface shape are large compared to R, $\kappa \approx (3/R)(n/R^2 + \pi/2)$.

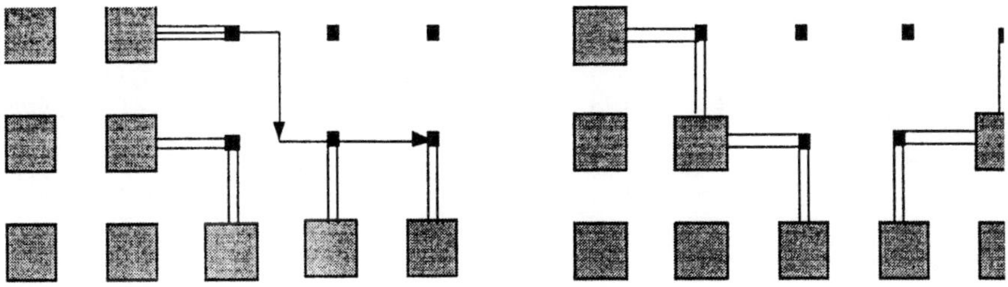

Figure 11.8 Schematic representation of a second-type walker (Kadanoff 1985).

If we move the interface at a given point, at each arrival or departure of a walker, the obtained interface is extremely fluctuating. Extending the method of noise reduction described in paragraph 11.2 to the walkers leaving the interface, Liang

(1986) can reproduce, with a value of m of the order of 10, evolutions of finger shapes perfectly analogous to the conventional numerical methods in paragraph 10.3 (figure 11.9a). We can also reproduce the first instabilities of the stationary finger already described in the paragraph mentioned above (figure 11.9b), as well as more branched patterns (figure 11.9c). This method allows us to obtain quantitative results, since the curve $\lambda(\kappa)$ of the stationary solutions of McLean and Saffman can be reproduced.

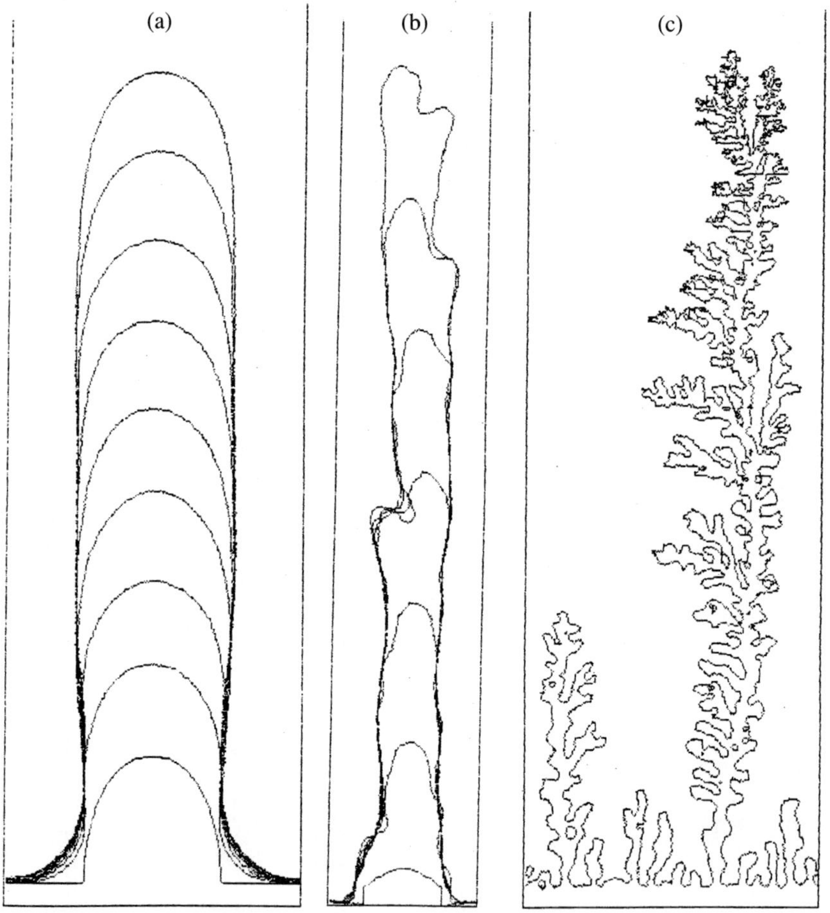

Figure 11.9 (a) $B = 5 \cdot 10^{-3}$, $m = 10$ (Liang 1986); (b) $B = 8 \cdot 10^{-4}$, $m = 10$; (c) $B = 5 \cdot 10^{-6}$, $m = 3$.

11.5 ANISOTROPY EFFECT

The DLA aggregate presented in figure 11.1 of paragraph 11.1 is anisotropic, since the particles move on a lattice of square symmetry. Nevertheless, the stochastic nature of the growth process weakens its anisotropic nature. Comparatively, a similar structure can be obtained without the lattice constraint in the following way. The centre of a particle moves in all directions with the same probability, over a distance equal to its diameter. If at a given time the particle and a part of the aggregate overlap, the particle attaches itself at the point where it touched the aggregate first. The obtained particle clusters appear to be very similar to those presented in figure 11.1. In particular, their fractal dimension $D_{h-r} = 1.7 \pm 0.07$ is very close to the case of square symmetry mentioned in the table 11.1. In three dimensions the fractal dimensions of the aggregates are also very close since $D_{h-r} = 2.50 \pm 0.08$.

For aggregates of similar sizes we can strenghten the anisotropy effects either by reducing the stochastic influence of the growth process or by introducing an anisotropic sticking rule.

As in paragraph 11.2, to reduce the influence of the fluctuations let us introduce the number m, defined as the number of times that a given site is reached by a particle before the particle attaches. The rule $m = 1$ generates the DLA. As shown by figure 11.10 the higher the value of m is, the more irregular and anisotropic the aggregate is, here on a square lattice. From $m = 5$ the irregular aggregate generated by frequent secondary branching acquires a dendritic morphology, that is with a branch in each direction of easy growth, each branch having a relatively stable tip, and secondary branches growing from the tip perpendicularly to the growth axis. For values of m of the order of 100 the branches get a regular, needle-like shape (Kertesz and Vicsek 1986).

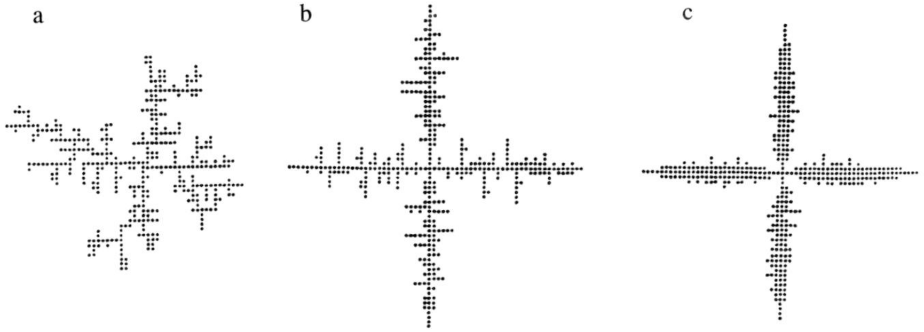

Figure 11.10 Aggregate of 400 particles on a square lattice. (a) $m = 2$, random fractal; (b) $m = 20$ dendritic growth; (c) $m = 400$, needle-like crystals (Kertesz and Vicsek 1986).

In a more in-depth study Eckmann and coworkers (1989) make small discotic particles diffuse outside a lattice. Each small disc features s antennas at angles $2\pi/s$. When a particle touches another particle on the border site it attaches to the closest antenna. In figure 11.11 we represent three aggregates obtained for the values $s = 3$, $s = 4$, $s = 5$, respectively, with the counter $m = 100$. The aggregate features s branches, which reflect the microscopic anisotropy.

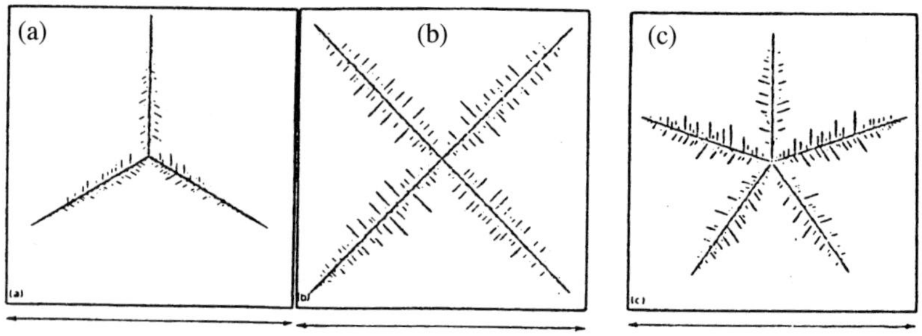

Figure 11.11 Dendritic aggregates for (a) $s = 3$, (b) $s = 4$, (c) $s = 5$ and $m = 100$ (Eckmann *et al.* 1989).

As shown on figure 11.12 the branch structure is well-determined. We notice a scaling invariance: a branch with a smaller number of particles, but which is stretched, presents the same aspect as a branch with a larger number of particles.

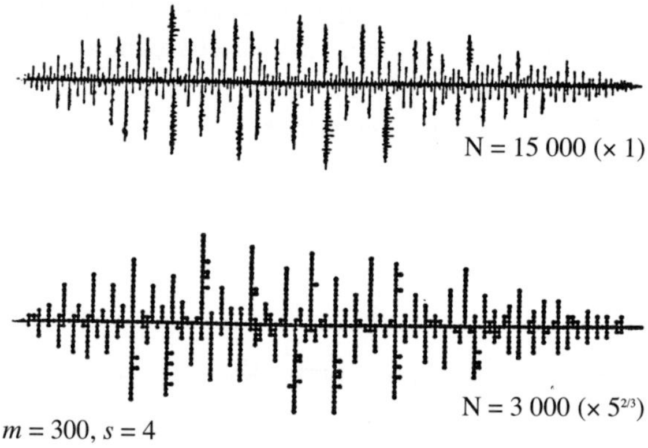

Figure 11.12 Scaling invariance of a branch, $s = 4$, $m = 300$. The upper and lower branches contain 15000 and 3000 particles respectively (Eckmann *et al.* 1989).

When m reaches large values, $m = 10000$ as in figure 11.13, the branches become compact.

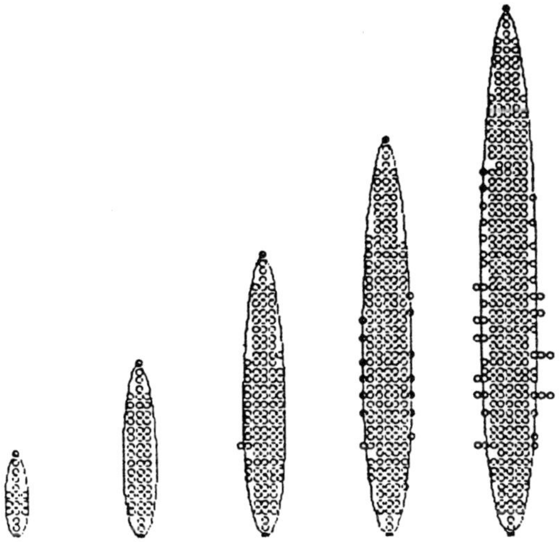

Figure 11.13 Comparison between the analytic form and the simulation for $s = 4$ and $m = 10000$ (Eckmann *et al.* 1989).

To determine the shape of these branches let us introduce f, the probability density of a particle attaching itself in the element dx. R is the length of the branch and x the longitudinal coordinate. Let us suppose that $s\Delta N$ particles attach themselves to the border sites of the branch. The number of particles which get attached at the abscissa x is therefore

$$\Delta n(x, R) = \frac{1}{R} f\left(\frac{x}{R}\right) \Delta N . \tag{11.1}$$

The scaling invariance of the branch structure implies that f does not explicitly depend on R. The number of particles getting attached at the abscissa x is therefore

$$n(x, R(N)) = \int_x^{R(N)} dR \frac{1}{R} f\left(\frac{x}{R}\right) \frac{dN}{dR} . \tag{11.2}$$

As in paragraph 11.2 the probability density f is proportional to the normal probability gradient u. If we approximate the shape of the aggregate by a cross, as in paragraph 10.4, the complex function W, with u as real part, has a derivative $dW/dz = z/(z^4 - R^4)^{1/2}$, since $u = 0$ at the border of the aggregate. Thus

$$f(y) = \frac{4}{\pi} \frac{y}{\sqrt{1 - y^4}} . \tag{11.3}$$

Writing that

$$\frac{dR}{dN} = \int_{(R-a)}^{R} dx \, \frac{f\left(\frac{x}{R}\right)}{R} = \int_{1-\frac{a}{R}}^{1} dy \; f(y) \tag{11.4}$$

we deduce $R(N) \approx aN^{3/2}$, and therefore, from equations (11.2) and (11.3),

$$n(x, R(N)) = \sqrt{\frac{x}{a}} \int_{\frac{x}{R}}^{1} dy \, \frac{1}{\sqrt{y}\sqrt{1-y^4}} . \tag{11.5}$$

As shown in figure 11.13 the shape determined by equation (11.5) is in good agreement with the numerical simulation. It only differs from the shape of the branch of Almgren and coworkers (1993) determined in paragraph 10.4 by some exponents. Kassner and Brener (1994) also follow this analysis, but instead of $\rho^2 V = \text{Cst}$ they consider $\rho = \text{Cst}$, of the order of the size of a particle. Then, as in paragraph 10.4, if $x \approx t^{\alpha}$, $\rho \approx t^{2-3\alpha}$, the new hypothesis implies that $\alpha = 2/3$. This new exponent allows us to find the shape of the branch given by equation (11.5).

The link between the growth of a DLA cluster confined to $m > 1$ and the growth of a crystalline dendrite in a capillary can be achieved by an operation similar to that in paragraph 11.2 (Arnéodo *et al.* 1991). We successively grow *N* clusters of DLA at $m > 1$, which contain the same number of particles *M* in a rectangular cell of width *w*. All these clusters are different from each other, but for a reason which is yet to be explained, their superposition allows us to find one of the stationary Saffman–Taylor finger forms determined in paragraph 7.5, of relative size λ smaller than $1/2$. After the superposition of all clusters let us define $r(x, y)$ as the number of particles occupying the site (x, y) divided by *N*. For a given *x* abscissa the $r(x, y)$ profile is at a maximum at the centre of the cell and equal to $r_{\max}$, and decreases towards the edges. We define the half-width of the finger as the average

$$\lambda = \frac{1}{r_{\max}} \int_{0}^{w/2} r(-\infty, y) dy \tag{11.6}$$

and we find that $\lambda = \Lambda(1/m^{3/2} w)$, where $\Lambda(u) \approx u^{1/2}$ at small values of *u*, and $\Lambda(u) \approx 1/2$ at high values. We thus find a behaviour similar to that of paragraph 8.10, where the smaller and smaller relative sizes can be obtained when the growth anisotropy is reinforced.

11.6 FINITE CONCENTRATION

The DLA aggregate grows particle by particle and its growth rate is determined by the number of particles injected per unit time. Its structure does not depend on this rate. A different problem is to consider the growth of such a cluster from a lattice covered by particles of density n_g. When a particle diffuses to a site next to the aggregate, it attaches itself there. Consider the case of growth in a linear geometry (Uwaha and Saito 1989). At low density its structure is similar to that of a DLA cluster. At higher density its structure becomes more compact (figure 11.14 a and b).

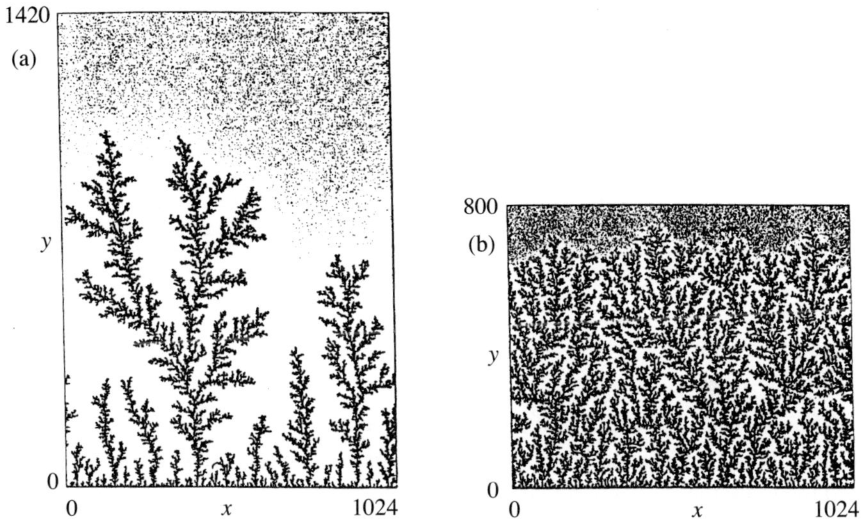

Figure 11.14 (a) $n_g = 0.08$ (Uwaha and Saito 1989); (b) $n_g = 0.2$ (Uwaha and Saito 1989).

In a stationary regime the envelope of the aggregate is almost planar and separates an upstream region where the density of free particles is n_g from a downstream region where the aggregate has a mean density $n_s = n_g$ because of the conservation of the number of particles. In the intermediate region of thickness of the order of the diffusion length D/V the density of free particles decreases from n_g to 0 whereas the density of the aggregate increases from 0 to $n_s = n_g$ (figure 11.15).

Contrary to the DLA cluster which is self-similar, the finite density aggregate is not self-similar because of the presence of the diffusion length D/V. However, the scales smaller than the diffusion length grow in the tail of the diffusion length where the free particles are sparse, as is the case in the DLA cluster. The corresponding structure is therefore fractal, of dimension D_f. If R is the scale under consideration, the local density of cluster particles is determined by the relation $n_s(R) \approx R^{D_f - d}$. If we extrapolate this relation to the scale D/V, where $n_s = n_g$ since the structure becomes compact, we obtain an expression for the growth rate of the aggregate as a function of its fractal dimension,

$$V = Dn_g^{1/(D_f - d)}. \qquad (11.1)$$

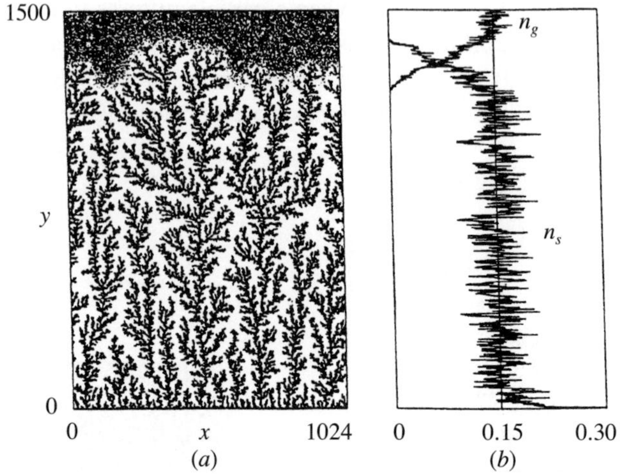

Figure 11.15 (a) $n_g = 0.15$; (b) density profile along the growth axis (Uwaha and Saito 1989).

Thus the fractal dimension of the aggregate controls its growth rate. This result is valid only if the diffusion length is far larger than the small scales of the aggregate, and therefore the spatial period of the lattice. Equation (11.1) imposes that the condition that the density of particles is sufficiently low. Note that the analysis developed here is conceptually similar to that developed in paragraph 7.7, concerning the form and the growth rate of a crystal in a small Péclet number capillary: upstream from the front there is a linear concentration profile on the scale of the diffusion length and around the interface a problem arises with the free boundary, similar to that in viscous fingering in a Hele–Shaw cell. In the case of the crystal in the capillary, since the external noise is low and the surface tension is high, the shape of the crystal is regular. In the case of the finite density aggregate it is the reverse, and its pattern is fractal. In both cases the growth rate of the interfaces is determined by a relation between the diffusion profile and the internal structure.

The growth rate V of the aggregate, a function of the particle density n_g can be obtained by numerical simulation. In a large domain of values of density n_g we determine the relation $V \approx 58 n_g^{3.5}$. Comparing with equation (11.1) we obtain $D_f = 1.71$, which is in good agreement with the fractal dimension of a DLA cluster.

11.7 ADVECTION EFFECT

In many configurations a flow or an electric field can sometimes change the Brownian motion of a particle into a given direction by advection. To model this effect on a lattice Meakin (1983) uses the following rule. An advection probability

p_0 between 0 and 1 is first chosen. If a randomly chosen second number is smaller than p_0 the particle moves in the advection direction by one spatial period of the lattice. If this number is larger than p_0 the particle is moved with equal probabilities in the four directions of the lattice. As shown on figure 11.16, we always obtain aggregates oriented in the advection direction but with the compactness increasing with p_0.

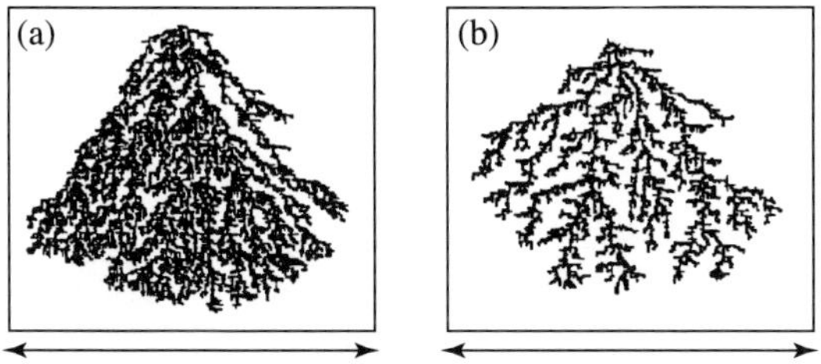

Figure 11.16 (a) Aggregate of 9582 particles, $p_0 = 0.5$; (b) 4629 particles, $p_0 = 0.1$ (Meakin 1983).

Whereas the purely diffusive aggregate does not feature any lengthscale, the advective effect introduces a diffusion length that can be evaluated in the following way: if n is the number of steps of the random walk, $l \approx \sqrt{n}$ by diffusion, and $l \approx p_0 n$ by advection, the two balance out for $n \approx 1/p_0^2$ or $l_D \approx l/p_0$. At distances smaller than l_D diffusion prevails and the structure of the aggregate is close to that of a diffusion-limited aggregate. This scaling invariance disappears on larger scales (Meakin 1983).

Another simulation (Erlebacher, Searson and Sieradski 1993) introduces the advective effect by an electric field, as follows: the particle is released at the border of the lattice and diffuses up to a distance L from the aggregate where it enters a region of space charge and where the electric field is more intense. Then, the particle is moved along a straight line to the closest point to the aggregate, where it is potentially attached. The operation is repeated JP times, where P is the perimeter of the aggregate and J the number of particles arriving at each iteration per unit perimeter. Then the shape of the cluster is modified, taking into account the number of particles arriving at each site. Changing the parameters L and J we generate a set of aggregates (figure 11.17), which we also find by electro-deposition (Grier *et al.* 1986, Sawada *et al.* 1986) in paragraph 11.8. We find for instance the typical form of a diffusion-limited aggregate for $J = 0$ and $L = 2$, and dense radial patterns obtained by electro-deposition in a stronger electric field ($J = 0.5$, $L = 50$).

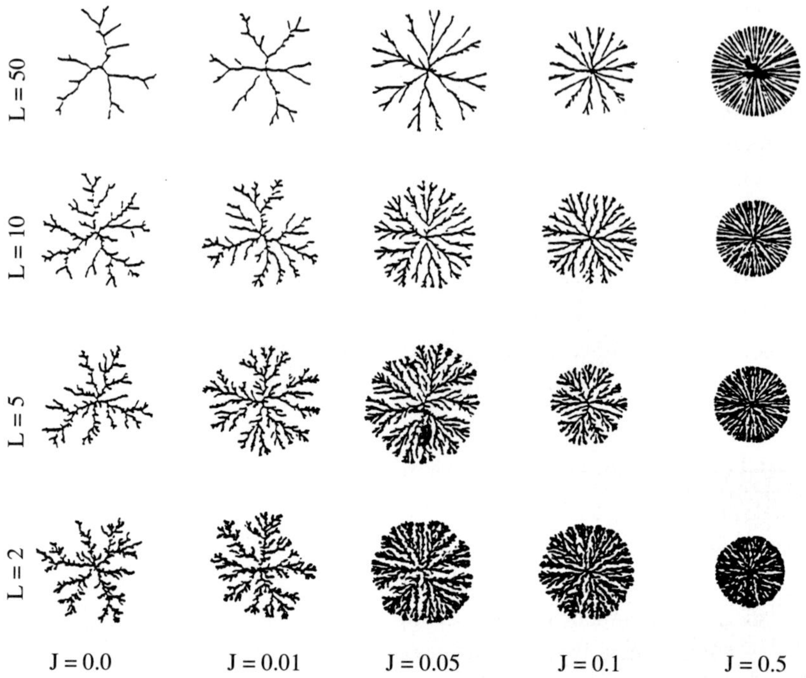

Figure 11.17 Aggregates for different values of *L* and *J*. The flux *J* increases from left to right. Migration lengths increase from the bottom to the top (Erlebacher *et al.* 1993).

11.8 EXPERIMENTAL GROWTH FORMS

Many growth forms similar to that of a DLA aggregate can be seen in nature. As will be shown here, some of them are the result of a growth mechanism very similar to that described in paragraph 11.1, whereas others only show a Laplacian dynamics analogy (viscous flow, quasi-stationary diffusion, Poisson law without space charge). In the first category we find: an aggregate of silver atoms deposited by epitaxy in the gas phase onto a platinum surface (Brune *et al.* 1994), or the sheets of zinc formed by electro-deposition already described in paragraph 1.1 (Matsushita *et al.* 1984). In the second category we find: viscous fingering developing in a miscible non-Newtonian fluid, fracture resulting from a disruption of a dielectric, or even a figure of chemical dissolution in a porous medium (Daccord and Lenormand 1986, Daccord 1987).

In the first experiment silver atoms are deposited by epitaxy in the gas phase onto a platinum surface. Once deposited these atoms migrate to the surface by Brownian motion until they get irreversibly attached on the perimeter of existing aggregates.

The parameters of the experiment are the temperature which controls the mobility of the atoms on the surface and the atom flux. The experiment is done under high vacuum and the observation of aggregates is achieved by scanning tunneling microscopy (Röder *et al.* 1993). The aggregates in figure 11.18a are generated at 110 K and at very small flux, of the order of 10^{-5} monolayer per second, a monolayer corresponding to 1.5×10^{15} atoms/cm^2. The largest ones contain about 3000 atoms, and develop very thin branches (one to two atoms wide) in all directions without an obvious correlation, just as those in paragraph 11.1 simulated by computer. The fractal analysis of an aggregate is done by counting the number of pixels N (of surface 2.3 Å^2) with greyscale values corresponding to the image, contained in a box of size ε. The linear relation $\log N$ as a function of $\log \varepsilon$ in figure 11.18b shows that the aggregates are fractal, of dimension $D \approx 1.78$, close to that of a DLA cluster.

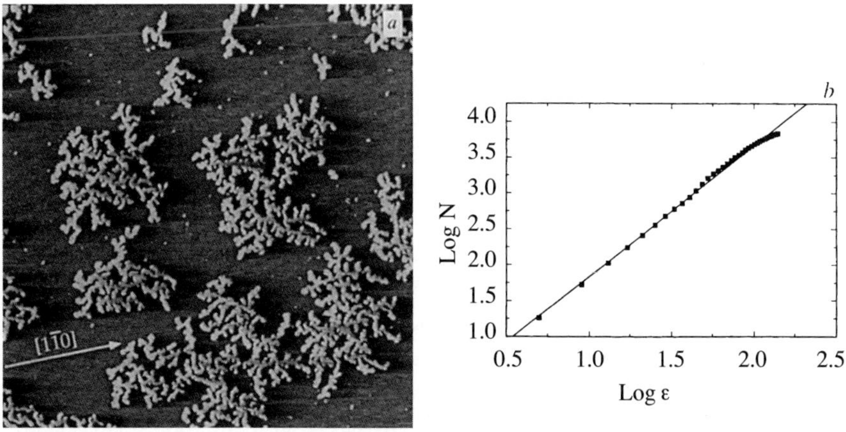

Figure 11.18 (a) Fractal aggregates of silver atoms (Brune *et al.* 1994) on a surface of platinum (111); (b) fractal analysis, Log N as a function of Log ε, of slope $D \approx 1.8$.

When the deposition flux of atoms is higher, for instance by a factor of 100, the aggregates become dendritic (figure 11.19a), with the crystalline symmetry of the substrate. The small increase in temperature, 130 K, implemented in order to obtain aggregates of size comparable to those in figure 11.19, does not account for the transition as shown in the insert of the upper part of the figure, where disordered clusters of the same temperature, yet with the preceding flux, are presented. The analysis of figure 11.19b shows that in spite of an important qualitative change in the structure, the structure remains fractal with a dimension close to the previous one, that is $D \approx 1.77$. This transition resembles the experiment by Ben-Jacob and coworkers (1985) shown in paragraph 10.4 (figure 10.8 b-c-d), in which the growth rate is increased, or the simulation by Kertesz and Viscek (1986) from figure 11.9 in para-

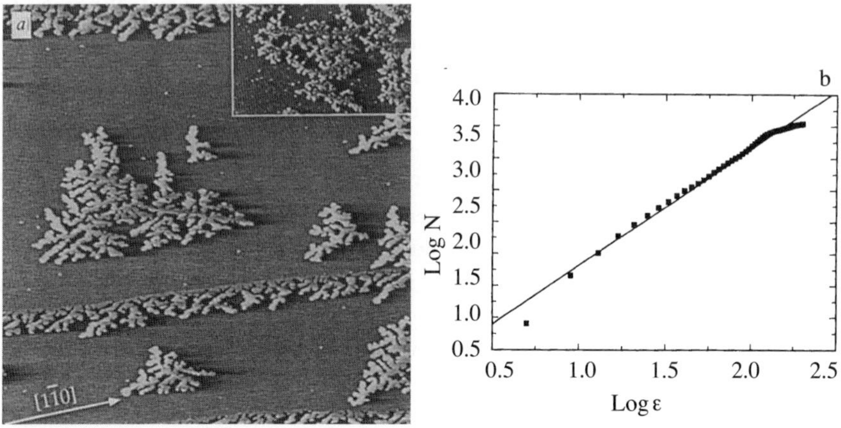

Figure 11.19 (a) Dendritic aggregates of silver atoms; (b) fractal analysis (Brune *et al.* 1994).

graph 11.5, in which the noise reducing factor m is increased. A computer simulation of this configuration (Hohage *et al.* 1996) reproduces some characteristics of these aggregates.

The second experiment deals with the formation of metallic zinc leaves by the electro-deposition process already described in paragraph 1.1 (Matsushita *et al.* 1984). Variants of this experiment have been done by Grier and coworkers (1986) and Sawada and coworkers (1986), in which the solution was confined in a very thin cell, of the order of 1 mm. The ionic concentration of the electrolyte ranges from 0.0025 M to 1 M and the potential difference between the two electrodes varies from 1 to 10 V. The observed aggregate structures are very diverse (figure 11.20). At low concentration the structure changes from a DLA-type structure for a small potential difference (figure 11.20a), to a dense structure for a larger potential difference (figure 11.20b). At higher concentration the aggregates form needles (figure 11.20c-d). This diversity of structures must be related to the higher complexity of the system (ionic concentration, electric potential), but also to the activation Butler–Volmer law already mentioned in paragraph 3.5. This law involves activation energies of a higher sensitivity of the deposition rates to the change in certain parameters, as is the case for a combustion reaction. As previously mentioned in paragraph 5.8, the growth dynamics is close to the dendritic growth and the structure of the aggregates is similar. At very low concentration and very low potential difference diffusion prevails and the structure is close to the DLA structure of paragraph 11.1.

The experiment by Nittmann and coworkers (1986) is similar to that by Saffman–Taylor in a Hele–Shaw cell, however, the two fluids are miscible (water, aqueous solution of polymer) but over a mixing time far longer than that of the development

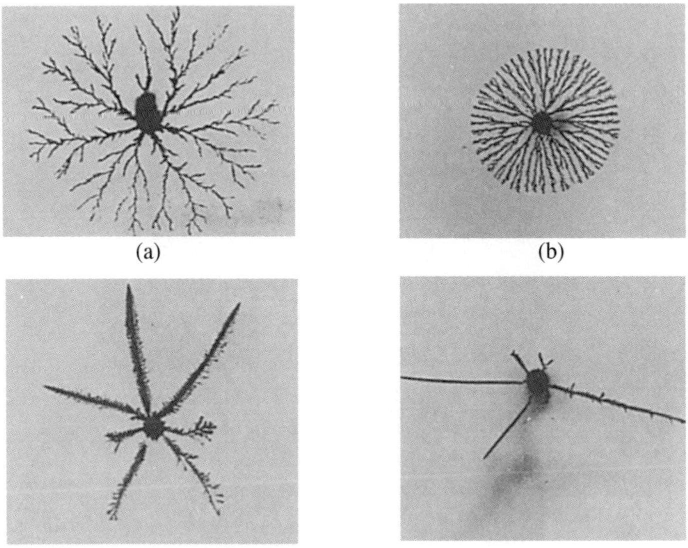

Figure 11.20 (a) DLA (0.01 M, 3 V), (b) dense aggregate (0.01 M, 10 V), (c) dendrites (0.5 M, 9 V), (d) dendrites (1 M, 10 V) (Grier *et al.* 1986).

of fingers. Moreover the more viscous fluid is non-Newtonian. Thus the surface tension between the two fluids (or the curvature effect) is neglected, and the development of fingers is increased by the non-Newtonian behaviour of the displaced fluid. As determined in paragraph 9.6, for a reasonable magnitude of external noise, the size of the fingers is of the same order of magnitude as the most unstable wavelength λ_c, and therefore all the smaller as the surface tension is reduced. We can thus obtain small fingering features with accessible fluid velocities. Moreover, since the viscosity of the non-Newtonian fluid is reduced as the pressure gradient increases, the formation of channels with a preferential displacement is increased. The fingering observed (figure 11.21) looks similar to that of a branch in a DLA cluster, although the fractal dimension of the object is $D = 1.39$ in this case. The same experiment done in a circular geometry (Daccord *et al.* 1986) produces a fractal structure of dimension $D = 1.7$, therefore similar to that of DLA. A model for this phenomenon is similar to the one for DLA since the velocity of a non-Newtonian fluid can be written as a function of the pressure gradient by a relation of the form $\vec{w} \approx \vec{\nabla} p \left|\vec{\nabla} p\right|^{m-1}$, $m > 1$, still with $\vec{\nabla} \cdot \vec{w} = 0$ since the boundary conditions at the interface are unchanged.

Another experiment deals with the breakdown of a dielectric, the most famous example being a flash of lightning. A simpler experiment consists of creating a two-

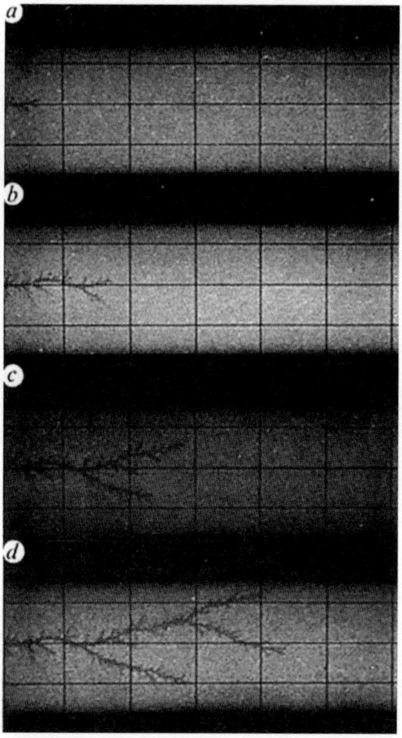

Figure 11.21 Fingering formed by water moving an aqueous solution of polymer with a viscosity ranging from 10^3 to 10^4 cP according to the velocity of the fluid (Nittmann *et al.* 1985).

dimensional radial discharge. An electrode is in contact with a 2 mm thick insulating glass plate. The other side of the plate is covered with a conductive material. The whole experiment is kept in a SF_6 atmosphere under pressure. We apply a 30 kV discharge during 1 μs and we obtain a Lichtenberg figure on the plate as shown in figure 11.22. It is again a fractal structure of dimension close to $D = 1.7$. The outside of the discharge figure is an insulating medium and the electric potential satisfies the Laplace equation. When the local electric field exceeds a critical value ions are released and the inside of the medium becomes conductive. A simple approximation consists of considering the inside as ideally conductive so that the discharge border is at a constant potential. The normal velocity of the discharge figure is assumed to be proportional to a power η of the local electric field (Niemeyer *et al.* 1984). The growth dynamics is simulated on a square lattice in the following way. At a given time the discharge figure consists of a set of connected links over which the potential is kept constant $(\Phi = 0)$. Outside, a circular electrode keeps the potential at $\Phi = 1$. We calculate the potential around the discharge figure solving the discretised Laplace equation

$$\Phi_{i-1,j} + \Phi_{i+1,j} + \Phi_{i,j-1} + \Phi_{i,j+1} = 0 \qquad (11.1)$$

where $\Phi_{i,j}$ is the value of the potential at the site (i, j). At the next time step a new link is added to the structure, weighted by the probability

$$p(i, j \rightarrow i', j') = \frac{(\Phi_{i',j'})^{\eta}}{\Sigma(\Phi_{i',j'})^{\eta}} \, . \qquad (11.2)$$

The discharge figures determined in this way are self-similar and have a structure similar to the DLA clusters. The fractal dimension of these structures depends on η, and decreases from $D = 2$ for $\eta = 0$ to $D = 1.6$ for $\eta = 2$. Thus, the higher values of η correspond to objects where secondary branching is less frequent. The value obtained for $\eta = 1$ is $D = 1.75$. This value is close to a DLA cluster since the probability distribution of a random walker satisfies the Laplace equation.

The dimensionality of a dielectric medium has great importance forthe structure and the fractal dimension of the obtained discharge figure. In fact with two-dimensional boundary conditions (Niemeyer *et al.* 1986) the structure obtained is far denser than that obtained on figure 11.22. This observation seems to be contradictory to the calculation by Satpathy (1986) who observed that at the same value of η the three-dimensionality of the potential creates denser figures. Owing to the complexity of the system, other factors must certainly intervene in the determination of the fractal dimension.

The last experiment which we consider is the chemical dissolution of a porous medium (Daccord and Lenormand 1987, Daccord 1987). Water is injected through

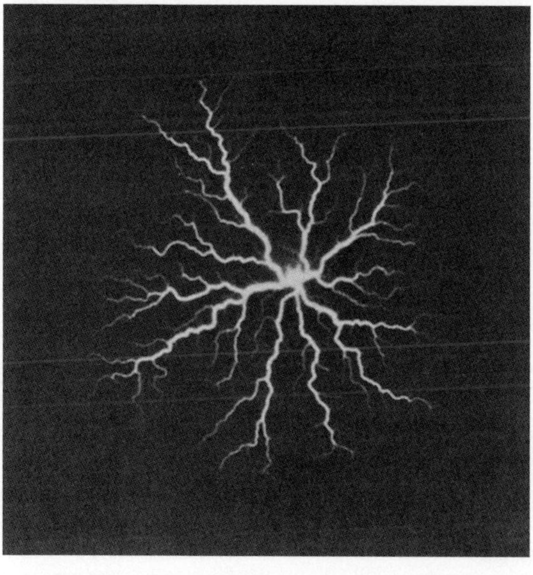

Figure 11.22 Lichtenberg figure. The geometry is three-dimensional since the external electrode is a conductive disc forming the bottom part of the cell (Niemeyer *et al.* 1984).

a point into a porous medium (mixture of calcium sulphate and water), soluble in water. As the water flows it dissolves the porous medium generating a figure of fractal dissolution. This figure is preserved by finally injecting a resin which fills the dissolution figure, then the external porous matrix is dissolved (figure 11.23). At a given time the dissolution figure separates a more permeable medium of smaller effective viscosity on the inside, from an outside medium which is less permeable since it is not dissolved by water. For a similar reason to viscous fingering in a Hele–Shaw cell the water flux crossing the interface between the two media is higher at the tips than in the dips. This results in a stronger dissolution at the tips, favouring the appearance of fingering. Contrary to a DLA cluster the branches widen with time. The model generating this structure is relatively complex since it involves a viscous flow, and a concentration of calcium sulphate which is coupled to the kinet-

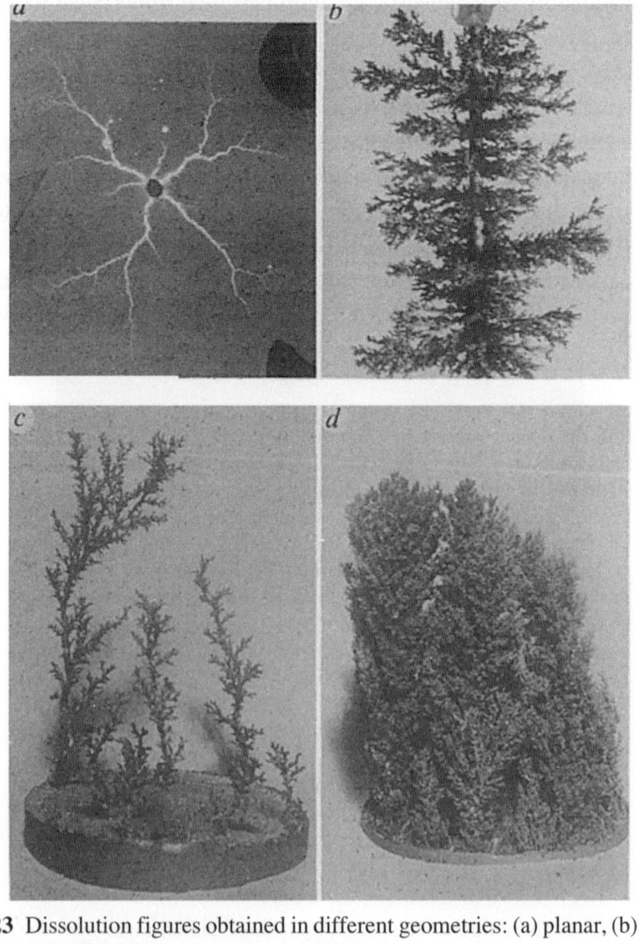

Figure 11.23 Dissolution figures obtained in different geometries: (a) planar, (b) cylindrical, (c)-(d) from a disc (Daccord and Lenormand 1986).

ics of interfacial dissolution. This novel example shows the universality of the growth forms, of which a simple representative is the DLA cluster.

BIBLIOGRAPHY

[1] C. Amitrano, A. Coniglio, P. Meakin, M. Zanetti, 'Multiscaling in Diffusion-Limited Aggregation', *Phys. Rev.*, **B, 44**, pp. 4974-4977, 1991.

[2] A. Arnéodo, Y. Couder, G. Grasseau, V. Hakim, M. Rabaud, 'Uncovering the Analytical Saffman–Taylor Finger in Unstable Viscous Fingering and Diffusion-Limited Aggregation', *Phys. Rev. Lett.*, **63**, p. 984, 1989.

[3] A. Arnéodo, F. Argoul, Y. Couder, M. Rabaud, 'Anisotropic Laplacian Growths: From Diffusion-Limited Aggregates to Dendritic Fractals', *Phys. Rev. Lett.*, **66**, p. 2332, 1991.

[4] A. Arnéodo, F. Argoul, E. Bacry, J.F. Muzy, M. Tabard, 'Golden mean Arithmetic in the Fractal Branching of Diffusion-Limited Aggregates', *Phys. Rev. Lett.*, **68**, p. 3456, 1992.

[5] D. Bensimon, L.P. Kadanoff, S. Liang, B.I. Shraiman, C. Tang, 'Viscous Flows in Two Dimensions', *Reviews of Modern Physics*, **58**, **4**, p. 977, 1986.

[6] R. Botet, R. Jullien, 'Diffusion Limited Aggregation with Disaggregation', *Phys. Rev. Lett.*, **55**, p. 1943, 1985.

[7] H. Brune, C. Romainczyk, H. Roder, K. Kern, 'Mechanism of the Transition from Fractal to Dendritic Growth of Surface Aggregates', *Nature*, **369**, p. 469, 1994.

[8] G. Daccord, 'Chemical Dissolution of a Porous Medium by a Reactive Fluid', *Phys. Rev. Lett.*, **58**, p. 479, 1987.

[9] G. Daccord, R. Lenormand, 'Fractal Patterns from Chemical Dissolution', *Nature*, **325**, p. 41, 1987.

[10] G. Daccord, J. Nittmann, H.E. Stanley, 'Radial Viscous Fingers and Diffusion-Limited Aggregation: Fractal Dimension and Growth Sites', *Phys. Rev. Lett.*, **56**, p. 336, 1986.

[11] J.P. Eckmann, P. Meakin, I. Procaccia, R. Zeitak, 'Growth and Form of Noise-Reduced Diffusion-Limited Aggregation', *Phys. Rev.*, **A 39**, p. 3185, 1989.

[12] J. Erlebacher, P.C. Searson, K. Sieradski, 'Computer Simulation of Dense-Branching Patterns', *Phys. Rev. Lett.*, **71**, p. 3311, 1993.

[13] D. Grier, E. Ben Jacob, R. Clarke, L.M. Sander, 'Morphology and Microstructure in Electrochemical Deposition of Zinc', *Phys. Rev. Lett.*, **56**, p. 1264, 1986.

[14] M. Hohage, M. Bott, M. Morgenstern, Z. Zhang, T. Michely, G. Comsa, 'Atomic Processes in Low Temperature Pt-Dendrite Growth on Pt (111)', *Phys. Rev. Lett.*, **76**, p. 2366, 1996. þ

[15] H.G.E. Hentschel, 'Fractal Dimension of Generalized Diffusion-Limited Aggregates', *Phys. Rev. Lett.*, **52**, p. 212, 1984.

[16] L.P. Kadanoff, 'Simulating Hydrodynamics: a Pedestrian Approach', *J. Stat. Phys.*, **39**, p. 267, 1985.

[17] K. Kassner, E. Brener, 'Continuum Description of Noiseless Diffusion-Limited Aggregation', *Phys. Rev. E*, **50**, p. 2161, 1994.

[18] J. Kertesz, T. Vicsek, 'Diffusion-Limited Aggregation and Regular Patterns: Fluctuations Versus Anisotropy', *J. Phys.*, **A19**, p. L257, 1986.

[19] C.H. Lam, 'Finite-Size Effects in Diffusion-Limited Aggregation', *Phys. Rev. E*, **52**, p. 2841, 1995.

[20] S. Liang, 'Random-Walk Simulations of Flow in Hele–Shaw Cells', *Phys. Rev.*, **A33**, p. 2663, 1986.

[21] B.B. Mandelbrot, *The Fractal Geometry of Nature*, Freeman, New York, 1982.

[22] B.B. Mandelbrot, 'Plane DLA is not Self-Similar; is it a Fractal that Becomes Increasingly Compact as it Grows?' *Physica*, **A 191**, p. 95, 1992.

[23] M. Matsushita, M. Sano, Y. Hayakawa, H. Honjo, Y. Sawada, 'Fractal Structures of Zinc Metal Leaves Grown by Electrodeposition', *Phys. Rev. Lett.*, **53**, p. 286, 1984.

[24] P. Meakin, 'Diffusion-Controlled Cluster Formation in 2-6 Dimensional Space', *Phys. Rev.*, **A27**, p. 1495, 1983.

[25] P. Meakin, 'Effects of Particle Drift on Diffusion-Limited Aggregation', *Phys. Rev.*, **B28**, p. 5221, 1983.

[26] M. Muthukumar, 'Mean Field theory for Diffusion-Limited Cluster Formation', *Phys. Rev. Lett.*, **50**, p. 839, 1983.

[27] L. Niemeyer, L. Pietronero, H.J. Wiesmann, 'Fractal Dimension of Dielectric Breakdown', *Phys. Rev. Lett.*, **52**, p. 1033, 1984.

[28] J. Nittmann, G. Daccord, H.E. Stanley, 'Fractal Growth of Viscous Fingers: Quantitative Characterization of a Fluid Instability Phenomenon', *Nature*, **314**, p. 141, 1985.

[29] M. Plischke, Z. Racz, 'Active Zone of Growing Clusters: Diffusion-Limited Aggregation and the Eden Model', *Phys. Rev. Lett.*, **53**, p. 415, 1984.

[30] Z. Racz, M. Plischke, 'Active Zone of Growing Clusters: Diffusion-Limited Aggregation and the Eden Model in Two and Three Dimensions', *Phys. Rev.*, **A31**, p. 985, 1985.

[31] P.A. Rikvold, 'Simulations of a Stochastic Model for Cluster Growth on a Square Lattice', *Phys. Rev.*, **A 26**, p. 647, 1982.

[32] H. Röder, E. Hahn, H. Brune, J.P. Bucher, K. Kern, 'Building One and Two-Dimensional Nanostructures by Diffusion-Controlled Aggregation at Surfaces', *Nature*, **366**, p. 141, 1993.

[33] S. Satpathy, 1986, Comment on 'Fractal Dimension of Dielectric Breakdown', *Phys. Rev. Lett.*, **57**, p. 649, respond by Niemeyer, Pietronero and Weismann.

[34] Y. Sawada, A. Dougherty, J.P. Gollub, 1Dendritic Fractal Patterns in Electrolytic Metal Deposits', *Phys. Rev. Lett.*, **56**, p. 1260, 1986.

[35] M. Uwaha, Y. Saito, 'Aggregation Growth in a Gas of Finite Density: Velocity Selection via Fractal Dimension of Diffusion-Limited Aggregation', *Phys. Rev.*, **A40**, p. 4716, 1989.

[36] C. Tang, 'Diffusion-Limited Aggregation and the Saffman–Taylor Problem', *Phys. Rev.*, **A31**, p. 1977, 1985.

[37] M. Tokuyama, K. Kawasaki, 'Fractal Dimensions for Diffusion-Limited Aggregation', *Physics Letters*, **100**, p. 337, 1984.

[38] T. Vicsek, 'Pattern Formation in Diffusion-Limited Aggregation', *Phys. Rev. Lett.*, **53**, p. 2281, 1984.

[39] T. Vicsek, Fractal Growth Phenomena, second edition, *World Scientific*, 1992.

[40] T.A. Witten, L.M. Sander, 'Diffusion-Limited Aggregation, a Kinetic Critical Phenomenon', *Phys. Rev. Lett.*, **47**, p. 1400, 1981.

[41] T.A. Witten, L.M. Sander, 'Diffusion-Limited Aggregation', *Phys. Rev.*, **B27**, p. 5686, 1983.

CHAPTER 12

Towards morphogenesis

12.1 A CERTAIN UNIVERSALITY

The patterns in growth presented in this book can be united into a single class, the *dynamics of curved fronts* (Pelcé 1986), or *fingered growth* (Kessler *et al.* 1988). They only constitute one class of growth patterns which exist in nature, but it is the class which is certainly the most interesting in terms of the richness of their growth forms. These ornamented patterns, resulting from instabilities which in general associate a transport phenomenon with interfacial dynamics, are shown in the table below.

Transport / Interfacial dynamics	Flow	Diffusion	Electro-diffusion
Surface tension	Viscous fingers	Dendrites	
		Electrode deposition	
Reaction-diffusion	Flames		

Thus, viscous fingers result from a dynamics involving a flow and a surface tension. Dendrites are found at the intersection of diffusion and surface tension, and flames between flow and reaction-diffusion. Electro-deposition exhibits a higher level of complexity, which is apparent in the diversity of forms which it can adopt. Since it results from a contribution of all the preceding growth form dynamics, it is situated at the centre of the above table.

It is interesting to note that each of the problems (stationary states, velocity of growth, stability, secondary branching) which naturally arose within each of the domains, was solved for one of them and then transposed to the others. Thus, the problem of forms and the growth velocities was resolved in the domain of viscous fingering because the velocity potential Φ which models that form satisfies a very simple equation, that is, the Laplace equation. The same problem arose in an equally simple way for dendritic growth, but it was technically more difficult as the anisotropy factor plays a role in the selection process. The stability of a curved form was feasibly taken into account for flames, where it appears in a paradoxical way. It was therefore necessary to consider all those patterns in turn.

If we now come back to morphogenesis itself, and therefore the growth of biological forms, before commencing our study it seems natural to ask if there exists an analogous pattern in nature, so that the problems raised for one of these patterns could be transposed to allow for a solution for the others. In the world of plant cells biologists observe two types of growth: diffuse growth and tip growth. The first mostly concerns multicellular organisms, that is, cell growth in tissues. It is the simplest growth model, by which the cell dimensions proportionally increase because of their initial inhomogeneity. The second concerns single cells, such as unicellular algae, filamenteous fungi, or neurons which have the property of growing at the tip of growth domes or cones, and give a tubular or dendritic morphology to the cell (Heath 1990). This growth model is all the more related to the material treated in this book, since the shape of these cells is greatly influenced by the conditions of the surrounding medium, in particular, by the presence of an electric field, the concentration gradient of different substances, and even flows. Following the method used here, it seems necessary to focus the study on some typical cells, for which numerous experimental data have already been obtained, and to ask some well-defined questions which could be solved starting from one of these forms. Note, however, that fingered growth is not exactly tip growth, and it could well be that this subtle difference in nomenclature reveals some important differences between the growth of the inanimate and the living forms.

12.2 TIP GROWTH

The most representative organism in this class is certainly the *Fucus* egg, a brown alga very widespread on the rocky coastlines of the northern hemisphere. The unfertilised egg is a very isotropic cell, without a visible axis of development. After fertilisation, a series of events leads to a polarisation of the cell which finally manifests itself by the germination of a rhizoid, giving the cell a characteristic pear shape (Kropf 1992). Several minutes after fertilisation, the membrane potential, of the order of – 20 mV, increases for several minutes up to a value close to 0 mV, then decreases on the time scale of several hours in order to stabilise at a value of the order of – 80 mV, which actually depends on the external concentration of potassium (Bentrup 1970, Weisenseel and Jaffe 1972, Taylor and Brownlee 1993). Approximately four hours after fertilisation, an ionic current loop appears in the extracellular matrix, the positive charges entering into the place of formation of the future rhizoid. The associated intensity of the electric current, which is at first weak, of the order of $0.05\ \mu A/cm^2$, grows until it is stabilised at $1\ \mu A/cm^2$ (Jaffe *et al.*

1974). The measurement of a current of such a small intensity could be carried out using a vibrating electrode (Jaffe and Nuccitelli 1974), with a spatial resolution of 30 μm, which is of the order of the size of the cell. The axis formed in this way is unstable at first, varying in direction, after which it becomes fixed by the cytoskeleton. Several hours later the germination commences. The cell growth therefore takes place essentially at the tip of the rhizoid, where a stronger intracellular calcium concentration (Berger and Brownlee 1993), and a more acidic pH (Gibbon and Kropf 1994) are found. In the absence of all external gradients, the growth axis is determined by the place of fertilisation. When an external gradient is applied (light flux, gradients of Ca^{2+}, K^{+}, potential, flow), the growth axis aligns itself according to the gradient. A collective effect results, in which different cells relatively close to one another influence the direction of their growth axis (Bentrup and Jaffe 1968).

As in the tubes of Pollen, certain filamentous fungi such as the *Neurospora crassa* or *Achlya bisexualis,* develop long tubes of diameter of the order of 30 μm, growing at a rate of 1 to 10 μm/min. In that case as well, the growth activity essentially takes place at the tip of the cell. Vesicles carrying different types of proteins (cellulose, polysaccharides, ionic channels) are synthesised using the Golgi apparatus and the endoplasmic reticulum, situated at a distance of several hundred microns behind the tip. The vesicles are transported towards the tip by a mechanism which is as yet unknown (molecular motors, electrophoresis), and they deliver their contents to the membrane, in a region of very limited extension close to the tip. In this case as well, a large current loop is observed in the extracellular matrix, essentially made up of protons which generally exit from the side of the cell and enter at the tip. The current characteristics are mostly reproducible from species to species : the intensity varying from 0.1 to 10 $\mu A/cm^2$, measured by the vibrating electrode which is sensitive to the presence of other substances, such as the amino acids or the sugars in the extracellular matrix (Kropf *et al.*1983, 1984, Gow 1984). A depolarisation of the tip of the order of 200 mV/cm, measured by the extra- and intra-cellular electrodes, therefore results (Slayman and Slayman 1962, Kropf 1986).

These cells exhibit chemotropism with certain substances. When an aminoacid source is applied several radii of curvature from the tip of a hypha of *Achlya bisexualis*, at the side, the tip of the cell changes its trajectory to align its growth axis with the axis of the applied gradient (Schreurs *et al.* 1989). These tubular forms often maintain an internal gradient of Ca^{2+}, the concentration of Ca^{2+} being higher at the tip (Reiss and Herth 1979), and sometimes even a pH gradient (Bachewich and Heath 1997, Robson *et al.* 1996). The shape of certain cells is very sensitive to the temperature of the medium. For example, the mutant cot-1 of *Neurospora crassa* transferred from 25 to 37 °C increases the number of secondary branches by an order of magnitude (Trinci 1979).

Neurons are, together with the glial cells and the Schwann cells, animal cells which grow by the tip growth mechanism. Deposited while in an embryonic state into a growing medium, at the beginning of growth they adopt the appearance of a fried egg : a membrane flattened onto a substrate with a protuberance towards the centre, i. e. the cell nucleus. After 30 minutes, the first neurites of similar morphologies develop at the beginning. After a day, one of the neurites is distinguished from the others by its fastest growth and finally becomes the axon. After several days, the neuron is essentially formed from a dendritic tree, with more or less dense ramifications according to the neuron under consideration, and from an axon of larger dimensions, structurally different to dendrites, essentially by the presence of microtubules (Lafont *et al.* 1993). On the same cover glass, where the conditions of uniformity have been thoroughly verified, the neurons have variable patterns, characterised by the number of their neurites and the relative length of the axon as compared to the neurites. These morphological characteristics depend significantly on the presence or the absence of the molecules of adhesion deposited on the cover glass, such as the laminin or the fibronectin (Chamak and Prochiantz 1989). Very reproducibly, the number of neurites grows and the relative length of the axon with respect to the neurites decreases regularly when the adhesion of the substrate is increasing, characterised for example by the concentration of the adhesion molecules (Bohin and Ternaux 1998). As for the fungi discussed above, certain neurons feature chemotropism with certain neurotransmitters, such as acetylcholine and glutamate (Zheng *et al.* 1994, Zheng *et al.* 1996). For example, when a source of acetylcholine (ACh) is applied at some distance from the growth cone of a motoneurone axon of Xenopus, the growth cone deviates towards the place where the concentration of ACh is the highest. In another experiment, raising the concentration of Ach plays the role of an inhibitor on the growth of neurites (Lipton and Kater 1989). In other experiments, it has been shown that certain motoneurons release some Ach (Young and Poo 1983). Thus, if a growth cone releases Ach, it can inhibit the growth of neurites close by (Layer 1990). There are therefore possibilities of autocrine interactions, by which the structure of the cell interacts with itself, a bit like the crystalline dendrites in the preceding chapters. This interaction is modulated by the presence of acetylcholinesterase (AChE), which is generally the case *in vivo*, which hydrolyses ACh. The neuron wraps itself around a layer of ACh, the local thickness of which depends on the concentration of ACh (Limozin *et al.* 1997). The resulting modulation of the concentration of ACh inside the membrane influences the local growth of the neurites. This growth dynamics becomes important when it is associated with the release of ACh at the synaptic sides, by the passage of an action potential. Could these variations in the local concentration not create a link between the perception of certain *stimuli* and a more long-term modification of the shape of certain neurons?

As the hyphae, the shape of the neurone grown *in vitro* strongly depends on the temperature of the medium. For example, a motoneuron of a rat develops a relatively short axon at 29 °C, whereas at 37 °C, it generates a longer axon surrounded by some primary neurites (Ternaux 1999, private communication). This observation is close to that of Cancalon (1983), who claims that the velocity of a growth cone increases, like a flame, exponentially with temperature.

12.3 IONIC CURRENT

E. Lund (1923) thought that the cellular electric field had an essential role in morphogenesis: '*it is proposed that the electric pattern is intimately related to the morphogenetic processes and polar or vector properties of cell and tissue functions. One function of this electric field or pattern is to act as a directive force in laying down new structures*'. Indeed, the numerous experiments indicate that the cells which grow by the tips, like algae (Peng and Jaffe 1976), hyphae (McGillivray and Gow 1986, Cho *et al.* 1991, De Silva *et al.* 1992), and neurons (Patel and Poo 1982), orient themselves in a sufficiently strong electric field, of the order of several V/cm, therefore, one order of magnitude larger than that of the typically measured intracellular electric field. However, the response is complicated, because the tips of the cells, or the neurites, can migrate towards the cathode or the anode by simply changing the nature of the acidic buffer keeping the extracellular pH constant (De Silva *et al.* 1992). Let us consider a spherical cell of radius $R \approx 10\ \mu\text{m}$, with a membrane which has a resistance of $r_m \approx 10^4 \Omega \cdot \text{cm}^2$, partially insulating the intracellular medium of resistivity $\rho_i = 200\ \Omega \cdot \text{cm}$ from the external medium of resistivity $\rho_e = 1\ 000\ \Omega \cdot \text{cm}$ (Kropf 1986). In an external electric field, E_e, the external potential is of the form

$$\Phi_e = E_e z + \frac{A_e}{r^2}\cos\theta\,, \quad \Phi_i = E_i z\,. \tag{12.1}$$

At the membrane, $\partial\Phi_e/\partial n = \rho_e i$, $\partial\Phi_i/\partial n = \rho_i i$, $\Phi_e - \Phi_i = r_m i$, such that due to the strong membrane resistance, $A_e \approx R^3 E_e/2$ (Robinson 1985), and $E_i \approx (3\rho_i R/2r_m)E_e \approx 3 \cdot 10^{-5} E_e$. Thus, the electric field practically does not penetrate inside the cell. It acts on the cell mainly by electrophoresis and electro-osmosis of membrane proteins, by its external tangential component which moves the external charges of the proteins (Poo and Robinson 1977, McLaughlin and Poo 1981).

However, as shown by Jaffe and coworkers (1974), certain ions such as H^+ and Ca^{2+}, can attach themselves to static and mobile ligands in the cytoplasm, and diminish the effective conductivity of the intracellular medium. Consider for instance a current mainly consisting of protons. When the protons enter into the cell, they attach themselves to fixed bases F^-, mobile bases M^- of total concentrations c_F and c_M and acidity constants K_F and K_M. From the law of mass action and the conservation of species F, we deduce

$$[F^-] = \frac{c_F K_F}{[H^+] + K_F} \tag{12.2}$$

and an analogous relation for species M. To simplify, suppose $[H^+] \ll [HM] \ll [M^-]$, then, $[H^+] = (K_M/c_M)[HM]$ such that, $i = -D_{HM}\Delta[HM]/FR = D_{HM}\Delta[F^-](c_M c_F K_F/K_M)/(FR[F^-]^2)$ where Δ here means the difference in the concentrations at the two poles of the cell. By electroneutrality, the charges F^-, of a higher concentration, are compensated by the majority of positive charges K^+. Since the latter are in equilibrium, $\Delta[K^+] = -[K^+](e/kT)\Delta\Phi$, such that $i = \sigma_{eff}\Delta\Phi/R$, with

$$\sigma_{eff} = \frac{eF}{kT}D_{HM}c_M\left(\frac{c_F}{K_M}\frac{K_F}{[F^-]}\right). \tag{12.3}$$

Take for example $D_{HM} = 10^{-5}\text{cm}^2/\text{s}$ and $c_M = 10^{-2}\text{mole/L}$, $K_M/K_F \approx 100$, $pH \approx pK_F$. Using $kT/e = 25$ mV and $F = 96\,500$, we deduce $\rho_{eff} = 5.2\times10^5\,\Omega\cdot\text{cm}$, that is, a resistivity 1 000 times larger than usual, which increases the induced intracellular field up to a value of the order of the natural field. Note also that, when the cellular shape is elongated, the current lines which are widely spaced on the outside, concentrate on the inside, thus increasing the intensity of the internal electric field (Kropf 1986). This can have an influence on the morphogenesis since the order of magnitude of electrophoretic mobility of particles in the cytoplasm is from 1 to 5 μm/s $(\text{V/cm})^{-1}$. Thus, in an internal electric field of 200 *m*V/cm, such a particle acquires a velocity from 12 to 60 μm/min and can thus feed an aggregate of vesicles near the cell tip before being incorporated into the cell wall.

Even in the absence of an external electric field, the tip growth is accompanied by ionic currents in the extracellular matrix. These currents are of very low intensities and could not be measured on cells of large size, such as those of unicellular algae (Nuccitelli 1988). The electrode first developed and used by Jaffe and Nuccitelli (1974) has one extremity which vibrates at a frequency of the order of 300 Hz, and explores a region of the order of 30 μm. It thus measures the difference in the potential which persists over that distance up to values of several nV which, using the resistivity of the medium, can be converted into current density by Ohm's law. At a given point, the electrode vibration in two perpendicular directions allows one to determine the direction of the current. By taking the measurement at several points around the cell, one can map the current lines.

These currents are observed around numerous cells and organisms, eggs and embryos, roots, muscles and nerves, algae, fungi and amoebas. In terms of tip growth they have been studied in detail, in the eggs of *Fucus* and *Pelvetia*, *Achlya bisexualis* and *Neurospora crassa* fungi, and tubes of *Pollen.* They organise in a polar fashion around the cell, in general entering at the tip and exiting at the sides. The order of magnitude of their intensity, $0.1 - 10\ \mu A/cm^2$, is reproducible among the considered organisms (Gow 1984, Kropf *et al.* 1984). These currents are generally associated with cell growth, their intensity being all the larger since their growth is fast. However, their relation is not always obvious. When a secondary branch emerges at some distance from the tip on the main trunk, the current entering this region exits at the main tip without significantly altering its growth rate. It is the same for a colony of relatively old hyphae, where the current therefore always exits at the tip. We can also observe these currents around an egg of *Fucus*, four hours after fertilisation, whereas its shape is still perfectly spherical (even if cytoplasmic gel is secreted at the same time by the cell wall in the region where the current enters the cell). A tube of pollen, the growth of which is inhibited by the addition of cytochalasin, equally continues to generate currents. Moreover, the ionic flux and the substances with which it is associated are characteristic of the organism: K^+, Ca^{2+} at the start, then K^+, Cl^- for the *Fucus*, H^+, K^+ for the *Pollen* tube, H^+, methionine (Met) for *Achlya Bisexualis*, H^+, sugar and phosphates for *Neurospora Crassa*.

By the intermediate of pumps and channels incorporated into the membrane, the cell maintains a potential difference between the inside and the outside. The inside potential goes from 80 mV for animal cells up to 200 mV for plant cells. The pumps, Na^+/K^+ or H^+, are powered by intracellular ATP, and generate an increase in the ion concentration which they transfer from either side the membrane. This increase in concentration is used by the cell to absorb different ions, substances by symport, mixed channels, such as in the case of H^+/Met for example. In the uniform stationary state, the sum of the ionic currents at the membrane must cancel out at the membrane by the conservation of charge, leading to a membrane potential, since, as in paragraph 3.5, the increases in concentration must be compensated by the potential difference. The ionic currents described above must clearly be associated with a non-uniform membrane potential and therefore either a non-uniform distribution of channels, pumps and symports, or a non-uniform functionality of those channels.

The channels are proteins of large molecular weight which go through the membrane many times, and contain numerous charged groups. One of the causes for a non-uniform distribution is therefore the electrophoresis of pumps and channels in the membrane, but here it is under the influence of an intracellular electric field. However, when the cell grows, the different channels are incorporated at the tip, the composition of which is assumed constant. As a result of a rapid cell growth, of the order of 1 to 10 μm/min. for the hyphae (Gow 1984) and the growth cones of neurons (Cancalon 1983) (which are in any case faster than the membrane electrophoretic phenomena which could lead to a segregation of the pumps and the symports), the relative proportion of these two types of channels must be assumed to be uniform along the cell. Indeed, by the influence of the intracellular electric field alone, a membrane protein of charge $z = 10$, with a diffusion coefficient $D = 10^{-9}\ \mathrm{cm}^2/\mathrm{s}$, moves by electrophoresis inside the electric field $E = 250\ \mathrm{mV/cm}$ at the velocity $v = zD(e/kT)E \approx 0.1\ \mu\mathrm{m/min}$. Inspired by the hypothesis of Jennings (1979), one of the causes for a non-uniforn functionality can therefore be deduced from the observed dependence of the current intensity on the amino acid concentration in *Achlya bisexualis* (Limozin *et al.* 1997). The membrane potential of the cell is maintained by a proton pump which ejects protons from the cytoplasm. The generated proton gradient can be used by the cell to absorb, using symports, different substances such as the aforementioned amino acids. Since the amino acid (aa) is absorbed by the cell using the symport H^+/aa, at a flux increasing with its concentration, the concentration decreases along the cell and generates a longitudinal gradient of absorption of that substance. Thus, at the tip, the proton flux entering by symport is larger than away from the tip, this difference being at the source of the electric field loop in the extracellular matrix, which enters at the tip and exits at the trunk (figure 12.1a). In two-dimensional growth, the order of magnitude of the current intensity calculated from the measured symport flux (Sanders *et al.* 1983), is of the same order as that measured in experiments (figure 12.1b). The three-dimensional geometry weakens this current by one order of magnitude (Limozin 1999).

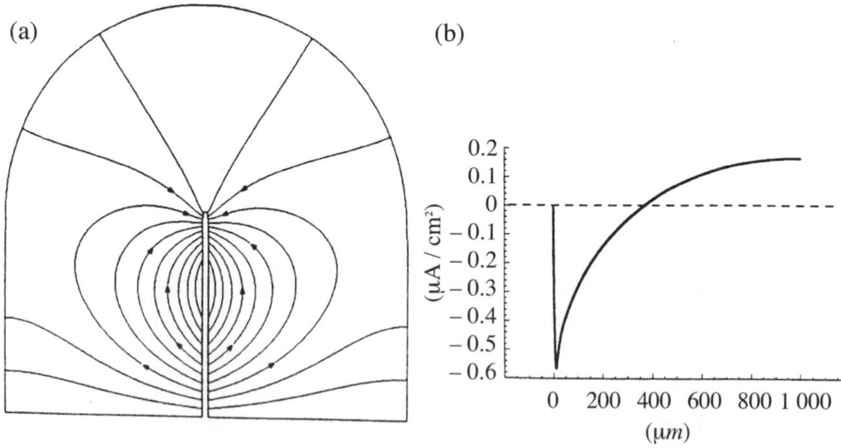

Figure 12.1 (a) Electric field lines generated by the cell, calculated by the model described above (Limozin *et al.* 1997); (b) the radial component of the current intensity as a function of the distance from the tip (Limozin *et al.* 1997).

Another cause of non-uniform functionality of the channels comes from the weakening of the pump activity at the tip (Potapova *et al.* 1988). Due to the growth being essentially localised at the tip, and therefore a larger energy consumption, less of the energy synthesised by the cell in the form of the ATP molecules is usable by the active transports at the tip, which decreases the intensity of the proton pumps in this region. This effect is quite similar to the kinetic effect of the growth of a dendrite at the interface temperature discussed in paragraph 3.2. It could be generalised to other active transports such as the Ca^{2+} pump maintaining a weak concentration of Ca^{2+} in the cytoplasm of a fertilised egg of *Fucus* or, of a neuron, where the Cl^- pump maintains a strong concentration of KCl inside the *Fucus* egg. At the tip of a *Fucus* rhizoid, where less energy is available to the active transports due to growth, the exclusion of Ca^{2+} ions from the cytoplasm as well as the uptake of Cl^- ions becomes weaker, sometimes causing a relative influx and outflux of Ca^{2+} and Cl^-, as observed. In the same way, a neuron growth cone where the energy consumption is larger, would become richer in Ca^{2+} by the weakening of the Ca^{2+} pumps.

It is also feasible that the cytoskeleton maintains a selective segregation of the pumps or the symports by attaching itself at different points on the plasma membrane. Indeed, a depolymerisation of actin filaments induced by the action of cytochalasin D decreases the current intensity around a *Fucus* egg after germination, as shown by Brawley and Robinson (1985). It is also possible that the unstable dynamics involving electro-diffusion in the intra and extra-cellular matrix, as discussed in the following paragraph, and the transmembrane ionic flux develop, generating non-uniform ionic structures.

The crucial question is therefore to find out whether the ionic currents interact with the form of the cell, a bit like a temperature field or a solute field interact with the dendritic structure. The external electric field measured by the vibrating electrode is much below the threshold of the response of an applied external electric field. The only remaining possibility therefore activating the internal electric field is the electrophoresis of vesicles towards the tips. In this case, the amplifying role played by the H^+ and Ca^{2+}, by attaching themselves to fixed and mobile negative charges, is essential. How can we then explain that, in certain cases, a current reversal does not affect the growth rate? One possibility could be that the internal electric field affects only the transport of vesicles in the cytoplasm, inducing them to regroup towards the tip to form an aggregate, the *Spitzenkörper*, very close to the plasma membrane. When the current is inverted in a transient way, this aggregate can still withstand the growth over a certain time, even if it is no longer fed, since the vesicles are no longer able to move towards the tip. If a clearer relation between currents and growth forms is established, the morphogenesis of a cell experiencing tip growth can have a direct physical analogue in electro-deposition, described in paragraph 3.5. Indeed, the ingredients (ions), their dynamics (electro-diffusion), as well as their transfer to the interface (Butler–Volmer kinetics) or at the membrane (kinetics of ionic channels) are perfectly analogous. Nevertheless, the local growth rate is directly proportional to the local ionic flux, for the electro-deposition process. Here, the transmembrane current can induce effects which do not manifest themselves until later, in another region of the cell, rendering the interaction between the form of the cell and the ionic structure non-local over time and space.

12.4 GROWTH MODELS

Due to the complexity of living organisms, the growth models must be simplified, meaning that the model feature a dynamics with few variables, which the first experiments bring to light. A first model is that of Turing (1952), who assumes that at the start of the morphogenesis of a living tissue a primordial structure with two concentration fields sets in, an activator and an inhibitor, reacting with one another and diffusing with different diffusion coefficients. As in paragraph 6.8, as soon as the activator diffuses faster than the inhibitor, the concentration field is spontaneously organised, with a well-defined wavelength function of the reaction rates and diffusion coefficients. Once this structure is set the cell exploits the differences in the concentration of the two substances to synthesise other substances in a non-uniform way, giving rise to the form.

More specifically to tip growth, the cell is deformed by the means of its membrane and its wall, necessary by the addition of new proteins (phospholipids, cellulose, polysaccharides), and by their extension. These new proteins are added by the deposit of vesicles, themselves synthesised by the Golgi apparatus and the endoplasmic reticulum. The dynamics of these vesicles towards the wall, the transfer kinetics, as well as the partial local elongation of the wall which the vesicles generate, determine the growth dynamics, since the other side of the elongation is due to the applied potentials.

After the model of Reinhardt (1892), the simplest model partially describing this dynamics is that of Bartnicki-Garcia and coworkers (1989). An aggregate of vesicles, the *Spitzenkörper*, is assumed to move at a constant velocity in a given direction. This aggregate sometimes releases vesicles in an isotropic way, which move along rectilinear trajectories. As soon as those vesicles arrive at the border of the already built up structure, they deposit there, increasing the volume of the intracellular space. As we shall see in paragraph 12.6, this model generates a rhizoid of a similar shape to the observed forms. The shape of the cell is just the shadow of the motion of an organelle, the cause of which is unknown.

Other models, conceptually more related to the physical models studied in the preceding chapters, have the advantage of generating tubular structures in a self-consistent way, without pushing away the problem to that of another structure. Thus, Picton and Steer (1982) develop a growth model for the *Pollen* tube, capable of explaining in particular the observed variation in growth rate as a function of the external concentration of Ca^{2+} (Picton and Steer 1983). Each of these processes can be modulated by the presence of Ca^{2+} ions, coming for example from the external medium through specific channels, which generates coupling between the wall dynamics and the diffusion of a morphogen.

In a conceptually similar way for the algae *Acetabularia*, Goodwin and Trainor (1985) assume that the wall is made up of a double layer: one layer is inside, made from the cytogel, the tensions of which are regulated by the concentration of Ca^{2+} ions, and the other is an external layer exhibiting ordinary elastic behaviour. A local decrease in the cytogel tension by a local increase in Ca^{2+} can therefore induce a stronger local curvature of the wall, which initiates the growth of a tip or a rhizoid.

To explain the phenomenon of chemotropism by amino acids, observed in the fungus *Achlya bisexualis*, Harold and Harold (1986) also consider the interaction of an external diffusion field, such as methionine, with the elongated cell wall. It is not clear why a higher concentration of the locally applied methionine would decrease the local rigidity of the wall and favour its extension.

In order to explain neuron morphogenesis, Hentschel and Fine (1994, 1996) consider that the local growth rate of the cell membrane depends on the internal concentration of Ca^{2+}, freely diffusing inside the cytoplasm. This concentration is controlled by the provision of external Ca^{2+}, and of transmembrane flux given as a function of its local internal concentration. In the same framework more complicated situations, but closer to experiments, can be envisaged. For example, a motoneuron of a rat continuously releases acetylcholine (ACh), at a surface flux of the order of 10^{-6} mole/m^2·s. This flux must generate a layer of ACh longing the neuron, the local thickness of which generally depends on the neuron geometry as well as the local concentration of ACh. This layer modulates the membrane concentration around a concentration at which the nicotinic receptors of ACh can be sensitive (Limozin *et al.* 1997). The flux of Ca^{2+} which they generate modulate the local actin polymerisation, and thus the local cell membrane elongation rate.

All these models have the common feature of coupling the dynamics of a wall, a membrane or a cell wall, to the diffusion of a morphogen in an intra- and extra-cellular matrix. In a way, these models exhibit a certain duality with the growth model of crystalline dendrites. In fact, in that case, from paragraph 5.3, the dynamics of the liquid–solid interface is coupled with the temperature field or the solute concentration field. The local growth rate is adjusted by the concentration flux, which is itself determined by the local shape of the interface. Thus, by associating the normal growth rate and the wall elongation as in paragraph 10.1, and by exchanging the roles of the concentrations and the flux, we establish a relation of duality between the two types of models. This correspondence is even clearer if the models are simplified in terms of the geometry, such that the growth dynamics is only regulated by the local shape of the wall (Pelcé and Pocheau 1992).

12.5 INSTABILITIES

In tip growth, a central question is the origin of the driving force which maintains the stationary tubular growth. In the model of the *Spitzenkörper*, the origin of the driving force is introduced *ad hoc* since the movement of the vesicle centres is imposed, which is somewhat unsatisfactory. Even though other models omit a large number of mechanisms of cellular functions, they have the advantage, just as those described in the preceding paragraphs, of including the morphological instabilities, spontaneously generating tubular forms. We must therefore distinguish between two types of instabilities: those which initially generate morphogen structures of a given form, in which the deformations of the cell only appear as a consequence of that primordial structure, and those which result from a direct coupling between the wall and morphogen dynamics.

In the first category, there is above all the Turing (1952) instability studied in paragraph 6.8. Biological systems should exhibit more examples of this instability than the reactive mixtures of inert matter, simply by the fact that the reagents and the reactions are much more diverse, in both their diffusion coefficients and the characteristic reaction times. As we can deduce from the cited paragraph, the typical wavelength of the unstable structure is of the order of $\lambda \approx \sqrt{D\tau}$, where D is a diffusion coefficient and τ is a reaction time. Thus, if $D \approx 10^{-6}\ \mathrm{cm^2/s}$, a time of the order of $\tau \approx 10^{-2}$ s leads to a wavelength of the order of 1 μm. The timescale over which the structure is established is of the order of $\sigma \approx D/\lambda^2 \approx 1$ min.

Other instabilities, which are more related to phenomena encountered in reality, can also lead to inhomogeneous structures. The cell maintains a uniform membrane potential between the inside and the outside, with the help of a set of ionic pumps and channels which transfer the ions from the inside to the outside of the cell, and *vice versa*. The membrane potential can become non-uniform giving rise to an electric field if, as in figure 12.2a, the channels bringing in a positively charged ion are more concentrated on one side of the cell than the other, or if, as shown in figure 12.2b, the channels remain uniformly distributed, but work more efficiently on one side than the other due to either the different value between the two concentration poles of certain substances, or the membrane potential. If those two mechanisms are coupled with the electro-diffusion dynamics in the intra and extra-cellular matrices, there may be amplifications.

For example, in the first case, the appearance of a weak electric field can move certain channels towards one of the poles by electrophoresis, and lead to a non-uniform repartitioning of those channels. The resulting charge current generates an electric field which is added to the initial field, and amplifies the initial cause (Jaffe 1977). The quantitative analysis of Larter and Ortoleva (1983) shows that this instability can be produced if the cell radius exceeds $R_c = Dc_0/z_e J_0$, where

$D \approx 10^{-6}\ \mathrm{cm^2/s}$ is the cytoplasmic diffusion coefficient, $c_0 \approx 10^{-3}\ \mathrm{mole/L}$ is the ionic concentration, $J_0 \approx 10^{-9}\ \mathrm{mole/m^2 \cdot s}$ is the flux of Ca^{2+} ions crossing the mobile channels, and z_e is the charge number carried by the external part of the ionic channel. Since the size of the cell is of the order of 100 μm, this condition is only satisfied if z_e is very large, of the order of 1000. For a more reasonable order of magnitude, we must take into account, as in paragraph 12.3, the fixed and mobile ligands for the Ca^{2+} ions, which act as amplifiers for the intracellular field. The growth rate of the instability is determined by the electrophoresis of the channels in the membrane, that is $\sigma \approx R^2/D_\gamma \approx$ a few hours, if the electrophoretic diffusion coefficient takes the minimum value, typically accepted as $D_\gamma \approx 10^{-9}\ \mathrm{cm^2/s}$.

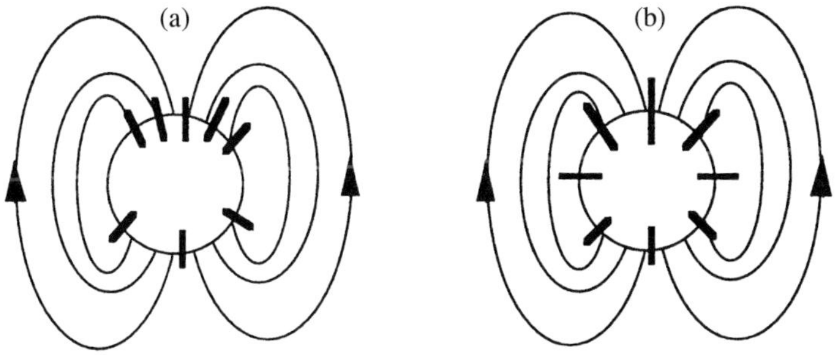

Figure 12.2 (a) Channel segregation; (b) non-uniform functioning of the channels.

In the other case, as in the non-uniform functioning of proton pumps in algae *Chara* (Toko *et al.* 1985), an initial electric field induces a potential difference between the membrane on opposite sides of the cell. If the uniformly distributed channels are voltage dependent, these generate different flux at the two poles and induce a charge current which amplifies the initial electric field (Pelcé 1993). The polar mode of that instability manifests itself if the radius of the cell exceeds $R_c \approx Dc_0/((kT/e)\partial J_0/\partial\Phi)$, or $R_c \approx D/(\partial J_0/\partial c)$, depending on whether we take into account the sensitivity of the channel to the membrane potential or ionic concentration (Pelcé 1993, Pelcé and Léonetti 1995). The typical timescale of the process is then $\sigma \approx R^2/D$, where $D \approx 10^{-6}\mathrm{cm^2/s}$ is the cytoplasmic diffusion coefficient, and therefore a factor of 10^3 shorter than the previous one, that is, several minutes (Léonetti and Dubois-Violette 1998). Thus, some minutes after the start of the calcium channel functionality, an asymmetric distribution of Ca^{2+} ions could be seen in the cell, subsequently amplified by other intracellular mechanisms. The characterisation of the channels at the start is therefore extremely important, allowing one to evaluate the critical cell radii, mentioned above. The range of applications of this second instability seems however limited, as the sensitivity of the channels to variations in the potential and the concentration should be the reverse to that of a typical passive channel.

The instabilities in the second category (Goodwin and Trainor 1985, Hentschel and Fine 1994, Kam and Levine 1997) couple the dynamics of a membrane or a wall to an intracellular diffusion field such as that of Ca^{2+}. Consider a locally flat part of the membrane (figure 12.3a). The cell adsorbs an activating substance such as methionine for *Achlya bisexualis*, or secretes an inhibiting substance such as ACh for the neuron, maintaining a gradient which we can assume to be uniform around the cell. Let us assume that the membrane points towards the exterior. Contrary to the Mullins–Sekerka instability (figure 6.4) in paragraph 6.5, the morphogen gradient is not significantly modified by the wall deformation, because the entering flux is not directly determined from the wall deformation. Thus the tip of the deformation explores a region where the morphogen concentration is lower, and amplifies its elongation. The equivalent of surface tension, which stabilises the amplification, will be a kind of membrane rigidity γ (Pelcé and Pocheau 1992), of unknown miscroscopic origin for the time being. The most unstable wavelength of this instability is therefore $\lambda \approx (\gamma R/(dv/dc))^{1/4}$, where R is the radius of the cell and dv/dc is the derivative of the local normal velocity with respect to the morphogen concentration.

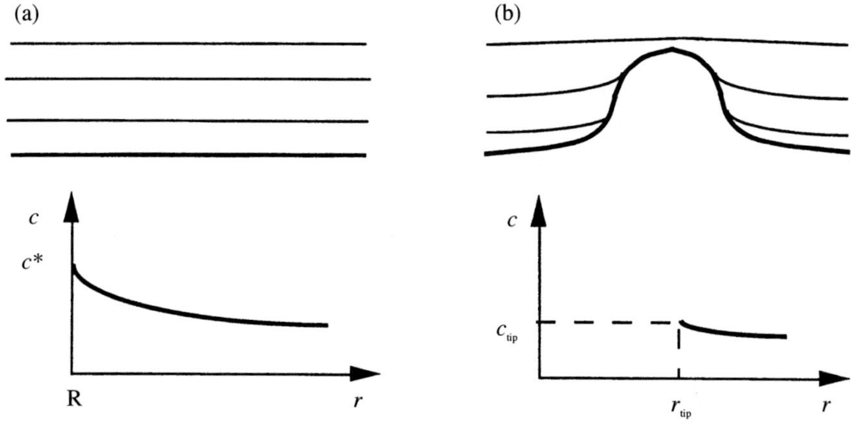

Figure 12.3 Schematic of the Hentschel–Fine instability.

12.6 RHIZOIDS

In the same way as in chapters 7, 8 and 10, the growth models shown in paragraph 12.3 should be able to generate tubular structures, enabling them to identify with the observed rhizoid shapes.

In the model of Bartnicki-Garcia and coworkers (1989), a source isotropically emits N vesicles per unit time into a plane, each of them increasing its surface area by a^2. If the source stays immobile, the cell is a disc of a radius which grows as $R \approx (Na^2t)^{1/2}$. Now assume that the source moves at a constant velocity V in a fixed direction y. We can therefore show that the cell forms a rhizoid of equation $y = x\cot(x/d)$, a hyphoid curve, where $d = Na^2/V$ is the distance from the tip to the source, shown in figure 12.4. This shape is tubular in the sense that it attains a constant width far away from the tip, $2\pi d$, just like a Saffman–Taylor finger. The faster the cell growth, the thinner its width, although this data does not have clear experimental evidence. Remarkably, as shown in figure 12.5, this curve perfectly reproduces the contour of numerous filamentous fungi such as *Polysticus versicolor* or *Pythium aphantdermatum*, the *Spitzenkörper* being clearly identified in the region indicated in the first case.

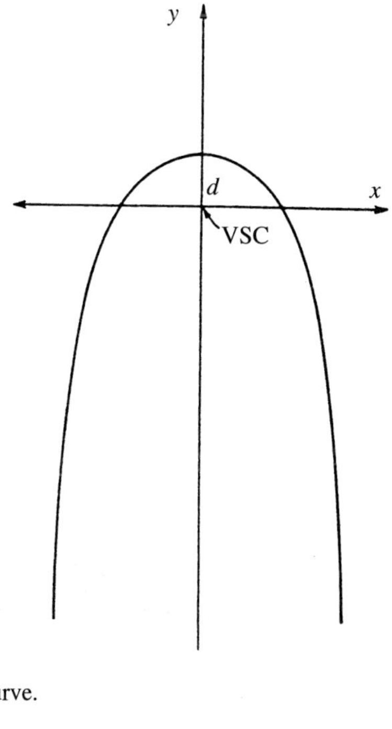

Figure 12.4 Hyphoid curve.

Even though simplified by for example assuming the incorporation kinetics of the vesicles uniform, this model at least has the advantage that it precisely derives the shape of the tip of a filamentous fungus.

In the second category of models, Denet (1996) simulates the evolution of an interface with a normal velocity depending on the external morphogen concentration diffusing in the extracellular matrix, diminished by the rigidity of the cell wall

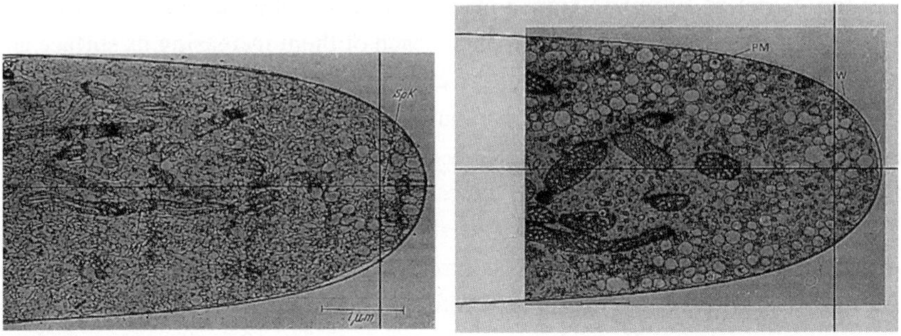

Figure 12.5 Relationship between a hyphoid and the contour of filamentous fungi *Polysticus versicolor* (on the left) and *Pythium aphantdermatum* (on the right) (Bartnicki-Garcia and al. 1989). For *Polysticus versicolor*, the presence of a *Spitzenkörper* is observed at the centre of the vesicle (VSC) of the corresponding hyphoid curve.

γ when the curvature of the interface varies too fast. The morphogen is adsorbed by the cell at constant flux. As previously mentioned in paragraph 12.3, this model is analogous to the crystalline growth model simulated in paragraph 10.5 by Saito *et al.* (1989). The difference can only be seen at the boundary conditions and an analogous simulation technique is therefore employed. It is observed that, as shown in figure 12.6, for initial conditions defined by experimental conditions, the interface develops a quasi-stationary tube of a well-defined width as soon as the rigidity γ exceeds a certain critical value γ_c. Contrary to the dendrite, for which the concentration of the solute is almost constant at the interface, the morphogen is absorbed by the cell at an almost constant normal flux. The morphogen concentration decreases along the trunk, freezing the local elongation of the cell wall at large distances from the tip.

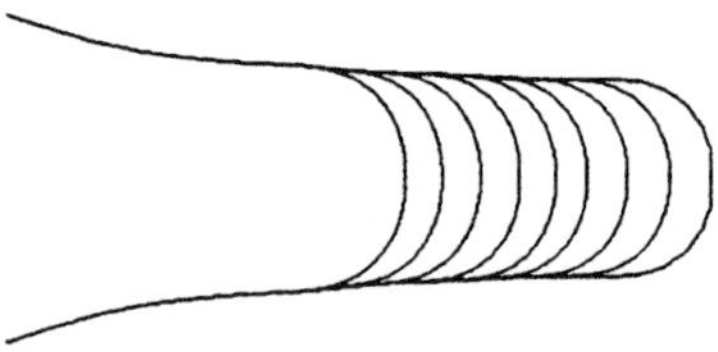

Figure 12.6 Growth of a rhizoid in the external morphogen diffusion model when $\gamma > \gamma_c$ (Denet 1996).

This same behaviour has already been observed by Pelcé and Sun (1993) in the geometric model of Pelcé and Pocheau (1992) (figure 12.7). In a different way to crystalline growth, for which the dendritic growth is only observed when the surface tension is anisotropic whatever its order of magnitude may be, the formation of a tubular structure takes place when the membrane rigidity exceeds a critical value.

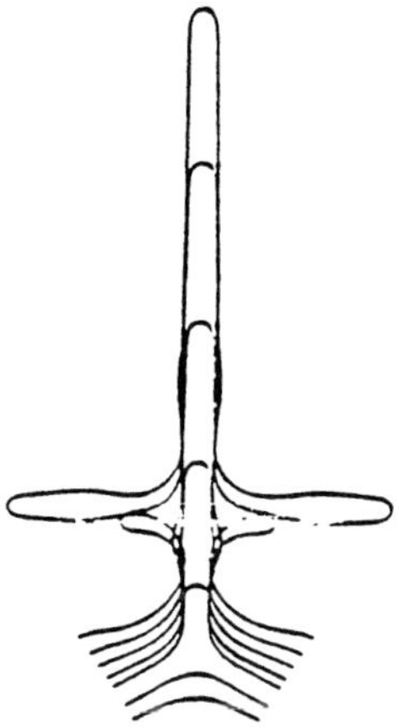

Figure 12.7 Rhizoid obtained by a simulation of the geometric model when $\gamma > \gamma_c$ (Pelcé and Sun 1993).

The problem of secondary branching is still open to debate. We will only give the results of the different observations. As for the growth forms of inert matter discussed in this book, secondary branching manifests itself as secondary branches of a similar morphology to primary branches, growing more or less regularly a certain distance away from the tip. For example Harold and Harold (1986) show how different substances (the cytochalasins affecting the dynamics of the cytoskeleton, or the ionophores eliminating the trans-membrane gradients of different substances such as protons or Ca^{2+} ions, or even the amino acid concentration) affect the density of secondary branching. A drop of methionine, placed near to the principal trunk, can also initiate a secondary branch in this region, even at large distances from the tip (Schreurs *et al.* 1989).

Other filamentous fungi are multinuclear in a sense that when the main tip elongates, a nuclear division occurs at more or less regular intervals. Soon after the division of a nucleus a new wall, the septum, is formed transverse to the principal direction of fungus growth, isolating the tip region from the rest of the cell. Along the principal branch, we thus observe a succession of compartments, each containing a nucleus. It is often observed that one secondary branch grows per compartment.

If the origin of secondary branching remains unclear, one can still wonder what the growth models in paragraph 12.4 can predict. In the model of vesicle aggregates secondary branching is only conceivable if the aggregate itself is split in two, or if another one is created at some distance from the tip. In the growth models with instabilities, if the morphogen concentration has not decreased too much at a certain distance from the tip, an analogous instability to the primary instability may be reproduced at the side of the primary branch, giving an analogous structure of the primary branch to the secondary branch. One can also conceive that the main tip splits in two, reproducing the phenomenon of tip splitting, already discussed in paragraph 9.1. Indeed, as shown by the models of Pelcé and Sun (1993) and of Denet (1996) in figure 12.8, the tip of a rhizoid splits in two when the wall rigidity γ is weaker than a critical value γ_c.

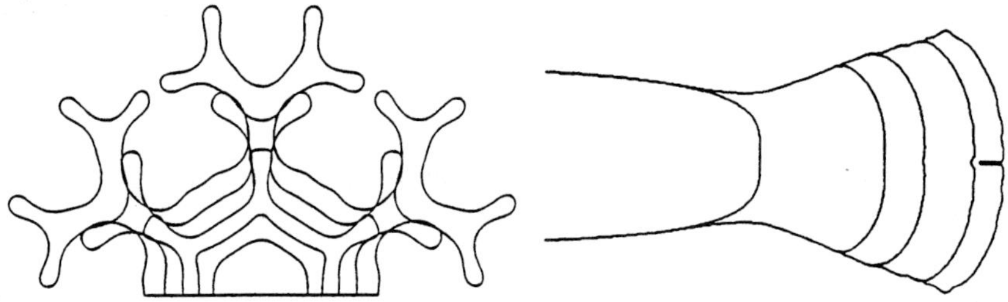

Figure 12.8 Tip splitting of rhizoids when $\gamma < \gamma_c$. Left: geometric model (Pelcé and Sun 1993). Right, model with external diffusion of morphogenes (Denet 1996).

This mode of instability could therefore be the origin of a very ornamented pattern of *Micrasterias* algae shown in figure 12.9, where the tip of a lobe is split in two many times.

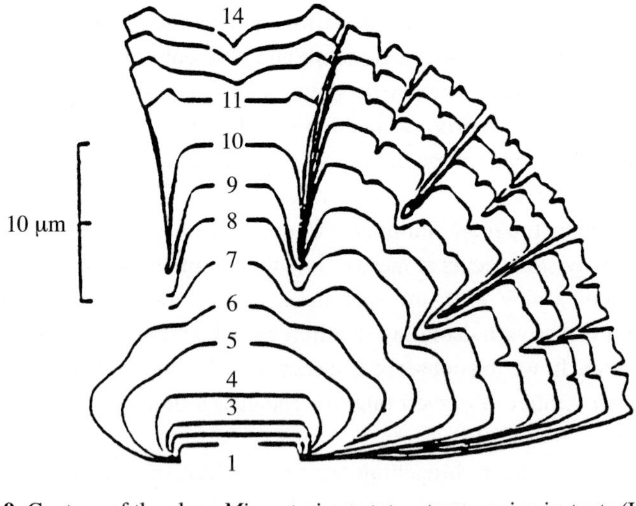

Figure 12.9 Contour of the algae *Micrasterias rotata*, at successive instants (Lacalli 1975).

The growth form showing the most secondary ramifications is certainly the neuron, particularly its dendritic tree. In order to characterise this complex structure, analogous fractal analyses to those in paragraphs 11.1 and 11.8 were carried out by Caserta and coworkers (1990) on the neurons of the retina and Takeda and coworkers (1992) on the cells of Purkinje. It has been observed that these structures are actually fractal, their dimension evolving with time. From this observation these authors have produced an analogy between the growth mechanisms of these cells and the branched patterns shown in paragraph 11.8. The problem is to reveal the diffuse substance from which the growth form originates, even if its existence has already been mentioned in the last century by Ramon y Cajal (1911). As previously mentioned in paragraph 12.2, the neurotransmitters could be good candidates. Another possibility is that morphogenesis is principally initiated from the inside of the neuron, the stochastic aspect of the growth being characterised by the probability of a branch emerging (Van Veen and Van Pelt 1994, Albinet and Pelcé 1996). To determine this probability, we must take into account the interaction between the cytoskeleton and the membrane in a more precise manner.

12.7 GROWTH BY ACTIVATION

The cell growth originates from the growth of the wall, the membrane and the cytoskeleton. Let us consider the easiest case where there is no wall, as is the case for animal cells and for neurons in particular. The membrane is made up of two layers of phospholipids and its stretching is a result of these molecules being incorporated. Its shape is maintained by the phytoskeleton which is made of actin filaments and microtubules (Alberts *et al.* 1994). The latter are rigid polymers coming from the polymerisation of actin and tubulin monomers, respectively. Since the monomers are dissymmetric, so are the polymers, making their assembly easier on the positive side than on the negative one. The speed of polymerisation of these filaments follows a simple kinetic law, the Wegner law (1968): $v = \delta(k_+ c - k_-)$, where c is the monomer concentration, δ the monomer length and the kinetic constants $k_+[\text{mole}^{-1}\text{s}^{-1}]$, and $k_-[\text{s}^{-1}]$ on each side. The k_- constant corresponds to the possible spontaneous breaking of a filament. One thus recovers a growth law analogous to that of electro-deposition, equation (3.2) in paragraph 3.5, with the important difference that the growth here is from the inside. The collective dynamics of these filaments coupled with diffusion can be determined in a way analogous to the one described in appendix 2 in paragraph 5.4 (Dogterom and Leibler 1993).

For usual monomer concentrations in the cytoplasm, $v \approx 20\ \mu\text{m/min}$, a value which is higher than the growth rate of the cell protuberance, such as the tip of a hypha or a growth cone, and much higher than the local growth rate of the wall of the grain or the soma. The elongation of the membrane or the wall thus limits the filament growth since the filament must wait for a sufficient space between its extremity and the membrane in order to attach a monomer. Hence, the filament extremity is subject to a force from the membrane directed towards the interior of the cell, thus slowing it down. In order to determine its velocity as a function of the applied force *f*, let us consider Peskin and coworkers' simple model (1993), in which a filament grows along a fixed axis against a body, applying a force *f* along the same axis but in a direction opposite to the growth. The body is subject to shocks from molecules of thermal energy kT, such that its relative position with respect to the filament extremity fluctuates, now and then leaving enough space for a monomer to attach there. The probability for such a space to appear is proportional to the Arrhenius factor $\exp(-f\delta/k\mathrm{T})$, so that $v = \delta(k_+ c\exp(-f\delta/kT) - k_-)$. Since at 298 K, $kT = 4.1\times10^{-21}\,\mathrm{J}$, and $\delta = 3$ nm for an actin monomer, a force of a few pN is enough to decrease the growth rate of a filament by an order of magnitude. Such a relation has been experimentally verified for example on microtubules by Dogterom and Yurke (1997). Since the filaments have an orientation with respect to the membrane, the normal growth rate of the membrane (or, indirectly, its local relative elongation) becomes a function of the interior force exerted on it. This function depends on an intermediary Arrhenius factor which may be very sensitive to temperature fluctuations, with a possible link with Trinci's (1979) observations of side branching of hyphae as well as observations on neurons by Cancalon (1983) and Ternaux (1999).

For many reasons which are yet to be defined (filaments stuck to the membrane, curvature energy, substrate adhesion, cytoplasmic flow), this interior force depends on the local curvature of the membrane or the wall. If this curvature is at a minimum for some characteristic protuberance sizes, the monomers attach themselves more easily to the filaments for these deformations, thus amplifying their growth.

Let us consider the example of figure 12.10 in which a membrane is attached to some filaments, thus exerting a force on the free filaments susceptible to growth, which can be approximated by an effective surface tension γ. Moreover, the membrane adheres to a substrate lowering this surface tension to $\mu = \gamma - w$ (D'Arcy Thompson 1917, Chamak and Prochiantz 1989). The tensions and adhesion coefficient values can be determined experimentally on some cells (Mege *et al.* 1987, Albersdörfer *et al.* 1997). By inserting new phospholipid molecules, the membrane elongates. However, its points of attachment prevent it from freeing enough space for the insertion of new actin monomers unless they are under favorable thermal fluctuations of the membrane, the amplitude of which has been determined (De Gennes and Papoular 1969 and Brochard and Lennon 1974). Taking into account the curvature energy of a membrane (Helfrich 1973), it is possible to show that the deformation energy of the membrane is at a minimum for a deformation size of a few μm, depending on the curvature parameter, the effective surface tension and the dimensions of the cell. As can be seen in figure 12.10, the membrane could wait until it has the excess length $2\pi\delta$, and homogeneously attach all the monomers at the same time. This event is rare since it requires too large an energy fluctuation. Favorised by a smaller excess of length and requiring less energy, the localised deformation is much more probable.

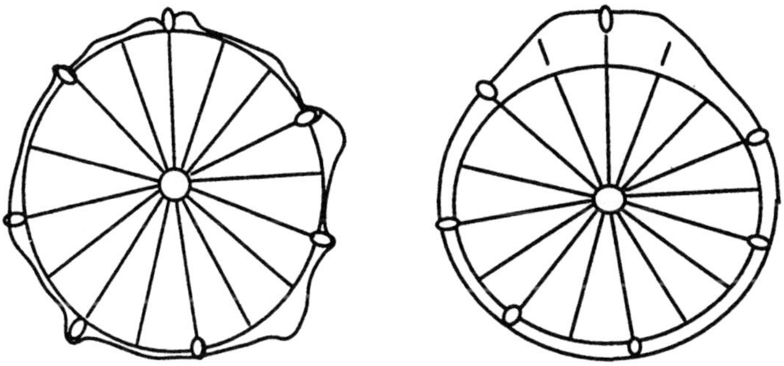

Figure 12.10 Membrane attached to filaments limiting the growth of other filaments. The membrane elongates but is held back by attached filaments (a). The thermal fluctuations excite some local minimal energy deformations and allow the insertion of new actin monomers (b).

Let us suppose that the growth is made more likely at the tip rather than the soma by an activation energy difference ΔE. Let $\alpha = \exp(\Delta E/kT)$ be the probability ratio of continuing a neurite and generating another one. Consider the following simple growth law (Pelcé *et al.* 1999): the perimeter of the soma, i.e. the cellular body, is divided into N possible growth sites. A first particle is attached with a probability $1/N$, then a second one with a probability p_2 on a free site and αp_2 on an occupied site, so that $(N-1)p_2 + \alpha p_2 = 1$. At the n-th step, the probability will be p_n for a free site and αp_n for an occupied one. So, if m is the number of branches at step t, $(N-m)p_t + m\alpha p_t = 1$. As can be seen in figure 12.11b, for $\alpha = 10^6$, and $N = 30$, the model generates several neurons of different branch lengths, and of total length $L = T\delta$, where T is the total number of elementary steps. Putting aside the fact that, the neurites generated by this model are assumed straight for simplicity, the shape of the neurons has some similarities with those obtained from cellular bodies initial placed on a cover glass in a culture medium (figure 12.11a). Let us divide T into intervals t_i, $T = t_1 + t_2 + \ldots + t_m$, where t_i corresponds to the time between the $(i-1)$-th branch and the i-th branch. Then $t_1 \approx 1/Np_1$, $t_2 \approx 1/(N-1)p_2 \ldots t_i \approx 1/(N-i+1)p_i$. Hence, if $1 << m << N$, $T \approx \alpha\, m^2/2N$ and

$$m \approx \left(2\frac{\pi D}{\lambda_c}\frac{L}{\delta}\exp\left(-\frac{\Delta E}{kT}\right)\right)^{1/2} \tag{12.1}$$

since $N \approx \pi D/\lambda_c$, where D is the diameter of the soma, λ_c the neurite size, and $T = L/\delta$. Note the similarities between this relation and that from equation (5.13) in paragraph 5.5 determining the velocity of a flame as a function of temperature.

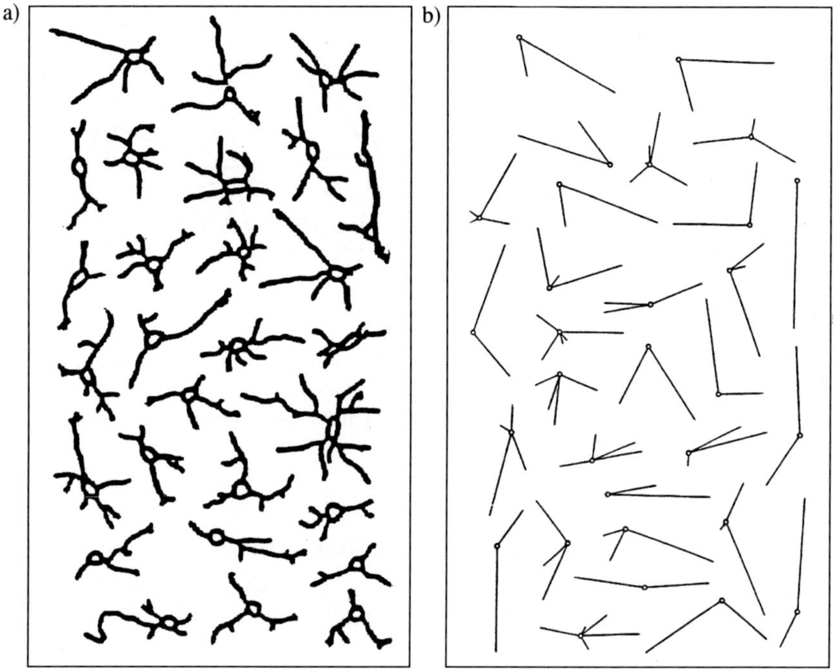

Figure 12.11 (a) Motoneurons of rats after one day in a growth medium (J. Bohin, J.P. Ternaux); (b) simulation of the model (C. Pelcé, L. Savtchenko).

For instance, if $D = 10\ \mu\text{m}$, $\lambda_c = 1\ \mu\text{m}$, $L = 300\ \mu\text{m}$, a total length generally obtained after a day of growth, $m = 3$ if $\Delta E/kT = 13.9$ or $\alpha \approx 10^6$, $m = 4$ for $\Delta E/kT = 13.2$ or $\alpha \approx 5 \cdot 10^5$.Thus, the larger α is, the smaller the number of branches and the larger the difference in length between branches. This observation is in qualitative agreement with the shape variations observed when the substrate adhesion or the temperature of the medium change. In the preceding model, $\Delta E \approx \mu\delta\lambda_c$, so that $\Delta E/kT \approx 10$ if $\mu \approx 10^{-5}\ \text{J/m}^2$, and a decrease in μ by 10^{-6} J/m^2 due to an increase in the concentration of adhesive molecules at the tip compared to the soma (Krivko *et al.* 1993), is enough to lower ΔE of kT and thus add another branch. Note however that such a reduced activation energy is also typical of enzymatic reactions. The presence of a catalyst in the growth cone and not in the soma could have the same effect. Moreover, from equation (12.1),

$$\frac{\Delta m}{m} = \frac{\Delta E}{2kT}\frac{\Delta T}{T} \tag{12.2}$$

such that an increase in temperature by 10.7 K is enough to generate one more branch when $\Delta E/kT = 14$.

Also note that the study of reduced systems such as vesicles containing actin filaments reveals the formation of tubular structures when the system temperature is increased by about ten degrees (Häckl *et al.* 1998).

BIBLIOGRAPHY

[1] A. Albersdörfer, T. Feder, E. Sackmann, 'Adhesion-Induced Domain Formation by Interplay of Long-Range Repulsion and Short-Range Attraction Force: A Model of Membrane Study', *Biophysical journal*, **73**, p. 245, 1997.

[2] B. Alberts, D. Bray, J. Lewis, M. Raff, K. Roberts, J.D. Watson, *Molecular Biology of the Cell*, third edition, Garland publishing inc, 1994.

[3] G. Albinet, P. Pelcé, 'Computer Simulation of Neurite Outgrowth', *Europhys. Lett.*, **33**, p. 569, 1996.

[4] C.L. Bachewich, I.B. Heath, 'The Cytoplasmic pH Influences Hyphal Tip Growth and Cytoskeleton-Related Organization', *Fungal genetics and biology*, **21**, p. 76, 1997.

[5] S. Bartnicki-Garcia, F. Hergert, G. Gierz, 'Computer Simulation of Fungal Morphogenesis', *Protoplasma*, **151**, p. 46, 1989.

[6] F.W. Bentrup, L.F. Jaffe, 'Analyzing the 'Group effect', Rheotropic Responses of Developing Fucus Eggs', *Protoplasma*, **65**, p. 25, 1968.

[7] F.W. Bentrup, 'Elektrophysiologische Untersuchungen am Ei von Fucus serratus: Das Membranpotential', *Planta*, **94**, p. 319, 1970.

[8] F. Berger, C. Brownlee, 'Ratio Confocal Imaging of Free Cytoplasmic Gradients in Polarising and Polarised Fucus Zygotes', *Zygote*, **1**, p. 9, 1993.

[9] J. Bohin, 'Étude préliminaire de l'effet d'adhésivité du substrat sur la morphogenèse d'un neurone', internship report UMR 138 CNRS, UPR 9041 CNRS.

[10] S.H. Brawley, K.R. Robinson, 'Cytochalasin Treatment Disrupts the Endogenous Currents Associated with Cell Polarization in Fucoid Zygotes: Studies of the Role of F-Actin in Embryogenesis', *The journal of cell biology*, **100**, p. 1173, 1985.

[11] F. Brochard, J.F. Lennon, 'Frequency Spectrum of the Flicker Phenomenon in Erythrocytes', *Journal de physique*, **36**, p. 1035, 1975.

[12] P. Cancalon, 'Influence of Temperature on Slow Flow in Populations of Regenerating Axons with Different Elongation Velocities', *Developmental brain research*, **9**, p. 279, 1983.

[13] F. Caserta, H.E. Stanley, W.D. Eldred, G. Daccord, R.E. Hausman, J. Nittman, 'Physical Mechanisms Underlying Neurite Outgrowth: a Quantitative Analysis of Neuronal Shape', *Phys. Rev. Lett.*, **64**, p. 95, 1990.

[14] B. Chamak, A. Prochiantz, 'Axones, dendrites et adhésion', *C.R. Acad. Sci. Paris*, **308**, p. 353, 1989; 'Influence of Extracellular Matrix Proteins on the Expression of Neuronal Polarity', *Development*, **106**, p. 483, 1989.

[15] C.W. Cho, F.M. Harold, W.J.A. Schreurs, 'Electric and Ionic Dimensions of Apical Growth in Achlya Hyphae', *Experimental Mycology*, **15**, p. 34, 1991.

[16] D'Arcy Wentworth Thompson, *On Growth and Form*, first published in 1917, abridged edition, Cambridge University Press, 1961. Translated into French by D. Teyssié, *Forme et croissance*, Seuil, 1994.

[17] P.G. de Gennes, M. Papoular, Vibrations de basse fréquence dans certaines structures biologiques, volume jubilaire en l'honneur de A. Kastler (P.U.F., Paris), 1969.

[18] B. Denet, 'Numerical Simulation of Cellular Tip Growth', *Phys. Rev.*, **E53**, p. 986, 1996.

[19] M. Dogterom, B. Yurke, 'Measurement of the Force–Velocity Relation for Growing Microtubules', *Science*, **278**, p. 856, 1997.

[20] M. Dogterom, S. Leibler, 'Physical Aspects of the Growth and Regulation of Microtubule Structures', *Phys. Rev. Lett.*, **70**, p. 1347, 1993.

[21] L.R. De Silva, J. Youatt, G.W. Gooday, N.A.R Gow, 'Inwardly Directed Ionic Currents of Allomyces Macrogynus and Other Water Moulds Indicate Sites of Proton-Driven Nutrient Transport but are Incidental to Tip Growth', *Mycol. Res.*, **96**, p. 925, 1992.

[22] B.C. Gibbon, D.L. Kropf, 'Cytosolic pH Gradients Associated with Tip Growth', *Science*, **263**, p. 1419, 1994.

[23] B.C. Goodwin, L.E.H. Trainor, 'Tip and Whorl Morphogenesis in Acetabularia by Calcium-Regulated Strain Fields', *J. Theor. Biol.*, **117**, p. 79, 1985.

[24] N.A.R. Gow, 'Transhyphal Electric Currents in Fungi', *Journal of general microbiology*, **130**, p. 3313, 1984.

[25] W. Häckl, M. Bärmann and E. Sackmann, 'Shape Changes of Self-Assembled Actin Bilayer Composite Membranes', *Phys. Rev. Lett.*, **80**, p. 1786, 1998.

[26] R.L. Harold, F.M. Harold, 'Ionophores and Cytochalasins Modulate Branching in Achlya Bisexualis', *Journal of general microbiology*, **132**, p. 213, 1986.

[27] I.B. Heath, *Tip Growth in Plant and Fungal Cells*, Academic Press, 1990.

[28] W. Helfrich, 'Elastic Properties of Lipid Bilayers: Theory and Possible Experiments', Z. Naturforsch, **28**, p. 693, 1973.

[29] H.G.E. Hentschel, A. Fine, 'Instabilities in a Cellular Dendritic Morphogenesis', *Phys. Rev. Lett.*, **73**, p. 3592, 1994.

[30] H.G.E. Hentschel, A. Fine, 'Diffusion-Related Control of Cellular Dendritic Morphogenesis', *Proc. Roy. Soc. Lond.*, **B263**, p. 1, 1996.

[31] L.F. Jaffe, R. Nuccitelli, 'An Ultrasensitive Vibrating probe for Measuring Steady Extracellular Currents', *J. Cell. Biol.*, **63**, p. 614, 1974.

[32] L.F. Jaffe, K.R. Robinson, R. Nuccitelli, 'Local Cation Entry and Self Electrophoresis as an Intracellular Localization Mechanism', *Ann. N.Y. Acad. Sci.*, **283**, p. 372, 1974.

[33] L.F. Jaffe, 'Electrophoresis along Cell Membranes', *Nature*, **265**, p. 600, 1977.

[34] D.H. Jennings, 'Membrane Transport and Hyphal Growth', in *Fungal Walls and Hyphal Growth*, J.H. Burnett and A.P.J. Trinci, eds. Cambridge University Press, 1979.

[35] R. Kam, H. Levine, 'Unicellular Algal Growth: A Biomechanical Approach to Cell Wall Dynamics', *Phys. Rev. Lett.*, **79**, p. 4290, 1997.

[36] D. Kessler, J. Koplik, H. Levine, 'Pattern Selection in Fingered Growth Phenomena', *Adv. Phys.*, **37**, p. 255, 1988.

[37] I.M. Krivko, D.A. Rusakov, S.V. Savina, G.G. Skibo, V.A. Berezin, 'Lateral Patterns of the Neural Cell Adhesion Molecule on the Surface of Hippocampal Cells Developing *in vitro'*, *Neuroscience*, **55**, p. 491, 1993.

[38] D.L. Kropf, 'Establishment and Expression of Cellular Polarity in Fucoid Zygotes', *Microbiological reviews*, **56**, p. 316, 1992.

[39] D.L. Kropf, Electrophysiological properties of Achlya Hyphae: Ionic currents studied by Intracellular Potential Recording, *The journal of cell biology*, **102**, p. 1209, 1986.

[40] D.L. Kropf, M.D.A. Lupa, J.H. Caldwell, F.M. Harold, *Science,* **220**, pp. 1385-1387, 1983.

[41] D.L. Kropf, J.H. Caldwell, N.A.R. Gow, F.M. Harold, 'Transcellular Ion Currents in the Water Mold Achlya. Amino Acid Proton Symport as a Mechanism of Current Entry', *The journal of cell biology*, **99**, p. 486, 1984.

[42] T.C. Lacalli, *J. Embryol. Exp. Morphol.*, **33**, p. 95, 1975.

[43] F. Lafont, M. Rouget, A. Rousselet, C. Valenza, A. Prochiantz, 'Specific Responses of Axons and Dendrites to Cytoskeleton Perturbations: an *in vitro* Study', *J. Cell. Sci.*, **104**, p. 433, 1993.

[44] R. Larter, P. Ortoleva, 'A Study of Instability to Electrical Symmetry Breaking in Unicellular Systems', *J. Theor. Biol.*, **96**, p. 175, 1982.

[45] P.G. Layer, 'Cholinesterases Preceding Major Tracts in Vertebrate Neurogenesis', *Bioessays*, **12**, p. 415, 1990.

[46] M. Léonetti, E. Dubois-Violette, 'Theory of Electrodynamic Instabilities in Biological Cells', *Phys. Rev. Lett.*, **81**, p. 1977, 1998.

[47] P.C. Letourneau, 'Cell-Substratum Adhesion of Neurite Growth Cones, and its Role in Neurite Elongation', *Experimental cell research*, **124**, p. 127, 1979.

[48] L. Limozin, B. Denet, P. Pelcé, 'Ionic Currents Generated by Tip Growing Cells', *Phys. Rev. Lett.*, **78**, 25, p. 4881, 1997.

[49] L. Limozin, P. Pelcé, L. Savtchenko, M.C. Chamoin, J.P. Ternaux, 'Predicted Acetylcholine Layer around Embryonic Motoneurones', *NeuroReport*, **8**, p. 2957, 1997.

[50] L. Limozin, « Étude des gradients cellulaires associés à la croissance par les pointes », PhD thesis held at Université de Provence, 1999.

[51] S.A. Lipton, S.B. Kater, 'Neurotransmitter Regulation of Neural Outgrowth, Plasticity and Survival', *TINS*, **7**, p. 265, 1989.

[52] E.J. Lund, 'Electric Control of Organic Polarity in the Egg of Fucus', *Botanical Gazette*, **76**, p. 288, 1923.

[53] A.M. McGillivray, N.A.R. Gow, 'Applied Electric Fields Polarize the Growth of Mycelial Fungi', *Journal of general microbiology*, **132**, p. 2515, 1986.

[54] S. McLaughlin, M.m. Poo, 'The Role of Electro-Osmosis in the Electric-Field-Induced Movement of Charged Macromolecules on the Surfaces of Cells', *Biophys. J.*, **34**, p. 85, 1981.

[55] J.L. Mege, C. Capo, A.M. Benoliel, P. Bongrand, 'Use of Cell Contour Analysis to Evaluate the Affinity between Macrophages and Glutaraldehyde-Treated Erythrocyes', *Biophys. J.*, p. 177, 1987.

[56] R. Nuccitelli, 'New Perspectives on Morphogenesis. Ionic Currents in Morpho-genesis', *Experientia*, **44**, p. 657, 1988.

[57] N. Patel, M-m Poo, 'Orientation of Neurite Growth by Extracellular Electric Fields', *The journal of neuroscience*, **2**, p. 483, 1982.

[58] P. Pelcé, *Dynamics of Curved Fronts,* Academic Press, 1988.

[59] P. Pelcé, 'Origin of Cellular Ionic Currents', *Phys. Rev. Lett.*, **71**, p. 1107, 1993.

[60] P. Pelcé, M. Leonetti, 'Ca^{2+} Circulations Generated by Ca^{2+} – Sensitive Ca^{2+} Channels', *Europhys. Lett.*, **30**, p. 221, 1995.

[61] P. Pelcé, A. Pocheau, 'Geometrical Approach to the Morphogenesis of Unicellular Algae', *J. Theor. Biol.*, **156**, p. 197, 1992.

[62] P. Pelcé, J. Sun, 'Geometrical Models for the Growth of Unicellular Algae', *J. Theor. Biol.*, **160**, p. 375, 1993.

[63] H.B. Peng, L.F. Jaffe, 'Polarization of Fucoid Eggs by Steady Electric Fields', *Developmental biology*, **53**, p. 277, 1976.

[64] C. Peskin, G.M. Odell, G.F. Oster, 'Cellular Motions and Thermal Fluctuations: The Brownian Ratchet', *Biophysical journal*, **65**, p. 316, 1993.

[65] J.M. Picton, M.W. Steer, 'A Model for the Mechanism of Tip Extension in Pollen Tubes', *J. Theor. Biol.*, **98**, p. 15, 1982.

[66] M.M. Poo, K.R. Robinson, 'Electrophoresis of Concanavalin A Receptors along Embryonic Muscle Cell Membrane', *Nature*, p. 602, 1977.

[67] T.V. Potapova, K.B. Aslanidi, T.A. Belozerskaia, N.N. Levina, 'Transcellular Ionic Currents Studied by Intracellular Potential Recordings in Neurospora Crassa Hyphae', **241**, p. 173, 1988.

[68] S. Ramon y Cajal, *Histologie du système nerveux de l'homme et des vertébrés*, Vol. I and II, A. Maloine, Paris, 1911.

[69] M.O. Reinhardt, 'Das Wachstum der Pilzhyphen. Ein Beitrag zur Kenntniss des Flächenwachstums Vegetabilischer Zellmembranenen'. *Jachbuch für wissenschaftlische Botanik*, **23**, p. 479, 1892.

[70] H.D. Reiss, W. Herth, 'Calcium Gradients in Tip Growing Plant Cells Visualized by Chlorotetrac cycline Fluorescence', *Planta*, **146**, p. 615, 1979.

[71] K.R. Robinson, 'The Responses of Cells to Electric Fields: a Review', *The journal of cell biology*, **101**, p. 2023, 1985.

[72] G.D. Robson, E. Prebble, A. Rickers, S. Hosking, D.W. Denning, A.P.J. Trinci and W. Robertson, 'Polarized Growth of Fungal Hyphae is Defined by an Alkaline pH Gradient', *Fungal genetics and biology*, **20**, p. 289, 1996.

[73] D. Sanders, C.L. Slayman, M.L. Pall, 'Stoichiometry of H^+ / Amino Acid Cotransport in *Neurospora Crassa* Revealed by Current-Voltage Analysis', *Biochimica et Biophysica acta*, **735**, p. 67, 1983.

[74] W.J.A. Schreurs, F.M. Harold, 'Transcellular Proton Current in Achlya Bisexualis Hyphae: Relationship to Polarized Growth', *Proc. Natl. Acad. Sci. USA*, **85**, p. 1534, 1988.

[75] W.J.A. Schreurs, R.L. Harold, F.M. Harold, 'Chemotropism and Branching as Alternative Responses of *Achlya Bisexualis* to Aminoacids', *J. Gen. Microbiol.*, **135**, p. 2519, 1989.

[76] C.L. Slayman, C.W. Slayman, 'Measurement of Membrane Potential in *Neurospora*', *Science*, **136**, p. 876, 1962.

[77] T. Takeda, A. Ishikawa, K. Ohtomo, Y. Kobayashi, T. Matsuoka, 'Fractal Dimension of Dendritic Tree of Cerebellar Cell during onto and Phylogenetic Development', *Neuroscience research*, **13**, p. 19, 1992.

[78] A. Taylor, C. Brownlee, 'Calcium and Potassium Currents in the Fucus Egg', *Planta*, **189**, p. 109, 1993.

[79] K. Toko, H. Chosa, K. Yamafuji, 'Dissipative Structure in the Characeae, Spatial Pattern of Proton flux as a Dissipative Structure in Characean Cells', *J. Theor. Biol.*, **114**, p. 125, 1985.

[80] A.P.J. Trinci, *The Duplication Cycle and Branching in Fungi, in Fungal Walls and Hyphal Growth*, J.H. Burnett and A.P.J. Trinci, eds., Cambridge University Press, 1979.

[81] A.M. Turing, 'The Chemical Basis of Morphogenesis', *Phil. Trans. Roy. Soc. Lon.*, **237**, p. 5, 1952.

[82] M.P. van Veen, J. van Pelt, 'Dynamic Mechanisms of Neural Outgrowth', *progress in brain research*, **102**, p. 95, 1994.

[83] A. Wegner, 'Head to Tail Polymerization of Actin', *J. Mol. Biol.*, **108**, p. 139, 1976.

[84] M.H. Weisenseel, L.F. Jaffe, 'Membrane Potential and Impedance of Developing Fucoid Eggs', *Developmental biology*, **27**, p. 555, 1972.

[85] S.H. Young, M-m. Poo, 'Spontaneous Release of Transmitter from Growth Cones of Embryonic Neurones', *Nature*, **305**, p. 634, 1983.

[86] J.Q. Zheng, M. Felder, J. A. Connor, M-m. Poo, 'Turning of Nerve Growth Cones Induced by Neurotransmitters', *Nature*, **368**, p. 140, 1994.

[87] J.G. Zheng, J.J. Wan, M.M. Poo, 'Essential Role of Filopodia in Chemotropic Turning of Nerve Growth Cone Induced by a Glutamate Gradient', *The journal of neuroscience*, **16**, p. 1140, 1996.

Index of authors

I

J

K

L

M

N

O

P

Q

R

S

T

U

V

W

X

Y

Z

Index of subjects

A

B

C

D

E

F

G

O

P

R

S

T

V

Z